ADVANCES IN
MOLTEN SALT CHEMISTRY
Volume 2

CONTRIBUTORS TO THIS VOLUME

A. Block-Bolten
Department of Metallurgy and Materials Science
University of Toronto
Toronto, Ontario, Canada

John R. Dickinson
Department of Chemistry
University of Virginia
Charlottesville, Virginia

S. N. Flengas
Department of Metallurgy and Materials Science
University of Toronto
Toronto, Ontario, Canada

K. W. Fung
Department of Chemistry
University of Tennessee
Knoxville, Tennessee

Keith E. Johnson
Department of Chemistry
University of Saskatchewan, Regina Campus
Regina, Saskatchewan, Canada

John E. Lind, Jr.
Department of Chemical Engineering
Stanford University
Menlo Park, California

G. Mamantov
Department of Chemistry
University of Tennessee
Knoxville, Tennessee

ADVANCES IN MOLTEN SALT CHEMISTRY
Volume 2

Edited by
J. BRAUNSTEIN
Oak Ridge National Laboratory
Oak Ridge, Tennessee

GLEB MAMANTOV
The University of Tennessee
Knoxville, Tennessee

and
G. P. SMITH
Oak Ridge National Laboratory
Oak Ridge, Tennessee

⚘ PLENUM PRESS • NEW YORK-LONDON • 1973

Library of Congress Catalog Card Number 78-131884
ISBN 0-306-39702-1

© 1973 Plenum Press, New York
A Division of Plenum Publishing Corporation
227 West 17th Street, New York, N.Y. 10011

United Kingdom edition published by Plenum Press, London
A Division of Plenum Publishing Company, Ltd.
Davis House (4th Floor), 8 Scrubs Lane, Harlesden, London, NW10 6SE, England

All rights reserved

No part of this publication may be reproduced in any form
without written permission from the publisher

Printed in the United States of America

ARTICLES PLANNED FOR FUTURE VOLUMES

Molecular Dynamics Calculations on Molten Salt Systems
L. V. Woodcock, School of Chemistry, Leeds University

Aluminum Chloride Containing Solvent Systems as Organic Reactions Media
R. A. Osteryoung and H. L. Jones, Department of Chemistry, Colorado State University

Chemistry of Thiocyanate Melts
D. H. Kerridge, Chemistry Department, University of Southampton

Magnetic Methods in Molten Salt Research
N. H. Nachtrieb, Department of Chemistry, University of Chicago

Experimental Techniques in Molten Fluoride Chemistry
C. E. Bamberger, Oak Ridge National Laboratory

Optical Interferometry Applied to the Study of Molten Salts
A. Lundin and S. E. Gustafsson, Chalmers Institute of Technology, Sweden

Phase Equilibria in Molten Salts
R. E. Thoma, Oak Ridge National Laboratory

Advances in Electrode Kinetics in Molten Salts
D. Inman and D. G. Lovering, Imperial College, London

Chemistry in Molten Nitrates
J. Jordan, Department of Chemistry, Pennsylvania State University

Relaxation Phenomena in Molten Salts
C. T. Moynihan, Catholic University of America

Studies of Dynamic Properties by Spectroscopic Techniques
J. H. R. Clarke, Chemistry Department, University of Southampton

Chemistry of Steel Making
A. R. Kay, Department of Metallurgy and Materials Science, McMaster University

Gas Solubility in Molten Salts
P. E. Field, Virginia Polytechnic Institute

The Chemistry and Thermodynamics of Molten Lif–BeF_2 Solutions
C. F. Baes, Jr., Oak Ridge National Laboratory

PREFACE

The first chapter in the second volume of this series deals with the physical properties of a novel group of melts, the organic molten salts as applied to the testing of models of molten salt behavior. In addition, there are three chapters on the chemistry of solute species in melts. One of these chapters deals with solubilities of reactive gases in melts. Another chapter treats coordination chemistry and electronic spectroscopy of Group VIII elements in fused salts. The last chapter is concerned with recent developments in electroanalytical chemistry in molten salt systems.

J. B., G. M., G. P. S.

CONTENTS

Chapter 1

MOLTEN ORGANIC SALTS—PHYSICAL PROPERTIES

    J. E. Lind, Jr.

1. Introduction 1
2. Equilibrium Thermodynamic Properties 2
    2.1. Phase Transitions 2
    2.2. PVT Properties 3
    2.3. Surface Tension and the Scaled Particle Approach . . . 7
3. Transport Phenomena 10
    3.1. Walden and the Walden Rule 10
    3.2. The Problem of Corresponding States 13
    3.3. The Second-Order Transition Model 14
    3.4. The Kinetic Model 15
    3.5. The Effect of the Coulomb Field on Transport Processes 19
    3.6. Salts with Added Nonelectrolytes 22
References 25

Chapter 2

SOLUBILITIES OF REACTIVE GASES IN MOLTEN SALTS

    S. N. Flengas and A. Block-Bolten

1. Introduction 27
    1.1. Thermodynamic Treatment of Solubility 30
    1.2. Reactive Systems 34

2. Experimental Techniques for Measuring Solubilities of Gases in Molten Salts 42
3. Solubility Data 54
4. Chemical Syntheses in Molten Salts 69
References 77

Chapter 3

HIGH-TEMPERATURE COORDINATION CHEMISTRY OF GROUP VIII

K. E. Johnson and J. R. Dickinson

1. Introduction 83
　1.1. Scope 83
　1.2. Methods of Studying Coordination at High Temperatures 85
　1.3. General Survey of the Coordination Chemistry of Group VIII 85
2. Electronic Spectroscopy and Bonding 87
　2.1. Atomic Spectra and Coupling Schemes 87
　2.2. Spectra of and Bonding in Diatomic Molecules 96
　2.3. Bonding in and Spectra of Polyatomic Species 103
3. Group VIII Species in the Gas Phase 123
　3.1. Diatomic Molecules 123
　3.2. Other Species 123
4. Group VIII Species in Molten Salts 129
　4.1. Review of Oxidation States 129
　4.2. Iron(III) $3d^5$ 130
　4.3. Cobalt(II) $3d^7$ 135
　4.4. Nickel(II) $3d^8$ 152
　4.5. Ruthenium(III) $4d^5$ 172
　4.6. Osmium(IV) $5d^4$ and Osmium(III) $5d^5$ 173
　4.7. Rhodium and Iridium 176
　4.8. Palladium and Platinum 179
5. Whither High-Temperature Coordination Chemistry? . . 181
　5.1. General Conclusions 181
　5.2. Suggestions for Further Study 184
Appendix A. The Coulomb (Direct) and Exchange Integrals . . 185
Appendix B. Effects of Electric and Magnetic Fields 186

Appendix C. The Correlation of Molecular and Atomic Electronic
States 187
Appendix D. Symbols Used in Sections 3–5 191
References 192

Chapter 4

ELECTROANALYTICAL CHEMISTRY IN MOLTEN SALTS—A REVIEW OF RECENT DEVELOPMENTS

K. W. Fung and G. Mamantov

1. Introduction 199
2. General Remarks 200
   2.1. Acid–Base Concepts 200
   2.2. Methodology 201
   2.3. Experimental Techniques 202
   2.4. EMF Series 203
3. Presently Important Solvent Systems 204
   3.1. LiCl–KCl Eutectic (58.8–41.2 mole %) 204
   3.2. NaCl–KCl (50–50 mole %) 212
   3.3. Chloroaluminates ($AlCl_3$–Alkali Chlorides) 218
   3.4. Other Chlorides 224
   3.5. Other Halides 230
   3.6. Nitrates 230
   3.7. Melts 241
References 246

Index 255

Chapter 1

MOLTEN ORGANIC SALTS—PHYSICAL PROPERTIES

John E. Lind, Jr.
Department of Chemical Engineering
Stanford University

1. INTRODUCTION

The following discussion will concern itself with the thermodynamic and transport properties of molten organic salts in so far as they shed light on the structure of simple ionic fluids. As a result several detailed studies of complex organic mixtures have been omitted and the interested reader should consult the comprehensive review articles by Gordon[1] and Reinsborough.[2]

The advantage of studying organic salts is their low melting points and the possibility of tailoring their size and shape. For a special class of salts which have isoelectronic anions and cations there are also nonelectrolytes isoelectronic to the salt ions which are fluid in the same temperature region. The direct comparison of nonelectrolytes to molten electrolytes is thus possible. The one detracting feature of these systems is that they are polyatomic with internal degrees of freedom which are not included in the simple theories often used to describe them. On the other hand, this problem raises the interesting question of how the rotational isomers are biased by the presence of the internal Coulomb field of the salt melt.

The differential thermal analysis of Janz and co-workers has been most helpful in showing the alterations in chain configurations which can occur before melting. We will first put these results in mind and then we can proceed to the equilibrium thermodynamic properties and finally the transport properties.

2. EQUILIBRIUM THERMODYNAMIC PROPERTIES

2.1. Phase Transitions

Phase diagrams have been summarized recently by Gordon[1] and will not be discussed here. Janz and co-workers[3,4] have made an extensive study of the solid–solid and solid–liquid phase transitions of 33 quaternary ammonium salts by differential thermal analysis. The results show changes of alkyl chain conformation in the cations at or below the melting point, especially for the longer chains. Thus in the use of simple models without internal degrees of freedom one must be wary of where these degrees of freedom will manifest themselves in the observable properties.

Coker et al.[3] have made a detailed study of $(n\text{-}C_5H_{11})_4\text{NSCN}$ which illustrates the changes of chain conformation. A solid transition occurs about 10° below the melting point of 49.5°C in which the volume of the solid increases some 5% to a value greater than that of the liquid. Upon heating to the melting point, the volume contracts sufficiently to permit an increase of volume upon fusion of 0.4%. They postulate a 2 *gauche–trans* kink occurring in each amyl chain, causing the extended chains to shorten without altering the long axis of the chain. This kink consists of two *gauche* bonds separated by one *trans*. From parafin results, 6 kcal/mole is estimated for this transition, in good agreement with the observed heat of the solid–solid transition of 5.4 kcal/mole. On the basis of the number of chain conformations, the entropy of the transition was calculated to lie between 12 and 18 eu, in agreement with the observed 17 eu. The entropy of fusion of 14 eu is not unreasonably higher than the values of 7–10 for the alkali thiocyanates, and therefore it is attributed primarily to the external degrees of freedom including rotation.

Since changes of chain conformation may occur at several temperatures as well as at the melting point, the total entropy of all transitions including fusion is tabulated for comparison in Table I for a few salts. In general the cations with short or branched chains exhibit lower entropy changes because the chains are hindered within the cation. Larger anions seem to permit disengagement of the chains in the liquid, permitting more freedom for the chains and a higher entropy.

The tetraalkylammonium iodides have also been examined by Levkov et al.[5] For this series with four identical alkyl chains on the cation they show that the total entropy of up to three premelting transitions is constant for the chains with odd numbers of carbon atoms at about 8 eu, while that for even numbers increases with chain length, starting at 11 eu for the ethyl

TABLE I. Total Entropies of Transition[4]

Salt	ΔS(total), eu/mole	Salt	ΔS(total), eu/mole
$(n\text{-}C_4H_9)_4NI$	22.6	$(C_4H_9)_4NCl$	15.6
$(n\text{-}C_5H_{11})_4NI$	33.1	$(C_4H_9)_4NBr$	18.8
$(i\text{-}C_5H_{11})_4NI$	23.7	$(C_4H_9)_4NI$	27.6
$(n\text{-}C_6H_{13})_4NI$	33.8	$(C_4H_9)_4NClO_3$	33.3
$(n\text{-}C_7H_{15})_4NI$	30.8	$(C_4H_9)_4NNO_3$	9.1

chain. In contrast the entropies of fusion alternate in the opposite fashion to make the total entropy of all transitions to be a more slowly increasing function of chain length. They ascribe the 8-eu premelting transitions of the chains with odd numbers of carbons to a crystal structure change plus a transition of the cation configuration from nordic cross to extended simple cross form. The formation of various rotational isomers of the chains like the kink-block mechanism described above is then left to the melting transition; whereas for chains with even numbers of carbon atoms this isomerization occurs in the premelting transitions. This explanation is in contrast to the thiocyanate process proposed above. These explorations should now be supported by X-ray studies.

The melting of the alkali acetates has also been studied volumetrically and spectroscopically by Hazelwood et al.[6] The sodium salt expands upon melting by 3.9%, which compares well with sodium nitrate, but the potassium salt actually contracts about 1% upon melting. The volumetric data indicate the possibility of transitions in the solid similar to the quaternary ammonium salts.

2.2. PVT Properties

The PVT properties of three salts have been measured[7] and directly compared to nonionic fluids containing molecules isoelectronic to the salt ions in an attempt to set bounds to the contribution of the internal Coulomb field of the electrolyte to the thermodynamic properties. The salts are tetra-n-butylammonium tetra-n-butylborate, tetraethylammonium tetra-n-propylborate, and tetra-n-propylammonium tetraethylborate. The thermodynamic properties of the latter two salts, which are isoelectronic to one another are essentially identical within 1–2% and this result suggests that the rotational isomers are independent of the electric field direction or the central atom for these asymmetric salts. The properties of these three salts were compared to their isoelectronic nonelectrolytes 5,5-dibutylnonane and the equimolar mixture of 3,3-diethylpentane and 4,4-dipropylheptane.

In order to relate the thermodynamic properties of the fluids to their microscopic structure the use of the independent variables temperature and volume is extremely useful. These are the independent variables of the cannonical ensemble which is most often used conceptually and in the development of theory. The pressure is then seen as a very sensitive function resulting from an integral whose integrand contains large positive and negative portions that nearly cancel each other at 1 atm pressure. Thus the thermodynamic properties of these systems were measured up to 8000 bars whenever possible and this permitted changes of volume as large as 30%. The temperature range for these salts, which melt near 100°C, was 40°C and was limited by decomposition.

Pressures of the order of 1000–2000 bars are required to bring the volume of the nonelectrolyte to that of the salt at the same temperature. Figure 1 shows the compressibility factors for the salt and nonelectrolyte for the asymmetric system. The compressibility factor for the salt is about eight less than that of the nonelectrolyte for all systems and at all temperatures and volumes. This difference should now be simply related to the Coulomb field in the salt.

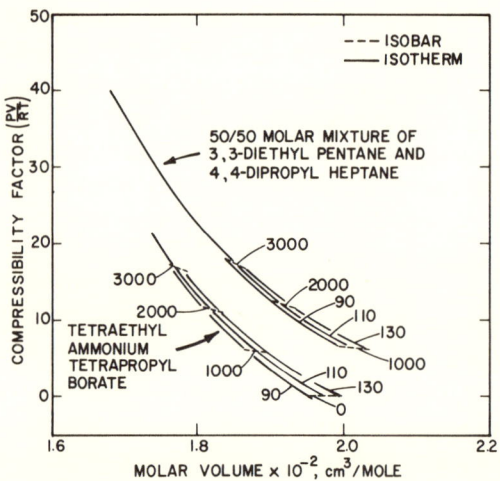

Fig. 1. Compressibility factor of tetraethylammonium tetrapropylborate and 50/50 molar mixture of 3,3-diethylpentane and 4,4-dipropilheptane.

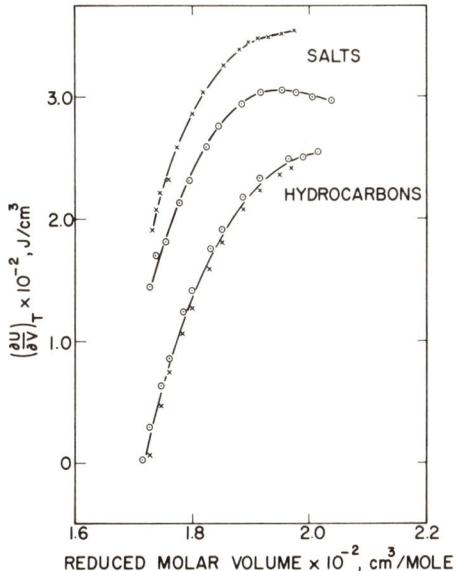

Fig. 2. The volume derivative of the energy scaled onto the asymmetric system data. Crosses: Eth$_4$NPr$_4$B. Dotted circles: Bu$_4$NBu$_4$B.

The equation of state for liquids can best be analyzed using the thermodynamic equation of state

$$P = T(\partial S/\partial V)_T - (\partial U/\partial V)_T \qquad (1)$$

The reason this equation is convenient is that the term $T(\partial S/\partial V)_T$ depends only upon the repulsive core of the molecules if a mean field is used for the attractive forces between the molecules. Such a separation permits immediate evaluation of core size from a fit of a hard- or soft-sphere model to $(\partial S/\partial V)_T$ without further specification of $(\partial U/\partial V)_T$. The examination of the thermodynamic equation of state shows that the major contribution to the lower pressure of the salt system arises from $(\partial U/\partial V)_T$, which increases because of the Coulomb field. Figure 2 shows $(\partial U/\partial V)_T$ for salts and nonelectrolytes at corresponding temperatures and volumes scaled onto that of the asymmetric salt using the law of corresponding states that applies to the hydrocarbons. The differences between $(\partial U/\partial V)_T$ for the salts and the hydrocarbons are roughly in correspondence with that predicted for crystalline salts with the same Madelung constant. The word "roughly" is used above because the values of $(\partial U/\partial V)_T$ are much more temperature dependent for the hydrocarbons than for the salt and therefore the differences of these derivatives have a significant temperature dependence.

Figure 3 shows that in most cases $(\partial S/\partial V)_T$ is lower for the salt than for the nonelectrolyte. Since the isotherms of $(\partial S/\partial V)_T$ for the different salts scale onto one another using the scaling parameters for the nonelectrolytes, this suggests that the decreased values of $(\partial S/\partial V)_T$ arise primarily from the distortions of the alkyl chains on the ions and not from the ordering effect of the Coulomb field upon the charge centers of the ions. The argument is that for the nonelectrolytes $(\partial S/\partial V)_T$ scales by the total number of degrees of freedom of the molecule and not merely by the center-of-mass degrees of freedom alone. The nonelectrolyte molecules at these densities can be represented by several soft spheres whose total number of degrees of freedom equals that of the polyatomic molecule. Chain distortions should scale with the total number of degrees of freedom, whereas the effects of charge ordering would involve the center-of-mass coordinates alone.

The total contribution to the decrease of the value of $(\partial S/\partial V)_T$ for the salts to the lower pressure of the salt is usually less than 15% except for the butyl salt at low densities and temperatures, where the contribution rises to 50%. Thus in general the increased value of $(\partial U/\partial V)_T$ for the salt

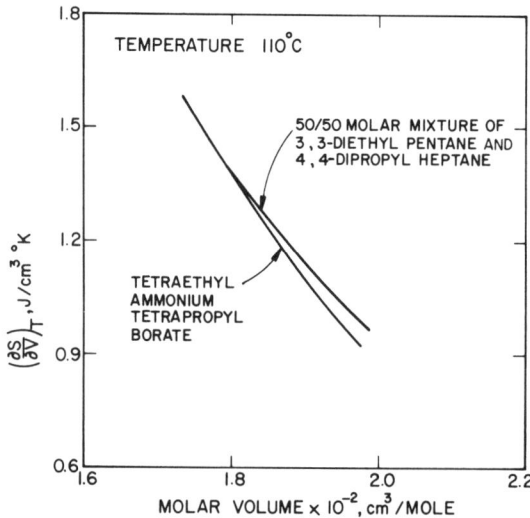

Fig. 3. Comparison of the volume derivatives of the entropy for salt and corresponding hydrocarbon mixture.

is the major difference between the salt and the nonelectrolyte and it produces a lower pressure for the salt than for the nonelectrolyte when they are compared at the same temperature and volume.

These results can be compared with a theoretical model to ask how close theory is to being able to predict the thermodynamic properties of molten salts. The simplist model is that of charging hard spheres to form a molten salt at constant volume and temperature. The experimental results presented above give a lower bound to the hard-sphere salt pressure because $(\partial U/\partial V)_T$ is an upper bound. The latter is an upper bound because any deformation of the real molecules by the Coulomb field will tend to lower the Coulomb energy and thereby tend to increase $(\partial U/\partial V)_T$.

Waisman and Lebowitz[8] have a solution to this model in the mean-spherical approximation, and their result is known to be equivalent to the first three terms of the mode expansion of Chandler and Andersen[9] and their results agree well with hypernetted chain calculations of Rasaiah and Friedman[10] for 1 M aqueous solutions. The fused salt is ten times the volume density of the 1 M aqueous solution. While the charge density of the salt is only three times as great as the solution, the dielectric constant is two rather than 78 and this results in a large increase in the electric field. The theory predicts for the salt a difference of one to two in the compressibility factor, leaving a residual difference of about seven. Looking at the equation of state, the theory predicts a small, 1% decrease of $(\partial S/\partial V)_T$ upon charging the hard-sphere system up into a molten salt. It also predicts a nearly temperature-independent increase of $(\partial U/\partial V)_T$ of only one-fifth to one-third of that observed. The theoretical difference is also much less volume dependent. These differences suggest that ion-induced dipole terms may be significant[11] since they would have greater temperature and volume dependence. Thus the theory predicts the correct sign of all the changes of the thermodynamic properties upon charging the system but it underestimates the effects. Part of this underestimate arises because the measured values are increased by effects peculiar to the real system, such as deformability of the ions and ion-induced dipole effects.

2.3. Surface Tension and the Scaled Particle Approach

About the earliest study of molten organic salts was the measurement of surface tensions by Walden[12] of a number of alkylammonium nitrates, thiocyanates, and halides. Most recently the chlorides have been studied by Kisza and Hawranek.[13] The surface tensions of these salts are similar to those for organic liquids and they range from 20 to 50 dyn/cm². These values are about half the values for the alkali halides, which vary from 60 to 120 dyn/cm².

Walden's early study[12,14] related the structure of the salt melts to the degree of alkylation. The negative of the temperature coefficient of the molar surface tension is the surface entropy and it had been shown empirically to be inversely related to the degree of association. Walden's surface tensions for salts with cations of the same molecular weight showed that the primary ammonium salts are the most highly associated, with the tertiary and quaternary the least associated.

Recently Bloom and Reinsborough[15] have measured the surface tension of pyrdinium chloride as well as transport properties and used it to interpret micelle formation of this salt with cetyl and myristyl cationic soaps. Earlier Coleman and Prideaux[16] studied mixtures of carboxylic acids with diethylamine and piperidine, but these are complicated systems which are too far afield of this discussion.

The scaled particle solution[17] to the hard-sphere model permits the calculation of the surface tension if the density and the size of the molecules are known. Mayer[18] has shown the theory to apply to both nonelectrolytes and salts. The same solution of the hard-sphere model permits the calculation of the various thermodynamic properties such as the thermal expansion coefficient. Since the thermal expansion coefficient is known for most salts for which there are surface tension data, the former can be used to calculate the latter.

The thermal expansion coefficient α is given by

$$\alpha = (1 - \eta^3)/T(1 + 2\eta)^2 \tag{2}$$

and the surface tension σ is approximately

$$\sigma = (kT/4\pi a^2)\{[(12\eta/(1 - \eta)] + 18[\eta/(1 - \eta)]^2\} \tag{3}$$

where η is the packing fraction $(\pi N a^3/6V)$ for spheres of diameter a. We know from Section 2.2 that the hard sphere should represent segments of large molecules rather than the molecule as a whole. Even so, very often the thermodynamic properties can be interrelated ignoring the fact and treating the molecule as a whole as a single sphere. This approximation normally leads to unusually high packing fractions. Fortunately the one tetraalkylammonium salt for which there are surface tension data is a very hindered salt, tetraisoamylammonium iodide, where there are relatively few internal degrees of freedom. Table II shows that the surface tension is predicted to within 10% from the expansion coefficient. The density is

TABLE II. Results of Scaled Particle Analysis

Compound	T, °C	$\alpha \times 10^4$	η	σ(calc)	σ(obs)	Ref.
(iso-amyl)$_4$NI	119.5	6.08	0.47	24.2	25.98	12
(iso-amyl)$_3$NHSCN	90	6.39	0.48	26.8	28.89	12
C$_6$H$_5$-N(CH$_3$)$_2$HBr	87.1	8.68	0.371	26.5	49.49	13
(CH$_3$)$_2$NH$_2$Cl	200	4.80	0.49	91.3	40.26	13

high, in fact the density is in a region where the scaled particle theory deviates significantly from the results of molecular dynamics. Table II also shows that even the tertiary ammonium thiocyanate fits the theory. However, the tertiary hydrobromide and the secondary hydrochloride deviate by factors of two in different directions from the theory.

Walden and Birr[21] showed that for isomeric picrates the thermal expansion coefficients of the primary, secondary, and tertiary ammonium salts are greater than for the quaternary salts. Presumably the larger expansion coefficients occur in sustems which are less ionic because of the dissociation of the salt into amine and picric acid. This explanation is in agreement with the comparison of the isoelectronic salts and hydrocarbons in Section 2.2. For them the nonelectrolyte has the greater thermal expansion coefficient when the comparison is made at constant pressure and temperature, in agreement with Walden and Birr, while for a comparison at constant volume and temperature the salt has a greater coefficient. Thus nonelectrolyte components are in equilibrium with the primary through tertiary ammonium salts and the shifts in this equilibrium will be reflected in the thermal expansion coefficient and produce unreasonable estimates of the surface tension.

Before concluding this discussion of surface tension let us make a more careful analysis of these intercorrelations of properties for which the scaled particle theory has been used. The approach assumes that the thermodynamic derivatives can be interrelated by the properties of a hard-sphere fluid. In particular, the compressibility, thermal expansion, and surface tension can be related to one another. The ability to relate the surface tension to the other properties is unique to the scaled particle theory and it is not possible to test the consistency of the approach to that relation. However, for the hard-sphere scaled particle solution to relate the thermal expansion coefficient and the compressibility we must have $(\partial^2 U/\partial V^2)_T \ll T(\partial^2 S/\partial V^2)_T$. This requirement is obvious when the thermal expansion coefficient α is expressed in terms of the thermodynamic equation of state

$$\alpha = -(\partial S/\partial V)_T [VT(\partial^2 S/\partial V^2)_T - V(\partial^2 U/\partial V^2)_T]^{-1} \qquad (4)$$

while the compressibility $\varkappa_T$ is

$$\varkappa_T = -[VT(\partial^2 S/\partial V^2)_T - V(\partial^2 U/\partial V^2)_T]^{-1} \qquad (5)$$

with $\alpha/\varkappa_T = (\partial S/\partial V)_T$.

We know from Section 2.2 that $(\partial S/\partial V)_T$ depends essentially upon the properties of the molecular cores but not upon the attractive field between the molecules. Therefore $(\partial S/\partial V)_T$ can be well represented by a hard-sphere model. Thus α and $\varkappa_T$ can be related by the hard-sphere model if the core diameter is known so $(\partial S/\partial V)_T$ can be evaluated. However, if the diameter is evaluated from either α or $\varkappa_T$, the derivative $(\partial^2 U/\partial V^2)_T$ must be negligible compared to $T(\partial^2 S/\partial V^2)_T$ or an incorrect diameter will result which compensates for this additional term.

Unfortunately the organic salts for which there are surface tension data have not had measurements made of these second derivatives of the energy and entropy with respect to volume. In fact these derivatives are known only for the salts in Section 2.2: tetrabutylammonium tetrabutylboride and tetraethylammonium tetrapropylboride. For the latter at 130° and 1 bar, $T(\partial^2 S/\partial V^2)_T = -5.2$ J/cm^6 and $(\partial^2 U/\partial V^2)_T = +2.7$ J/cm^6. Their values differ by only a factor of two and this is reflected by the fact that the compressibility computed from the thermal expansion coefficient using the scaled particle approach is only one-fifth the observed value.

Whether these results are typical of the organic salts for which there are surface tension data will have to await further measurements. Suffice it to say that it is not at all clear that these techniques of using the scaled particle theory to interrelate properties are valid when applied to these organic salts with large ions for which there are surface tension data.

3. TRANSPORT PHENOMENA

3.1. Walden and the Walden Rule

Some of the earliest and certainly the most extensive measurements were made by P. Walden and co-workers. The density, electrical conductivity, and viscosity of primary through quaternary picrates were studied.[20] They have melting points between 80 and 200°C. Alkyl chains varied from methyl to pentyl with also primary and secondary cetyl ammonium salts.[21]

The surface tensions of quaternary ammonium salts suggested that they were little associated and primarily ionic. Similarly, the molar volumes of the quaternary salts were smaller than the volumes of other ammonium ions with equal numbers of carbon atoms on the cation and their thermal expansion coefficients were also smaller. These data supported the idea that the quaternary salts were completely dissociated since the Coulomb field in the ionized melt should cause a contraction of the volume. The quaternary ammonium salts also had the lowest viscosities and the highest electrical conductivities. Walden then proposed the use of his rule for these completely dissociated electrolytes at their melting points[14]:

$$\Lambda \eta \sqrt{M} = 10.5 \tag{6}$$

where Λ is the equivalent conductance, η is the viscosity in poise, and M is the molecular weight of the salt. This equation holds not only for quaternary ammonium picrates but also alkali nitrates and chlorides, which have a variation of 1000° in their melting points. It is also closely obeyed by the quaternary ammonium picrates at infinite dilution in both organic and aqueous solvents at room temperature. At temperatures above the melting point the measured product decreases from this value.

From Eq. (6) and the measured viscosity, a theoretical conductivity at the melting point for primary through tertiary ammonium salts could be calculated. This conductivity represented a fully ionized melt of the same viscosity. The ratio of the measured conductance to the theoretical value represented the degree of dissociation. The results were not in accord with the analysis of the surface tension. The tertiary ammonium picrate turned out to be less dissociated than the primary salt by the conductance criterion. The inconsistency in the results from the two physical properties suggested a much more complicated structure for the primary and secondary ammonium salts.

Since the Walden product proved unsatisfactory for the primary and secondary ammonium salts, the use of the melting point as a corresponding state was rejected and comparisons of the conductance and viscosity were made at the same temperature between salts having the same number of carbon atoms on the cation. The conductivity as shown in Table III uniformly increases with number of alkyl chains. The viscosity decreases to the tertiary salt and then increases slightly for the quaternary salt. Walden suggested that the tertiary salt underwent simple association which decreased the conductance without building up high-molecular-weight species which would increase the viscosity. The viscosity could even drop because of the

TABLE III. Walden's Results for Alkylammonium Picrates at 150°C for Salts Having the Same Number of Carbon Atoms on the Cation

Salt	Relative viscosity	Relative conductivity
Primary	3.5	0.15
Secondary	1.5	0.30
Ternary	0.8	0.25
Quaternary	1	1

smaller Coulomb field. The high viscosities of the primary and secondary ammonium picrates suggested high-molecular-weight species, perhaps ions solvated by associated molecules.

To attempt to further clarify the situation not only were several iodides and perchlorates studied but mixtures of fused picrates[19,22] were studied. The quaternary amyl iodides and perchlorates obeyed Walden's rule but the butyl salts were as much as 25% low. The mixtures of picrates were nearly ideal, with slightly negative excess volumes of less than 1% for the mixtures of quaternary salts. Slightly negative excess volumes of mixing are reasonable for completely ionized melts with a common anion since they might be capable of being represented by a two-component hard-sphere model which shows negative excess volumes of mixing. For mixtures of a quaternary salt with a primary or secondary ammonium salt the excess volumes are less negative and may become positive at high concentrations of primary or secondary ammonium salt. For these mixtures the excess volume is positive by no more than 0.4% at any concentration.

Unfortunately the ions of these picrate salts are not very spherically symmetric so that the application of the conformal solution theory of Davis and McDonald[23] does not warrant an attempt. A quick analysis agrees with Walden that the quaternary salts are ionic while those with lower degrees of alkylation are less ionic. If the effects of size and intermolecular attractive forces are separable, one can say that mixing of hard-sphere molecules of different size leads to small negative excess volumes. Then Prigogine's theory of regular solutions of molecules of the same size[24] can be used. When a liquid of strong intermolecular forces such as an ionic salt mixes with a weakly interacting fluid of associated salt molecules to form a mixture with weak interactions, the excess volume of mixing tends to be positive compared to mixtures of associated salts where all the intermolecular interactions are comparable. The mixture of an ionic salt with a molecular fluid should yield a fluid with weak interactions since the dipolar molecules would screen the ions from one another by providing a solvent of higher dielectric constant. The result is thus in agreement with Walden's data.

For the mixture of quaternary ammonium picrates the equivalent conductance and viscosity vary linearly with composition. As salts with lower degrees of alkylation are mixed, both transport properties tend to have minima. The deviation from the linear rule is greatest for the mixture of primary and secondary ammonium salts. This latter system behaves much like the $PbCl_2$–KCl system,[25] having a minimum in the transport property and a positive excess volume of mixing. Complexes such as $PbCl_3^{-3}$, $PbCl_4^{-2}$, and $PbCl_6^-$ have been postulated to explain the inorganic case. A spectroscopic study of the picrates would be very helpful in understanding their structure.

3.2. The Problem of Corresponding States

Walden found that the Walden products of completely iodized melts could be predicted for a wide range of salts at the melting point. However, when he attempted to analyze the data for associated systems he found it easier to compare at the same temperature individual properties for salts having about the same volume but different degrees of alkylation.

Since then the use of activation energies and Eyring's transition-state theory has obviated to some extent the need for a theory of corresponding states so long as the Arrhenius plots of the transport properties are linear. Table IV lists a number of activation energies for conduction and viscosity. The ratios of the two activation energies are near unity for most of the organic salts and Janz et al.[26] have suggested that the interlocking of the chains of the cations prevents the smaller anion from moving in conduction in the melt without cooperative motion of the large cations which must move in the shearing motion of viscous flow. In contrast the ratio of activation energies is much greater than unity for the alkali halides. The motion of one species of ion is not nearly so constrained by the other and in LiI the Li^+ can move in the interstices between the I^- ions. The one organic salt which behaves like the alkali halides, having a high ratio of activation energies, is tetrapropylammonium tetraphenylboride.[27] In this salt the structure may be dominated by the anion which has some space between the phenyl rings for the smaller cation to pass through.

Such an approach then leads into various hole and jump theories as these activation energies and entropies can be related to thermodynamic

TABLE IV. Activation Energies

Salt	E_Λ	E_η	E_η/E_Λ	T, °C	Ref.
KCl	3.42	6.59	1.93	790–930	47
LiI	1.30	4.42	3.41	460–650	47
KNO_3	3.58	4.30	1.20	410–540	47
AgI	1.48	5.26	3.57	530–830	47
CH_3COOK	5.71	7.31	1.28	305–350	6
$(C_3H_7)_4N(C_6H_5)_4B$	3.6	9.5	2.6	210	27
	2.6	8.9	3.4	232	27
$(C_4Hg)_4N(C_6H_5)_4B$	6.97	8.24	1.19	240–270	27
$(C_3H_7)_4NBF_4$	5.00	6.04	1.23	250–275	27
$(C_4H_9)_4NBF_4$	7.5	8.3	1.10	163	27
	5.6	6.3	1.12	256	27
CH_3NH_3Cl	2.94	3.99	1.35	230–260	13, 49
$(i\text{-}C_5H_{11})NH_3$ picrate	11.3	12.5	1.11	150	20, 48
$(i\text{-}C_5H_{11})_3NH$ picrate	9.8	8.9	0.91	150	20, 48
$(i\text{-}C_5H_{11})_4N$ picrate	7.1	9.5	1.34	150	20, 48

properties. These theories have not been applied to organic melts to any extent. In recent years two other approaches have been taken to the problem of the comparison of the properties of different salts. A frontal attack has been made by Angell and Moynihan[28] on the corresponding states problem and their approach appears to be useful for a wide range of complicated systems. The second approach, which is based on a detailed mechanical theory, is limited to simple ions but gives a more detailed picture of the dynamics in the melt. These approaches will be discussed in the next two sections.

3.3. The Second-Order Transition Model

Angell and Moynihan[28] have used the ideas of the quasilattice model of Gibbs and DiMarzio for the glassy state to define a characteristic temperature for transport phenomena. They and Easteal have applied this approach in the low-temperature region of inorganic melts as well as to $ZnCl_2$ and its solutions with pyridinium chloride, α- and β-picolinium hydrochloride, and quinolinium hydrochloride.

This quasilattice model predicts an Ehrenfest second-order transition at a temperature T_0 below which the fluid remains frozen in one of its configurations of lowest energy for amorphous packing. Angell and Moynihan identify this temperature with the temperature at which the metastable supercooled liquid has lost entropy, below the normal freezing point in excess of the normal solid, equal to the entropy of fusion at the normal first-order freezing point. The expression for the transport properties are obtained by analogy from the probability of a configurational fluctuation given by the model. The transport properties are exponential functions of the reciprocal of $(T - T_0)$.

In general the measured glass transition temperature lies above T_0. Easteal and Angell[29] have shown that the measured glass temperature and T_0 from conductances parallel one another in the $ZnCl_2$–pyridinium chloride system. This agreement supports the method of analysis and suggests that the easily measured glass transition temperature can be used in the same fashion as the transport properties to interpret the structure of these complicated melts.

3.4. The Kinetic Model

Perhaps the most elegant and exact but therefore the most limited approach to the transport properties is through a general equation such as the Boltzmann or Fokker–Planck equation describing the time-dependent distribution functions of the system. The results of the zeroth-order theories can be conveniently cast in the form of the friction constants of irreversible thermodynamics.

The friction formalism of irreversible thermodynamics for 1–1 electrolytes yields the following expressions for the electrical conductivity Λ and the self-diffusion coefficient D of an ion[30]:

$$(\Lambda/e^2)^{-1} = C_-\xi_{+-} \tag{7}$$

$$(D_+/kT)^{-1} = C_-\xi_{+-} + C_+\xi_{++} \tag{8}$$

$$(D_-/kT)^{-1} = C_+\xi_{-+} + C_-\xi_{--} = C_+\xi_{+-} + C_-\xi_{--} \tag{9}$$

where ξ is the concentration-dependent friction constant, C is the concentration in ions or molecules per cubic centimeter, and e is the electronic charge in appropriate units. The Onsager relations state that $\xi_{+-} = \xi_{-+}$. We see that the Nernst–Einstein equation which is useful for solutions at infinite dilution,

$$\Lambda/e^2 = (D_+/kT) + (D_-/kT) \tag{10}$$

holds only if $\xi_{+-} = \xi_{++} = \xi_{--}$. For KCl, $\xi_{+-}/2 \simeq \xi_{++} \simeq \xi_{--}$. The first approximate equality for KCl is also known to be valid for tetrapropylammonium tetrafluoroborate. Thus the Nernst–Einstein equation does not hold for the fused salts.

Is there then a basis for Walden's rule? First statistical mechanics must furnish a theory for the transport phenomena in order to relate the friction coefficients to other transport phenomena. The development of a theory for uncharged molecules in a dense fluid is difficult and still very incomplete. The addition of a Coulomb field would appear to make the solution even more difficult. Fortunately at zero order the Coulomb field may only affect the equilibrium properties but not the transport properties. We will therefore analyze the transport data of fused salts by using the equilibrium radial distribution function determined by the Coulomb field but in the calculation of the transport coefficients the Coulomb field will be omitted wherever the pair potential is needed. This process assures that nearest neighbors are of opposite sign. A calculation on this basis using an experimental distribution function from X-ray data and the pair potentials obtained from crystal data has predicted[31] the electrical conductivity of KCl within 20% using the Brownian diffusion model of Kirkwood and Rice. Thus the assumption appears sound. A heuristic argument supporting the neglect of the Coulomb field merely points out that if the net charge surrounding a given ion were smeared uniformly on a conducting sphere, the potential would be uniform within the sphere and the reference ion could move within the sphere with no net electrical force exerted on it. Thus the Coulomb field contributes to the transport phenomena to the degree that this spherical symmetry is lost in the real systems. A further discussion of this problem is given in Section 3.5.

The electrical conductance is inversely related to ξ_{+-}, and in this order of approximation all ions of the same sign travel with the same mean drift velocity, so that the only ions of opposite sign exert a net drag in the conduction process. In self-diffusion the reference ion is suffering a fluctuation relative to all other ions, so ions of both signs exert a drag.

For simplicity of explanation the friction constant between molecules α and β for a hard-sphere system[32,33] are given below rather than those for a system of molecules interacting with continuous pair potentials:

$$C_{\alpha\beta}\xi_{\alpha\beta} = (8/3)\varrho_\alpha \sigma_{\alpha\beta}^2 g(\sigma_{\alpha\beta})(mkT)^{1/2} \tag{11}$$

The collision diameter is $\sigma_{\alpha\beta}$, ϱ_α is the particle density, and $g(\sigma_{\alpha\beta})$ is the radial distribution function. If the ions are not of the same mass m, twice the reduced mass may be used as a first approximation. The zeroth-order viscosity η for a 1–1 electrolyte can then be expressed in terms of the friction constants:

$$\eta = \tfrac{1}{5}\varrho_\alpha{}^2[\sigma_{+-}^2\xi_{+-} + \tfrac{1}{2}(\sigma_{++}^2\xi_{++} + \sigma_{--}^2\xi_{--})] \tag{12}$$

If ξ_{++} and ξ_{--} are small enough to neglect, this equation reduces to

$$\Lambda\eta = \tfrac{1}{5}\varrho e^2\sigma_{+-}^2 \tag{13}$$

which is similar in form to Walden's rule. The right-hand side is roughly proportional to the reciprocal of the cube root of the molar volume, which for organic salts is also about proportional to the reciprocal of the cube root of their mass. The difference between Walden's square root and the cube root of mass is not detectable over the limited range of molecule weights available. In general the like-ion friction constants are not negligible but probably vary sufficiently in proportion to ξ_{+-} that the Walden rule is reasonably adequate. In a previous analysis,[27] we used the Nernst–Einstein relation and obtained reasonable mean ion size parameters for the salts from the Walden product.

The use of the Brownian diffusion model will give similar results for the Walden product. All the present theories are inadequate for the computation of the viscosity and yield values about half of the observed value when the viscosity is computed from the observed friction constants.

At present the unlike-ion friction constants are the most adequately understood and they should be correlated at corresponding states. If a two-parameter potential is sufficient, Helfand and Rice[34] have shown that the reduced temperature and volume would be kT/ε and V/σ^3, where ε characterizes the depth of the potential well and σ is a length scaling the potential. Unfortunately, for polyatomic ions the thermodynamic scaling parameters refer to subunits or segments of the molecule which have three degrees of freedom, not distinguishing internal and external degrees of freedom. However, the transport properties depend primarily upon the external degrees of freedom and thus there is no simple relation between the effective pair potentials needed to correlate thermodynamic properties and those needed to correlate transport properties.

To understand either transport or equilibrium properties the independent variables to use are temperature and volume as pointed out in Section 2.2, for these are independent variables of the canonical ensemble which is most often used and which is the most simply conceived. To use these independent variables measurements must be made at high pressure to uncouple the temperature and the volume. Pressure should be avoided as an independent variable since it is a very sensitive function of temperature and volume.

The only data that permit such an analysis on organic salts are viscosity coefficients for tetrapropylammonium, phosphonium, and arsonium tetrafluoroborates.[35] The hardsphere theory would suggest that the arsonium salt has the greatest viscosity because of its large size and mass. At 1 atm the measured arsonium viscosity is lowest; however, if the salts are compared at the same volume and temperature, the predicted order is observed as shown in Fig. 4.

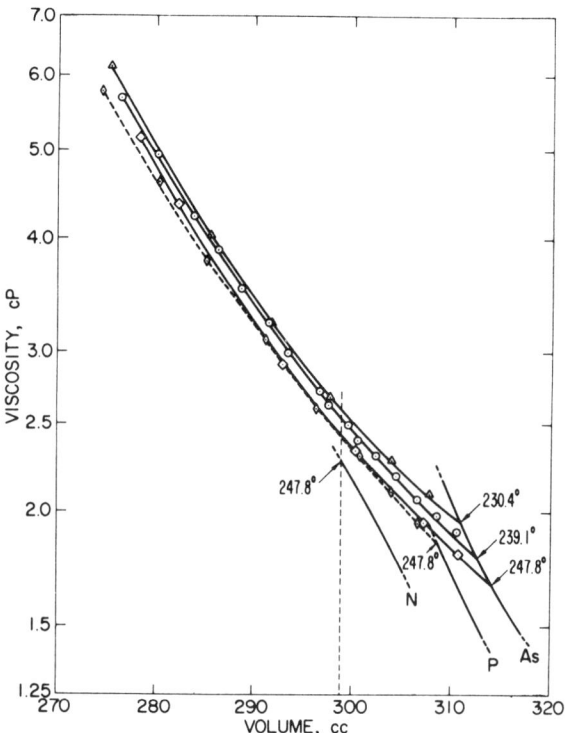

Fig. 4. Viscosity as a function of molar volume for the tetra-*n*-propylammonium (N), phosphonium (P), and arsonium (As) tetrafluoroborates. The isotherms intersect the sharply rising 1-atm isobars at the right. Reprinted from Ref. 35, p. 3037, by courtesy of the American Institute of Physics.

The volumetric data on these salts are not sufficiently accurate to analyze by the thermodynamic equation of state. A rough analysis in that paper using a hard-sphere theory and assuming that one hard sphere was equivalent to an ion yielded packing densities of 0.53–0.56, which are rather high. From our present knowledge presented in Section 2.2 we know that the cation has slightly more than three degrees of freedom and that this would cause the apparently high density if only three degrees of freedom were assumed. At any fixed volume and temperature the pressure is expected to be greatest for the arsonium salt and least for the ammonium salt, as is found to be the case. The arsonium salt should have the highest pressure because the As—C bonds are relatively unhindered to rotation compared to the N—C bonds and thus the arsonium ion will have effectively more degrees of freedom than the ammonium. At a fixed temperature and volume the salt with the highest pressure should have the lowest melting point at 1 atm and the highest volume at the same temperature and equal pressures. These results are in agreement with these systems and explain that the unusual relation of their viscosities at 1 atm arises from the equation of state and not from the transport phenomena.

The only other organic system for which pressure measurements are available is tetrahexylammonium tetrafluoroborate, the conductance of which has been measured up to 1000 bars by Barton and Speedy,[36] but no volumetric data are available. They do, however, estimate that the activation volume is about half that of the smaller ion. This value is small compared to the activation volume for self-diffusion in the rare gases and suggests that the scaled repulsive potential of the organic salt is steep compared to that of the rare gases.[37] This result is reasonable since the penetration of the surface atoms of the polyatomic molecule is comparable to that for a similar monatomic but the size of the polyatomic is much greater.

3.5. The Effect of the Coulomb Field on Transport Processes

The first question to be asked is why the Brownian diffusion model of Kirkwood should give reasonable results for the unlike-ion friction constants, as mentioned in Section 3.4, when the Coulomb potential is ignored and the experimental radial distribution function used. The assumptions in the Brownian diffusion model are difficult to evaluate but Douglass et al.[38] have shown it to be a factor of $\pi/2$ greater than their result using a Gaussian autocorrelation function. Now from molecular dynamics Alder et al.[39] have shown for hard spheres at high densities that the autocorrelation function crosses the zero axis and becomes asymptotically zero from the negative side. This suggests that the Brownian approximation is invalid for hard spheres, as one might expect since the Brownian model assumes many small diffusive steps rather than straight-line trajectories between impacts which reverse the relative velocity vector. However, let us go on to consider the hard-sphere model because of the surprising result that it yields.

To calculate the unlike-ion friction constant from the hard-sphere theory a hard-sphere diameter is required. Section 2.2 has shown that $(\partial S/\partial V)_T$ is essentially independent of the Coulomb potential and can be used conveniently to estimate the packing fraction and thus the sphere diameter. Because of the sorting of charge types by the Coulomb field KCl can be treated as a one-component fluid with nearest neighbors being always of opposite sign. The hard-sphere packing fraction computed from $(\partial S/\partial V)_T$ at about 800°C for KCl is 0.246, or $V/V_0 = 3$. This is a rather low density and the molecular dynamics shows that at this density the autocorrelation function is always positive and somewhat greater than the Gausssian given by the Enskog theory for gases. The hard-sphere diameter for KCl is 2.76 Å and the unlike-ion friction becomes 0.4×10^{-9} after a 16% decrease is made on the Enskog result based on the molecular dynamics findings. Thus the KCl is found to be at a density where the friction is less than the Enskog result rather than at the higher densities where it is greater than the Enskog result because of the back-scattering represented by the negative portions of the autocorrelation function. The hard-sphere result is only one-third of the experimentally measured friction constant.

Thus the low density of KCl does tend to suggest that a Gaussian autocorrelation function is not inappropriate although any hard-sphere model seems to understimate the friction. The Brownian diffusion model with experimentally determined pair distribution function and a pair repulsive potential evaluated from the crystal data seems most adequate. This result may occur because the coupling of the ions by the nondissipative Coulomb field may cause high-frequency motions of the ions in the presence of one another's repulsive potentials, thus fulfilling the requirements of the Brownian diffusion model where the ions suffer Brownian diffusion between collisions. Molecular dynamics calculations, like the Monte Carlo results of Woodcock and Singer[40] on the thermodynamic properties, are needed on the velocity autocorrelation function to determine if the agreement of the Brownian model is just fortuitous or not.

Tetra-*n*-propylammonium tetrafluoroborate can be compared with KCl. This organic salt is compared since it is the only one for which a like-ion friction constant is known. The ion sizes of these two salts differ by a factor of two and the properties of the salts can be compared at absolute temperatures which differ by nearly a factor of two. Note that the charge density of the organic salt is about one-sixth that of KCl. All of the friction constants for the organic salt are about four times those for KCl. Since the anion–cation friction constant has been predicted for KCl by theory without the Coulomb field being introduced into the nonequilibrium part of the calculation, the Coulomb field should not contribute significantly to the anion–cation friction in the organic salt. The fact that the like-ion friction constants behave similarly to the anion–cation friction constants tempts the suggestion that the Coulomb forces are also unimportant to the nonequilibrium part of the calculation of like-ion friction. Caution is required, however, because the like-ion friction constants for KCl are three times larger than any reasonable estimates from theory. This discrepancy may arise from the inadequacy of the theory for nonelectrolytes or from the neglected Coulomb forces as explained below.

Since the transport properties depend primarily upon the repulsive potential, the Coulomb field may play a role through the repulsive force between like ions. It appears to keep like ions apart in the melt in second-neighbor shells even though there is sufficient space for like ions to pack into the nearest-neighbor shells. This repulsive potential must drop off rapidly because of the screening effect of other ions. Thus it would drop off more rapidly in KCl because of its higher charge density, making the effective like-ion diameter smaller for KCl and resulting in the smaller like-ion friction constant for the inorganic salt. Otherwise these salts are similar, for at their respective melting points and charge densities the ratio of Coulomb to thermal energies for the two salts agrees within 10%. To summarize we can say that the inability to predict like-ion friction may be the fault of the theory for nonelectrolytes, or it may be caused by the repulsive Coulomb force between like ions which may scale proportional to the ion size because of the screening effect of the other ions at these high charge densities.

Second, a direct comparison has been made of the viscosities of the salt tetrabutylammonium tetrabutylboride with the isoelectronic 5,5-dibutylnonane at the same temperature and volume.[42] From the arguments in Section 3.4 there should be little difference between the salt and nonelectrolyte. At the highest temperature of comparison of 180°C, the ratio of salt to nonelectrolyte viscosity varied from 5.5 to 4.5 as the density decreased. At lower temperatures and higher densities the ratio increased to over ten. A theoretical estimate was made form the Brownian diffusion model for anion–cation interactions. This theory is convenient since it involves only the ratio of integrals but not their sums. Thus pair potentials could be roughly estimated with the hopes that their inaccuracies would tend to cancel in the ratio and that the perturbations made on the functions characterizing the nonelectrolyte to obtain those of the electrolyte would dominate the expressions. The Coulomb field was ignored except for the shift it causes in the position of the first peak of the radial distribution function. The shift caused by the unshielded Coulomb pair interaction was used to calculate the shift of position of the radial distribution function from that of the nonelectrolyte. The nonelectrolyte was assumed to have 8.2 nearest neighbors positioned about the minimum of the pair potential and this shift of position of the minimum of the potential for the salt required 6.9 nearest neighbors in order to maintain the same mean density. Using the nonelectrolyte pair potential with this shift of the distribution function resulted in a viscosity ratio of two, or about half the experimental value. Thus just the shift in position of nearest neighbors can account for a sizable fraction of the difference between electrolyte and nonelectrolyte viscosities. A forthcoming comparison of friction constants should permit a less ambiguous analysis since the theory for the friction constants is in much better shape than that for the viscosity.

3.6. Salts with Added Nonelectrolytes

Of the few scattered studies of concentrated solutions of organic salts, Kenausis et al.[43] were the first to propose an explanation in some detail on the basis of the system $(n\text{-}C_5H_{11})_4\text{NSCN}$ with p-xylene added. By use of the Walden product they showed that approximately one pair of ions was removed from conduction per molecule of xylene added at the two diverse temperatures of 52° and 90° and at concentrations as high as 0.3 mole fraction of xylene. They argue that the nonelectrolyte causes an asymmetry in the Coulomb field about the ions near the nonelectrolyte and induces pairing. From Seward's data[44] on tetrabutylammonium picrate with butanol added, the amount of ion pair formation per molecule of added nonelectrolyte decreases with increasing dielectric constant of the bulk nonelectrolyte.

Shortly thereafter Longo et al.[45] published a note showing that the addition of nitrobenzene to this same salt caused a maximum to occur in the Walden product of the mixture. Since these data would imply greater dissociation in solution than in the salt, a different explanation was sought.

Recently Lind and Sageman,[46] using the friction formalism of Section 2.4, examined a number of systems near the limit of the pure salt. The formalism of Laity for 1–1 electrolytes yields[31]

$$(\Lambda/e)^{-1} = \xi_{+-}C_- + [(\xi_{+0}\xi_{-0}/(\xi_{+0} + \xi_{-0})]C_0 \qquad (14)$$

$$(D_+/kT)^{-1} = \xi_{+-}C_- + \xi_{++}C_+ + \xi_{+0}C_0 \qquad (15)$$

where ξ is the friction constant, C is the concentration of ions or molecules per cubic centimeter, and the subscripts minus, plus, and zero refer to anion, cation, and nonelectrolyte. Since no self-diffusion data are available, only the conductance can be analyzed. The second term on the right-hand side of Eq. (14) is the drag caused by the nonelectrolyte molecules and includes the classical electrophoretic effect. Since this term cannot be measured independently, limits were set of $\xi_{+0} = \xi_{-0} = 0$ and $\xi_{+0} = \xi_{-0} = \xi_{+-}$. With these assumptions ξ_{+-} can be calculated. The difference between no nonelectrolyte-ion friction and equating it to the anion–cation friction produces little ambiguity in the results. An explanation is now sought for the variation of ξ_{+-}.

Figure 5 shows a plot of the derivative of the friction constant with respect to mole fraction of nonelectrolyte against the ratio of the volume of the nonelectrolyte to the volume of the larger ion of the salt. The assumption of negligible ion–nonelectrolyte friction was used. If all friction constants were made equal, the derivatives would increase by about 0.5. If the volume of the nonelectrolyte is larger than the larger ion of the salt, this derivative is positive. If the volume is smaller, the derivative is negative. The graph appears more consistent when the denominator in the absicca is the volume of the larger ion rather than the total volume of the salt. Thus if the nonelectrolyte is of the same size as the larger ion, the anion–cation friction is unaltered. If the size of the nonelectrolyte differs from that of the larger ion, it appears that the anion–cation radial distribution function is unaltered. This statement is supported by the small deviations from ideal mixing of hard spheres. For the latter we see that the distribution function is only slightly perturbed upon mixing. Since zeroth-order theories of transport only depend on the radial distribution function and the pair potential, neither of which vary upon addition of a nonelectrolyte, the explanation lies outside the zeroth-order theory. A heuristic explanation can be given, however. The zeroth-order theories do not account for multibody correlations and the coupling of two-body correlation functions. It is just here that size may be very important. If the nonelectrolyte is larger than either ion, then several ions must move cooperatively to permit motion of the

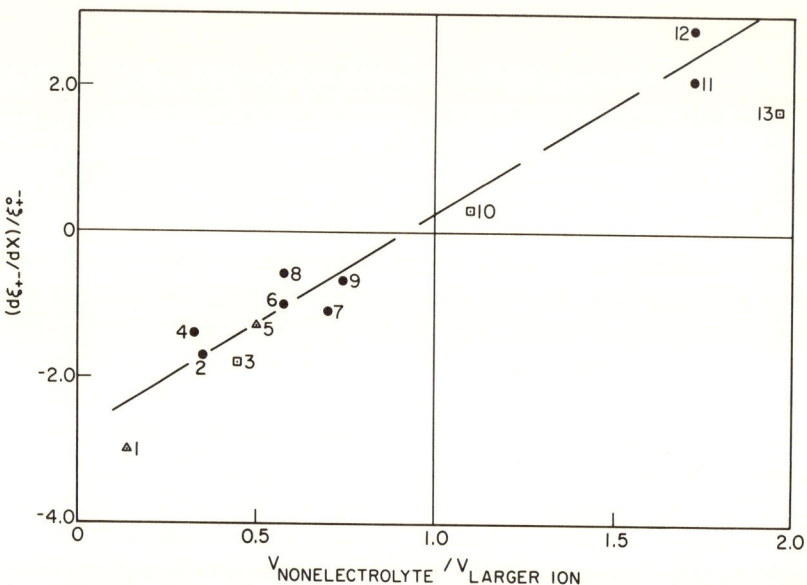

Fig. 5. Effect of added nonelectrolyte on anion–cation friction near the limit of a pure fused salt. Systems: 1. (i-C_5H_{11})$_4$NBF$_4$–CH$_2$OHCH$_2$OH (114.7°).[45] 2. (C$_4$H$_9$)$_4$NPi–C$_4$H$_9$OH (91°).[44] 3. LiClO$_3$–H$_2$O (131.8°).[50] 4. (n-C$_5$H$_{11}$)$_4$NSCN–xylene (92°).[43] 5. (i-C$_5$H$_{11}$)$_4$NBF$_4$–p-diisopropylbenzene (114.7°).[45] 6. N$_2$H$_5$Cl–H$_2$O (95°).[51] 7. NH$_4$NO$_3$–H$_2$O (180°).[52] 8. N$_2$H$_5$NO$_3$–H$_2$O (95°).[51] 9. AgNO$_3$–H$_2$O (221.7°).[52] 10. LiClO$_3$–H$_2$O (131.8°).[50] 11. TlNO$_3$–HgCl$_2$ (282°).[53] 12. TlNO$_3$–HgCl$_2$ (182°).[50] 13. LiClO$_3$–C$_3$H$_7$OH (131.8°).[50]

nonelectrolyte. This phenomenon may then be coupled to the anion–cation friction. If the nonelectrolyte is much smaller than the ions, it can move about easily in the melt and may, through its rapid motions, decrease the coupling of motions between ions. The above discussion in terms of the size of the larger ion could have been equally well given in terms of the volume of a neutral ion pair. The only reason for choosing the former is the smaller scatter produced in Fig. 5. In this figure the problem of corresponding states raises its head. The available experimental data are usually limited to one arbitrary temperature, and this leads to some of the scatter. The variation with temperature can be seen for the TlNO$_3$–HgCl$_2$ system.

Because these variations in the anion–cation friction arise from correlations for which the present theory is inadequate, conductance measurements may prove a useful means for probing these very interesting correlational phenomena.

REFERENCES

1. J. E. Gordon, in: *Techniques and Methods of Organic and Organometallic Chemistry* (D. B. Denney, ed.), Vol. 1, pp. 51–188, Marcel Dekker, New York (1969).
2. V. C. Reinsborough, *Rev. Pure Appl. Chem.* **18**:281 (1968).
3. T. G. Coker, B. Wunderlich, and G. J. Janz, *Trans. Faraday Soc.* **65**:3361 (1969).
4. T. G. Coker, J. Ambrose, and G. J. Janz, *J. Am. Chem. Soc.* **92**:5293 (1970).
5. J. Levkov, W. Kohr, and R. A. Mackay, *J. Phys. Chem.* **75**:2066 (1971).
6. F. J. Hazlewood, E. Rhodes, and A. R. Ubbelohde, *Trans. Faraday Soc.* **62**:310 (1966).
7. T. Grindley and J. E. Lind, Jr., *J. Chem. Phys.*, **56**:3602 (1972).
8. E. Waisman and J. L. Lebowitz, *J. Chem. Phys.* **52**:4307 (1970).
9. D. Chandler and H. C. Andersen, *J. Chem. Phys.* **54**:26 (1971).
10. J. C. Rasaiah and H. L. Friedman, *J. Chem. Phys.* **48**:2742 (1968); **50**:2965 (1969).
11. F. H. Stillinger, Jr., in: *Molten Salt Chemistry* (M. Blander, ed.), pp. 1–108, Interscience Publishers, New York (1964).
12. P. Walden, *Bull. Acad. Sci. (St. Petersburg)* **8**:405 (1914).
13. A. Kisza and J. Hawranek, *Z. physik. Chem. (Leipzig)* **237**:210 (1968).
14. P. Walden, *Z. physik. Chem.* **157**:389 (1931).
15. H. Bloom and V. C. Reinsborough, *Aust. J. Chem.* **21**:1525 (1968); **22**:519 (1969).
16. R. N. Coleman and E. B. R. Prideaux, *J. Chem. Soc.* **1936**:1350; **1937**:1022.
17. H. Reiss, in: *Advances in Chemical Physics* (I. Prigogine, ed.), Vol. 9, pp. 1–84, Interscience Publishers, New York (1965).
18. S. W. Mayer, *J. Phys. Chem.* **67**:2160 (1963).
19. P. Walden and E. J. Birr, *Z. physik. Chem.* **160**:45 (1932).
20. P. Walden, H. Ulrich, and E. J. Birr, *Z. physik. Chem.* **130**:494 (1927); **131**:1, 21, 31 (1928).
21. P. Walden and E. J. Birr, *Z. physik. Chem.* **160**:57 (1932).
22. P. Walden and E. J. Birr, *Z. physik. Chem.* **160**:61 (1932).
23. H. T. Davis and J. McDonald, *J. Chem. Phys.* **48**:1644 (1968).
24. I. Prigogine and V. Mathol, *J. Chem. Phys.* **20**:49 (1952).
25. H. Bloom and E. Heymann, *Proc. Roy. Soc.* **A188**:392 (1947); H. Bloom, *Rev. Pure Appl. Chem.* **7**:389 (1963).
26. G. J. Janz, R. D. Reeves, and A. T. Ward, *Nature* **204**:1188 (1964).
27. J. E. Lind, Jr., H. A. A. Abdel-Rehim, and S. W. Rudich, *J. Phys. Chem.* **70**:3610 (1966).
28. C. A. Angell and C. T. Moynihan, in: *Molten Salts, Characterization and Analysis* (G. Mamantov, ed.), pp. 315–375, Marcel Dekker, New York (1969).
29. A. J. Easteal and C. A. Angell, *Aust. J. Chem.* **23**:929 (1970).
30. R. W. Laity, *J. Chem. Phys.* **30**:682 (1959).
31. G. Morrison and J. E. Lind, Jr., *J. Phys. Chem.* **72**:3001 (1968).
32. E. Helfand, *Phys. Fluids* **4**:681 (1961).
33. H. C. Longuet-Higgins and J. A. People, *J. Chem. Phys.* **25**:884 (1956).
34. E. Helfand and S. A. Rice, *J. Chem. Phys.* **32**:1642 (1960).
35. S. W. Rudich and J. E. Lind, Jr., *J. Chem. Phys.* **50**:3035 (1969).
36. A. F. M. Barton and R. J. Speedy, *Chem. Comm.* **1969**:1197.
37. J. W. Tomlinson, *Rev. Pure Appl. Chem.* **18**:187 (1968).
38. D. C. Douglass, D. W. McCall, and E. W. Anderson, *J. Chem. Phys.* **34**: 152 (1961).
39. B. J. Alder, D. M. Gass, and T. E. Wainwright, *J. Chem. Phys.* **53**:3813 (1970).

40. L. V. Woodcock and K. Singer, *Trans. Faraday Soc.* **67**:12 (1971).
41. S. W. Rudich and J. E. Lind, Jr., *J. Phys. Chem.* **73**:2099 (1969).
42. G. Morrison and J. E. Lind, Jr., *J. Chem. Phys.* **49**:5310 (1968).
43. L. C. Kenausis, E. C. Evers, and C. A. Kraus, *Proc. Natl. Acad. Sci.* **48**:121 (1962); **49**:14 (1963).
44. R. P. Seward, *J. Am. Chem. Soc.* **73**:515 (1951).
45. F. R. Longo, P. H. Daum, R. Chapman, and W. G. Thomas, *J. Phys. Chem.* **71**:2755 (1967).
46. J. E. Lind, Jr., and D. R. Sageman, *J. Phys. Chem.* **74**:3269 (1970).
47. G. J. Janz, *Molten Salts Handbook*, Academic Press, New York (1967).
48. N. N. Greenwood and R. L. Martin, *J. Chem. Soc.* **1953**:1431.
49. K. Gatner and A. Kisza, *Z. physik. Chem. (Leipzig)* **241**:1 (1969).
50. A. N. Campbell and D. F. Williams, *Can. J. Chem.* **42**:1984 (1964).
51. R. P. Seward, *J. Am. Chem. Soc.* **77**:905 (1955).
52. A. N. Campbell, E. M. Kartzmark, M. E. Bednas, and J. T. Herron, *Can J. Chem.* **32**:1051 (1954).
53. G. J. Janz, A. Timidei, and F. W. Dampier, *Electrochimica Acta* **15**:609 (1970).

Chapter 2

SOLUBILITIES OF REACTIVE GASES IN MOLTEN SALTS

S. N. Flengas and A. Block-Bolten

Department of Metallurgy and Materials Science
University of Toronto
Toronto, Ontario, Canada

1. INTRODUCTION

It is well known that gases are soluble in molten salts at high temperatures. Solubilities are usually relatively high for the reactive gases and become less with the inert and noble gases.

It is also possible to distinguish between reactive and inert gas solubility by their temperature dependence. For the former, in general, solubilities decrease with increasing temperature, while for the latter the solubilities increase.

Thus gaseous solubility in molten salts, by analogy with the aqueous solutions, should be best described as "chemical" or "physical." In some systems both types may be present.

Physical solubility appears to arise from the concept that a molten salt structure contains voids or so-called "holes." Reiss et al.,[1] applying concepts from fluid mechanics, calculated an expression for the work necessary to create a spherical cavity in a real fluid. This work was then taken to represent the energy for the dissolution of a gas molecule in a liquid. Solubilities of noble gases, such as helium in benzene, appeared to satisfy the predictions from the model.

Uhlig[2] considered a two-step solubility process. At first a cavity is created at the expense of work done against the macroscopic surface tension of the solvent and then a gaseous molecule is fitted into the cavity with an additional energy term representing the interactions of the gas with the surrounding solvent molecules. The model is applicable to the dissolution of inert gases in aqueous and organic solvents.

Blander et al.[3] investigated solubilities of the noble gases He, Ne, and Ar in the eutectic mixture LiF–NaF–KF and interpreted the observed solubility in terms of a cavity model similar to that of Uhlig.[2] According to the authors, if the gas does not interact with liquid, the free energy change upon mixing is related to the surface energy of the "hole" created by the gas. The cavities are considered to be of molecular or atomic size and the surface tension of the solvent along the walls of the cavity is taken to be equal to the macroscopic value. In spite of the many assumptions, the equations appear to predict the observed solubilities. The solubility relationship given by these authors is

$$-RT \ln(C_d/C_g) = KA\gamma \qquad (1)$$

where C_g is the concentration of gas in the gas phase in equilibrium with the gas in the liquid at a concentration C_d, and A is the total surface area of the spherical cavity. For spherical atoms $KA = 18.08r^2$, where r is the atomic radius in angstroms.

It should be realized that the "hole theory" assumes the existence of atomic or molecular size cavities in a liquid which otherwise has the properties of a free surface.

A simplified version of a "hole theory" proposed by Altar[4] and Fürth[5] has also been applied to molten salts by Bockris and co-workers[6] without too much success.

However, the concept of a "void" where such voids are not limited to monatomic type vacancies appears to be fruitful and is supported by measurements of volume changes on melting for several pure chloride salts[7-9] and also by studies of radial distribution functions by X-ray and neutron diffraction.[10] For example, it has been shown that the volume increase which cannot be accounted for during melting of NaCl is 23% and that of AgCl is 5.7%.[11] There are salts, however, where melting is accompanied by a decrease in volume, such as $Rb(NO_3)$ and CH_3COOK.[12] The latter salts are characterized by large anions which probably dominate the volume of the phase and create large interstitial-like positions. Melting in this case may be thought of as a decomplexing reaction where larger ionic assemblies are broken down into smaller ionic species of better packing capacity. Where multiatomic voids exist they may be thought to be convenient sites for the physical dissolution of gases. As with any type of vacancy, such voids are expected to represent distortions in the symmetry of the electrostatic fields of the ions surrounding a void and therefore the voids themselves should be expected to be polarized and behave like electrified cavities. Within this electrified space the dissolution of a neutral gaseous atom or molecule should not be expected to proceed without some type of interaction, probably of the induced-polarization type. In this sense physical dissolution of a gas may also be described as the effect of the mutual interaction between polarized gaseous atoms or molecules and the electrified cavities.

Since the number of voids is probably increasing with temperature, the solubility of the gas is also expected to increase accordingly. Predictions of solubilities in molten salts based on the scaled particle theory and the theory of corresponding states were published by Lee and Johnson.[13]

Copeland et al.[14-21] made the further assumption that the dissolved gas creates its own "holes" and that the originally existing voids remain virtually unfilled unless high pressures are applied.

Ubbelohde[12] points out that the interaction of ionic melts with nonionic molecules must first occur at the surface of the melt. Further insertion of large molecules into the ionic melt is expected to increase the total volume and thus a large energy input for the separation of the ions is required. If the molecules inserted are polarizable, especially if they behave like dipoles, part of this energy will be recovered. But the overall process will be endothermic, indicating that the solubility of large nonionic molecules will be small and increase with increasing temperature.

Systems which may be described by "physical" solubility are the solutions of the noble and the so-called "inert gases" in molten salts. It should be emphasized that a given gas which is inert with respect to a given melt is not necessarily inert toward another system. For example, the solubility of CO_2 in molten alkali halides,[22-24] sodium borate melts,[25] or soda-silica melts[26] demonstrates this behavior.

Physical solubilities are usually low in magnitude and are accompanied by endothermic partial molar heats of mixing. However, the overall heat effect for dissolution could still be exothermic because of the predominance of the heat contribution due to condensation.

Chemical solubilities of "reactive gases," such as the halogens,[27-29] CO_2, SO_2, SO_3, HF, and HCl, as well as the volatile chlorides of the transition metals, such as $TiCl_4$, $ZrCl_4$, $HfCl_4$,[30] and $NbCl_5$, may be explained by chemical reactions leading to the formation of one or more complex ionic species which are taken to exist in the molten phase and whose concentrations are determined by an equilibrium constant. In this case, solubility is defined by the proper equilibrium conditions with respect to both the pressure of the reactive vapor and the activities of the species in solution. All concentrations, from the pure liquid salt phase to the corresponding pure gas, taken in its form as a pure compressed liquid, may be achieved.

Partial molar heat effects accompanying this type of solubility are usually exothermic and the excess entropies of mixing are expected to be negative.

Thus the study of gaseous solubilities in fused salts is of interest because it is closely related to the structure of liquids. In addition, from the practical viewpoint the solubilities of reactive gases are of interest for understanding corrosion reactions, for the purpose of purifying melts by degassing techniques, and for the preparation of electrolytic solutions by dissolving the vapor $TiCl_4$, $ZrCl_4$ in alkali halide and alkaline earth halide type melts.

A recent important development is related to the realization of chemical reactions between gases using a molten salt phase as the reaction medium or as a catalyst. For example, vinyl chloride has been prepared from gaseous ethylene in a melt containing cuprous chloride and potassium chloride.[31–33]

The present review article shall be concerned with the treatment of solubility and of chemical syntheses in reactive systems which present more interest for industrial applications. The solubility of reactive gases in fused salts has not been reviewed recently except in short articles on solubilities by Blander,[34] Battino and Clever,[35] Bloom and Hastie,[36] and Janz.[37]

1.1. Thermodynamic Treatment of Solubility

The solubilities of gases in liquids may be expressed in various ways. In the definition of the Bunsen coefficient the solubility is expressed as the volume of the ideal gas reduced to 0°C and 1 atm pressure soluble in a unit volume of liquid under the gas pressure of 1 atm and at the temperature of measurement. The Ostwald solubility $k_1 = V_g/V_{solv}$ is defined as the ratio of the volume of absorbed gas to the volume of the absorbing liquid, measured at the same temperature. The Bunsen solubility coefficient α and the Ostwald coefficient L on the assumption of ideal gas behavior are related to the simple g-moles/liter concentration scale c by the relationships which are applicable when $P = 1$ atm:

$$\alpha = 22.414c$$
$$L = (T_g/273.16)\alpha = RT_g c$$

Solubilities are also expressed as mole fractions, as gram-moles of gas per gram-mole of liquid, or as weight percent.

Considering the solubility of a simple gaseous molecule Y in a molten salt, the equilibrium may be described by the equation

$$Y(\text{in solution at } P, T) \rightleftharpoons Y(\text{gas at } P, T) \tag{2}$$

For a rigorous thermodynamic treatment it is necessary to choose the reference state of the gas as the pure gas at unit fugacity and any temperature T, and for the gas in solution its state as a pure liquified gas A under atmospheric pressure and any temperature T. This method may be followed for systems which may be liquified at high temperatures and low pressures.

For reaction (2) at equilibrium the following relationship is valid:

$$G^\circ_{Y(\text{gas}; 1\text{ atm}, T)} + RT \ln P_Y = G^\circ_{Y(\text{liquid}; 1\text{ atm}, T)} + RT \ln a_Y \tag{3}$$

where a_Y is the activity of dissolved gas. It follows that

$$\Delta G_V^\circ = RT \ln(a_Y/P_Y) \tag{4}$$

where the G°'s refer to the components in the chosen standard states. The pressure dependence of the free energy of the liquid has been ignored in this calculation as negligibly small. Accordingly, for a solubility experiment conducted with $P_Y = 1$ atm,

$$d(\ln a_Y)/d(1/T) = \Delta H_V^\circ/R \tag{5}$$

where ΔH_V° and ΔG_V° are, respectively, the enthalpy and the free energy of vaporization of liquid Y at any temperature. Using the mole fraction as the concentration unit and assuming that in the dilute solutions the activity coefficients are independent of composition and depend only on temperature (Henrian behavior), this equation may be written as

$$d(\ln N_Y)/d(1/T) = (\Delta H_V^\circ - \overline{\Delta H_Y})/R \tag{6}$$

where $\overline{\Delta H_Y}$ is the limiting enthalpy of mixing of the liquified gas Y. Hence the plot of $\ln N_Y$ versus $1/T$ should yield a curve having a slope equal to $\Delta H_T/R$, where ΔH_T is the total enthalpy obtained by this method, given as

$$\Delta H_T = \Delta H_V^\circ - \overline{\Delta H_Y} \tag{7}$$

ΔH_T also represents the enthalpy change during removal of 1 g-mole of dissolved gas from the solution.

To calculate the partial molal enthalpies of mixing $\overline{\Delta H_Y}$, the enthalpy of vaporization at each temperature should be obtained from tabulated data and then substracted from ΔH_T. In the case of reactive gases the enthalpies of mixing are in general exothermic.

Considering the mixing reaction to proceed in two steps,

$$Y(\text{pure liquid}; 1 \text{ atm}, T) \rightarrow Y(\text{gas}; 1 \text{ atm}, T) \qquad \Delta S = \Delta S_V$$
$$Y(\text{in solution}; 1 \text{ atm}, T) \rightarrow Y(\text{pure liquid}; 1 \text{ atm}, T) \qquad \Delta S = -\overline{\Delta S_Y} \tag{8}$$

the total entropy change for the process given by reaction (2) is

$$\Delta S_T = \Delta S_V^\circ - \overline{\Delta S_Y} \tag{9}$$

where ΔS_V° is the standard entropy of vaporization of Y at the temperature T and $\overline{\Delta S_Y}$ is the partial molar entropy of mixing of liquefied gas. If reaction (2) is conducted with the gas at P atmospheres, then the expression for the total entropy becomes

$$\Delta S_T = \Delta S_V^\circ - R \ln P - \overline{\Delta S_Y} \tag{10}$$

In general exothermic heats and negative partial molar excess entropies of solution indicate that the solution process involves strong interactions between Y and the ions in the melt.[38,39]

For gaseous systems which liquefy at low temperatures and under high pressures the foregoing choice of standard states is no longer practical as it involves long extrapolations in the calculation of the enthalpies and entropies of vaporization.

Accordingly, the treatment of solubility chosen by most authors is based on the following two choices of reference states.

In the first choice the reference state for the gas Y in the gas phase is chosen to be the hypothetical state of Y at a concentration of 1 g-mole/ml of gas and any temperature T. From the ideal gas law it follows that this state can only be realized if the gas is pressurized to a pressure P_Y' given by the expression

$$P_Y' = P_Y / 1000 C_g \tag{11}$$

where P_Y is the actual pressure of Y in reaction (2) and C_g is the concentration of gas in the gas phase given as gram-moles per milliliter.

For Y in solution the concentration scale is similarly given as gram-moles of Y dissolved per milliliter of solution (C_d) and therefore the reference state for the dissolved gas is a solution containing Y at a concentration of 1 g-mole/ml. In the second choice of reference states Y in the gas phase is taken with respect to pure gas Y at 1 atm pressure and any temperature T as the reference state while that of the dissolved gas remains the same as above.

Depending upon the choice of frame of reference, the equilibrium constant for the solubility reaction (2) is either

$$K_C = C_d/C_g \tag{12}$$

or

$$K_P = C_d/P_Y \tag{13}$$

From the relationship

$$P_Y = C_g 1000 RT \tag{14}$$

it follows that

$$K_C/K_P = 1000 RT \tag{15}$$

From Eq. (15) and the well-known van't Hoff relationship,

$$d(\ln K_C)/d(1/T) = -\Delta H_C^\circ/R \tag{16}$$

and

$$d(\ln K_P)/d(1/T) = -\Delta H_P^\circ/R \tag{17}$$

From Eq. (15)–(17), it is readily shown that

$$\Delta H_C^\circ = \Delta H_P^\circ + RT \tag{18}$$

where ΔH_P° and ΔH_C°, respectively, are the enthalpies derived from the corresponding plots of $\ln K_C$ and $\ln K_P$ versus $1/T$.

The temperature dependence of solubility must also obey the Le Châtelier principle. Thus if the solution process is exothermic, the solubilities are expected to decrease at higher temperatures. If, however, the dissolution of the gas is accompanied by the absorption of heat, then higher temperatures should promote increased solubilities.

Enthalpies of mixing ΔH_C° and ΔH_P° obtained through Eq. (16) and (17), respectively, refer to the solubility process and therefore have signs which are opposite to the enthalpy ΔH_T which is applicable to gas evolution.

The relationship between ΔH_C° or ΔH_P° and the true partial molar property $\overline{\Delta H_Y}$ derived by Eq. (7) may be readily obtained from a consideration of the reference states involved in each calculation. For example, $\overline{\Delta H_Y}$ is defined as

$$\overline{\Delta H_Y} = \bar{H}_Y - H_{Y(\text{pure liquid}; P=1\,\text{atm}, T)}^\circ \tag{19}$$

and

$$\Delta H_P^\circ = \bar{H}_Y - H_{Y(\text{pure gas}, P=1\,\text{atm})}^\circ \tag{20}$$

Because the limiting partial molar enthalpy content $\bar{H}_Y$ is independent of the choice of standard states from Eq. (18)–(20), it is readily shown that

$$\overline{\Delta H}_Y = \Delta H_P^\circ + \Delta H_V^\circ \tag{21}$$

or

$$\overline{\Delta H}_Y = \Delta H_C^\circ + RT + \Delta H_V^\circ \tag{22}$$

where $R = 1.987$ cal deg abs./mole.

Comparing Eq. (7) and (21), it also follows that

$$\Delta H_T = -\Delta H_P^\circ \tag{23}$$

$$\Delta H_P^\circ = -\Delta H_V^\circ + \overline{\Delta H}_Y \tag{24}$$

These considerations should be taken into account when one attempts to compare heats of mixing reported by various authors as given in Table III. So-called "entropies of mixing" reported by various authors are even more complicated. For example, the entropy defined from the plot of $\ln K_P$ versus $1/T$ has the significance of

$$\Delta S_P^\circ = -\Delta S_T \tag{25}$$

where ΔS_T is defined in Eq. (10). Entropy changes due to the pressure effect on the liquid have been neglected.

Equation (24) offers an interesting insight into the factors which determine the endothermic or exothermic nature of gaseous solubility. Thus for a nonreactive gas $\overline{\Delta H}_Y$ is expected to have positive values and the mixing process can still be exothermic if $\Delta H_V^\circ > \overline{\Delta H}_Y$. For reactive gases $\overline{\Delta H}_Y$ is always negative and the overall heat of solution ΔH_P° is expected to have large negative values.

1.2. Reactive Systems

The solutions of the volatile transition-metal chlorides in alkali metal chlorides may be considered typical of the reaction-type solubility.

Systems of this kind are represented by the solutions of volatile chlorides $TiCl_4$ (b.p. 136.4°C), $ZrCl_4$ (subl. p. 331°C), $HfCl_4$ (subl. p. 317°C), etc. The phase diagrams[30,40–50] of these systems indicate the presence of congruently melting compounds of the general formula M_2XCl_6, where M and X are an alkaline and a transition metal, respectively.

The concept of complex species in molten salts is significantly different from that in aqueous solutions. In dilute aqueous systems complex ionic species of the type $A_n B_m^{2-}$ are surrounded by the dielectric water solvent molecules and essentially do not interact with each other. The local electrical neutrality requirement is satisfied by ion–dipole type interaction and by the resulting solution effects. In an ionic melt the electrical neutrality requirement is satisfied by the existence of the positive and negative quasi-lattices and for a given ion the nearest neighbors are always ions of opposite sign. The overall balance of the attractive and repulsive forces implies that anions and cations are characterized by an average nearest-neighbor coordination number which is proportional to the valence of the species present. Thus, in a melt containing excess MCl and the complex M_2XCl_6 species it is convenient for the purpose of thermodynamic calculations to describe the composition in terms of "free" M^+, Cl^-, and XCl_6^{2-} ions, where X = Ti, Zr, Hf. However, in view of the continuity of the medium, it is also evident that the XCl_6^{2-} configurations in an environment of M^+ and Cl^- ions cannot occupy separate lattice sites and represent a continuous three-dimensional ionic network within which the X^{4+} species are always surrounded by Cl^- anions which in turn are interacting with the M^+ cations in the order

$$M^+, (Cl^-, X^{4+}, Cl^-), M^+$$

The interaction of X^{4+} species with other cations is expected to occur through bridging Cl^- ions. In this sense, the complex XCl_6^{2-} species may be described by X–Cl bond distances which should be shorter than the X–Cl bond distances in a hypothetical totally dissociated ionic melt.

It is evident that within the continuous ionic network configurations which are characterized by specific bond distances are statistical entities and may be described as discrete complex species.

The compounds K_2TiCl_6,[51] Rb_2TiCl_6,[52] Cs_2TiCl_6,[45–52] Li_2ZrCl_6,[53] Na_2ZrCl_6,[54] K_2ZrCl_6,[54] Rb_2ZrCl_6,[53] Cs_2ZrCl_6,[44] $BaZrCl_6$,[53] $SrZrCl_6$,[53] Li_2HfCl_6,[53] and Na_2HfCl_6[44] have been prepared in the pure state from reactions involving the gaseous XCl_4 vapor and the finely divided solid alkali chloride salt.

A typical phase diagram representing the MCl–XCl_4 systems is depicted in Fig. 1(a). The phases indicated in the MCl–XCl_4 type systems can only exist under the equilibrium pressure of the XCl_4 reactive vapor and should be investigated in a totally sealed cryoscopic apparatus. The predicted shape of the pressure–temperature curves for various selected compositions is shown in Fig. 1(b). The correspondence between these two diagrams is obtained from the following considerations. For the thermodynamic treatment the system shall be divided into two subsystems: MCl–M_2XCl_6 and M_2XCl_6–XCl_4 (Fig. 1a and 1b).

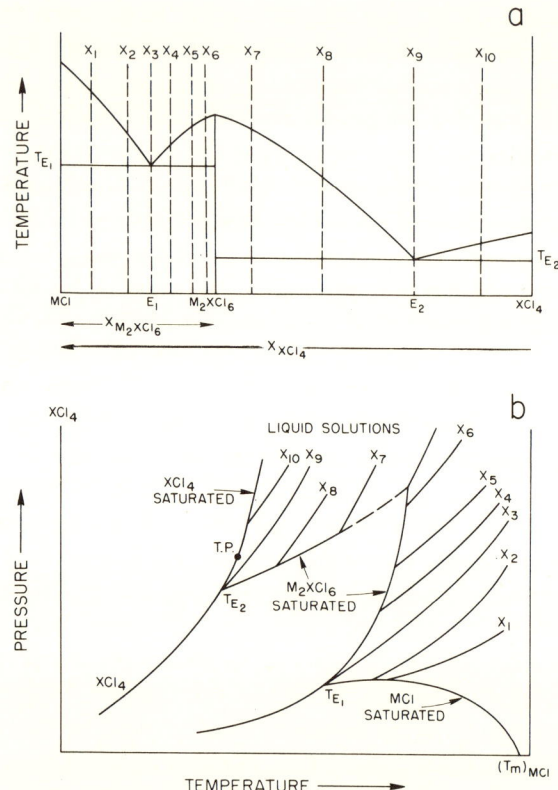

Fig. 1. Predicted reaction pressure versus temperature curves for the systems MCl–XCl_4, where M is an alkaline earth metal and X is one of the reactive metals Ti, Zr, or Hf. The analysis is only applicable to a binary system of this type which is described by a simple eutectic-type phase diagram and in the absence of solid solubility.

Within subsystem MCl–M_2XCl_6 the equilibrium pressure of XCl_4 vapor is determined by the reaction

$$M_2XCl_6 \rightleftarrows 2MCl + XCl_4 \qquad (26)$$
$$\text{(solid, liquid,} \quad \text{(solid, liquid,} \quad \text{(vapor)}$$
$$\text{or solution)} \quad \text{or solution)}$$

The equilibrium constant can be written as

$$K_{eq} = a_{MCl}^2 f_{XCl_4} / a_{M_2XCl_6} \qquad (27)$$

where a_{MCl} and $a_{M_2XCl_6}$ are the activities of MCl and M_2XCl_6, respectively, and f_{XCl_4} is the fugacity of XCl_4 vapor. At low pressures and high temperatures the fugacity may be taken as being equal to the pressure of the tetrachloride vapor. The states of unit activity for the condensed phases are selected as the pure solids at all temperatures. For the vapor the standard state is chosen as the metal tetrachloride vapor at 1 atm pressure.

At temperatures below the temperature of the eutectic T_{E_1} the phases present are the pure solids MCl and M_2XCl_6 and XCl_4 vapor. Because of the choice of standard states in this range the activities of a_{MCl} and $a_{M_2XCl_6}$ are unity. Thus

$$K_{eq} = P_{XCl_4} \tag{28}$$

The variation of the equilibrium pressure with temperature is given by the van't Hoff relationship

$$d(\ln P_{XCl_4})/d(1/T) = -\Delta H_R^\circ/R \tag{29}$$

That is, over a limited temperature range the plot of $\ln P_{XCl_4}$ versus $1/T$ (°K^{-1}) should be linear with slope $-\Delta H_R^\circ/R$, where ΔH_R° is the heat for the decomposition reaction

$$M_2XCl_6 \to 2MCl(s) + XCl_4(1 \text{ atm}) \tag{30}$$

Pressure measurements at temperatures below the eutectic temperature are independent of compositon and should be represented by a single curve. Equation (28) is only valid when the pure materials do not form solid solutions. If appreciable solid solubility exists, the equilibrium constant would have to include the activities of the solids.

The second region of interest includes all compositions on the M_2XCl_6 side of the eutectic T_{E_1} restricted by the liquidus tie line. In this region the solution of MCl and M_3XCl are saturated with respect to the solid M_2XCl_6. Accordingly, $a_{M_2XCl_6(\text{solution})} = 1$ and thus

$$P_{XCl_4} = K_{eq}/a_{MCl}^2 \tag{31}$$

For comparison, the equilibrium constant obtained from measurements below the eutectic may be expressed as a function of temperature and then extrapolated to any desired temperature.

From Eq. (31) it may be predicted that the temperature–pressure curve will change slope at the eutectic temperature since as the temperature is increased beyond this point the activity of MCl becomes less than unity. Thus the pressure of the vapor would be expected to increase as shown in Fig. 1(b).

In the range between the eutectic and the liquidus temperatures for a $MCl-M_2XCl_6$ mixture rich in M_2XCl_6, the composition of the solutions in equilibrium with the pure solid changes following the liquid solubility line. Thus the activity of MCl will be changing continuously because of temperature and composition changes. At any specific temperature in this range the pressure of XCl_4 exerted by any two mixtures of initially different composition will be identical up to the temperature of the lower-melting mixtures. The liquidus solubility line will therefore be represented by a common pressure curve. When the temperature is increased beyond the corresponding liquidus temperature the system becomes a homogeneous solution of MCl and M_2XCl_6 in equilibrium with XCl_4 vapor. The equilibrium pressure is now given by

$$P_{XCl_4} = K_{eq} a_{M_2XCl_6} / a_{MCl}^2 \tag{32}$$

Therefore at a temperature $T = T_L$ (liquidus) the pressure curve will undergo a decrease of slope because of the introduction of the activity of M_2XCl_6, which is now less than unity, as shown in Eq. (32). This is illustrated by curves 3–6 in Fig. 1(b), where each of these curves represents a different composition.

Considering compositions in the MCl saturated side of the eutectic at $T_L > T > T_{E_1}$, where T_L is a liquidus temperature, the phases present are pure solid MCl in equilibrium with a solution of M_2XCl_6 and MCl, and XCl_4 vapor.

In this range, as defined before,

$$a_{MCl} = 1, \qquad P_{XCl_4} = K_{eq} a_{M_2XCl_6} \tag{33}$$

For a given mixture, increasing the temperature above the eutectic would cause a decrease in the slope of the pressure curve. The predicted shape of the pressure curve representing the liquidus tie line in the phase diagram is again given in Fig. 1(b). The pressures of XCl_4 vapor in equilibrium with the saturated solutions decrease as the solutions become more dilute in M_2XCl_6 and eventually reach zero value at the melting point of pure MCl, as shown in Fig. 1(b).

For temperatures higher than the liquidus, Eq.(32) becomes again applicable and the pressure curves for various compositions slope upward at the corresponding liquidus temperatures, as shown by curves 1–3 in Fig. 1(b).

The pressure–temperature diagram for the $M_2XCl_6-XCl_4$ subsystem can be predicted using different considerations. At temperatures below the eutectic temperature T_{E_2} the system consists of the two immiscible pure solids M_2XCl_6 and XCl_4. However, as the XCl_4 pure solid or liquid ($TiCl_4$, $ZrCl_4$, or $HfCl_4$) usually has a sublimation pressure which is several orders of magnitude greater[55,56] than the XCl_4 vapor pressure produced by the thermal decomposition of the compound M_2XCl_6, it follows that the decomposition reaction of M_2XCl_6 is totally suppressed.

The equilibrium pressure P_{XCl_4} over the system is entirely due to the sublimation of XCl_4, as shown by the single curve in the PT diagram in Fig. 1(b).

Considering compositions representing the XCl_4 side of the eutectic E_2, as the temperature becomes greater than the eutectic temperature T_{E_2}, the phases present include solid XCl_4 in equilibrium with a solution of M_2XCl_6 and XCl_4, saturated with respect to the latter. Hence the pressure over the system is still determined by the pressure of pure XCl_4. All compositions on this side of the eutectic will be characterized by a common pressure curve which should have only one point of inflection at the triple point of solid XCl_4. The change of slope should then only represent the change from pure solid to pure liquid.

For temperatures higher than the liquidus, all XCl_4 has now reacted to form a homogeneous solution and the equilibrium pressure should be given by

$$P_{XCl_4} = a_{XCl_4} P^\circ_{XCl_4} \tag{34}$$

where a_{XCl_4} is the activity of XCl_4 in this solution. Since $a_{XCl_4} = 1$, the pressure curve representing the solutions should have a lower slope and intersect the sublimation curve for XCl_4 as shown by curves 10 and 9.

For mixtures representing the M_2XCl_6 side of the eutectic E_2 the pressure curve follows the previously described common path up to the eutectic temperature. At higher temperatures the curves follow the common curve describing the M_2XCl_6 saturated solutions in which the XCl_4 component predominates, and change slope upward at the corresponding liquidus temperature, as shown in Fig. 1(b). Equation (24) also describes the all-liquid range.

The curves representing solutions saturated with M_2XCl_6 in the two subsystems should intersect at the melting temperature of pure M_2XCl_6, as shown in Fig. 1(b).

From this discussion it is evident that the pressure–temperature relationships in reactive systems of this type are complicated and are described by a family of curves which inflect at the characteristic temperatures of the phase diagram.

The problem of solubility of the reactive vapors $TiCl_4$, $ZrCl_4$, and $HfCl_4$ in molten alkali and alkaline earth chlorides is directly related to the phase diagrams of the binary systems MCl–XCl_4 and any desired solubility may be obtained, providing that the pressure of the XCl_4 vapor is maintained at the value required by composition and temperature.[57]

The thermodynamic treatment for the solubility of transition metal chloride vapors in molten salts may be readily extended to other reactive systems. For example, the chlorination reactions of metal sulfides at high temperatures may be thought of as a mechanism describing the solubility of Cl_2 in sulfide melts.

Considering the exchange reaction at high temperatures

$$Cl_2(gas) + MS \rightleftarrows MCl_2 + \tfrac{1}{2}S_2(gas) \tag{35}$$

the equilibrium constant is a function of the measured P_{S_2} and P_{Cl_2} pressures only if the reactants MS and MCl_2 are present as pure solids at temperatures below the eutectic point in the binary mixture MS–MCl_2. For these temperatures

$$K_{eq} = P_{S_2}^{1/2}/P_{Cl_2} \tag{36}$$

At temperatures higher than the eutectic temperature

$$(P_{S_2}^{1/2}/P_{Cl_2})_{eq} = K(a_{MS}/a_{MCl_2}) \tag{37}$$

where a_{MS} and a_{MCl_2} are the activities of MS and MCl_2, respectively in the binary MS–MCl_2 mixtures.

The plot of $\log(P_{S_2}^{1/2}/P_{Cl_2})_{eq}$ against $1/T$ should be a function only of temperature at temperatures below the eutectic. At higher temperatures the curves should change slope at the characteristic liquidus temperatures in exactly the same manner as for the XCl_4–M_2XCl_6 subsystems discussed earlier.

The solubility of a halogen gas Y_2 in a fused salt MX may be understood in terms of reactions of the type

$$Y_2 + MY(\text{melt}) \rightleftarrows Y_3^- \tag{38}$$

where Y_3^- is a complex anion.

Although the solubility mechanism is given by a reaction, the MY_3 species cannot be considered as strong complexes of the type described previously. This is indicated by the relatively low solubility in these systems at normal pressures. For example, for the solubility of gaseous Br_2 in molten AgBr[29] the mixing reaction may be written as

$$Br_2(liq) + [(AgBr)_{m-n} \cdot (AgBr_3)_n] \rightarrow$$
$$[(AgBr)_{m-n-1} \cdot (AgBr_3)_{n+1}], \quad m > n \tag{39}$$

Since the partial molar heat of mixing for AgBr in the AgBr concentrated solution should be almost zero, then from Eq. (39) it follows that

$$\overline{\Delta H}_{Br_2} = \overline{\Delta H}_{AgBr_3} + \Delta H_R° \tag{40}$$

where $\Delta H_R°$ is the heat for formation of $AgBr_3$ from pure liquid bromine at 1 atm and molten AgBr.

The formation of a Br_3^- species may be thought of as the result of attachment of the nonpolar bromine molecule to a Br^- ion in the melt by ion-induced dipole type interactions. Such species as Br_3^- are usually described as $Br^-(Br^+ \cdots Br^-)$ triplets. The energy accompanying the ion-induced dipole interaction is expected to be of rather low magnitude, and therefore the term $\Delta H_R°$ should not constitute a significant contribution to the total energy of mixing. The dissolved bromine, which is closely associated with the Br^- ions in the form of the Br_3^- triplet, is not expected to be sufficiently polarized as to occupy distinctive anionic and cationic positions in the liquid sublattices. In this sense the dissolved bromine should occupy the same lattice site as a simple Br^- anion. Such host Br^- ions are shielded from the surrounding Ag^+ cations by the attached bromine molecules and a volume disturbance in the local symmetry of the ionic distributions is expected to occur.

The net result should be that next-nearest-neighbor repulsions of the type

$$Br^--Ag^+-Br^-, \quad Br^--Ag^+-Br_3^-, \quad \text{and} \quad Br_3^--Ag^+-Br_3^-$$

are expected to be significantly different and a pronounced heat effect due to differences in these interactions should be the predominant factor in the heat of mixing.

The structural properties of these solutions may be understood by a quasichemical treatment based on local order and on the predominance of next-nearest-neighbor anion repulsions. Heat effects accompanying the reaction part of the heat of mixing are neglected.

In a previous publication[58] it has been shown that a local order parameter σ may be defined as

$$\sigma = (P_{XY} - N_X N_Y)/N_X N_Y \tag{41}$$

where P_{XY} is the probability of finding an unlike anion pair XY. The fractions of unlike (XY) and like (XX, YY) pairs are given as[59]

$$P_{XY} = \lambda = N_X N_Y (1 + \sigma);$$
$$\varkappa = N_X^2 - N_X N_Y \sigma; \qquad (42)$$
$$\nu = N_Y^2 - N_X N_Y \sigma$$

In these expressions Y and ν refer to the solute species and X and $\varkappa$ to the solvent; N_X and N_Y are the mole fractions. With n_{tot} as the total number of moles and n_Y the number of moles Y, b an interaction parameter of the system, γ the activity coefficient, and γ^H a Henrian activity coefficient, the equation for the partial molar heat of mixing is derived as

$$\Delta \bar{H}_{AY} = [\lambda + n_{tot}(\partial \lambda / \partial n_Y)]RTb \qquad (43)$$

and, for dilute solutions of AY in a solvent AX,

$$\Delta \bar{H}_{AY} = (1 - 2\gamma_{AY}^H N_Y + \lambda_{AY}^H N_Y^2)RTb \qquad (44)$$

The partial molar entropy for dilute solutions becomes

$$\Delta \bar{S}_{AY} = -R \ln \gamma_{AY}^H N_Y + Rb(1 - 2\gamma_{AY}^H N_Y + \gamma_{AY}^H N_Y^2) \qquad (45)$$

and the activity coefficient which is given as

$$\gamma_{AY} = \nu / N_Y^2 \qquad (46)$$

represents the ratio between the real and the ideal numbers of pairs.

The method has been applied[58] in the calculation of the structural properties of the solutions of gaseous Br_2 in molten silver bromide and indicates that the partial molar properties are determined by the predominance of the Br_3^-–Br_3^- repulsions.

2. EXPERIMENTAL TECHNIQUES FOR MEASURING SOLUBILITIES OF GASES IN MOLTEN SALTS

Table I gives a review of experimental techniques used in the measurements of the solubility of reactive gases in molten salts.

Quenching techniques have been widely applied. However, subsequent analysis of such quenched solutions is usually based on assumptions regarding the amount of gas escaping during the cooling period. Pearce[25] measured the solubility of CO_2 in sodium borate melts by this method and analyzed the amount of CO_2 evolved under vacuum. Førland et al.[60]

TABLE I. Reactive Gases in Molten Salts. Experimental Techniques and Solubility Data

Gas[a]	Melt	Ref.	Experimental technique	Solubility data
F_2	$CuCl_2$–KCl (0.75)	75	Weight change, X-ray, chemical analysis	0.75 mole/mole melt, 250°C; reaction solubility
F_2	$NiCl_2$–KCl (0.66)	75		1 mole/mole melt, 275°C; reaction solubility
F_2	$CoCl_2$–KCl (0.75)	75		7/8 mole/mole melt, 325°C; reaction solubility
Cl_2	NaCl	61	Analysis of the halogen content in a quenched sample equilibrated with the halogen	5.63×10^{-5} ml/ml melt or 16×10^{-5} g/g melt at 820°C
Cl_2	KCl	61		23.8×10^{-5} ml/ml melt or 5.03×10^{-5} g/g melt at 820°C
Cl_2	LiCl	61		32.7×10^{-5} ml/ml melt or 6.98×10^{-5} g/g melt at 620°C
Cl_2	$CaCl_2$	61		6.96×10^{-5} ml/ml melt or 2.20×10^{-5} g/g melt at 800°C
Br_2	NaBr	61		26.9×10^{-5} g/g melt at 800°C
Br_2	KBr	61		98.2×10^{-5} g/g melt at 800°C
Br_2	$CaBr_2$	61		18.4×10^{-5} g/g melt at 800°C
I_2	NaI	61		287×10^{-5} g/g melt at 700°C
I_2	KI	61		629×10^{-5} g/g melt at 700°C
Cl_2	LiCl–KCl (eut) + UCl_4 (1–3 wt%)	76	Equilibrium studies, kinetic studies, stripping method, wetted rod contactor	Cl_2 soluble mainly due to $UCl_4 + Cl_2 = UCl_6$ reaction; its equil. constant $\ln K = 3.11 - 3250/T$, 400–700°C
Cl_2	LiCl	77	Stripping method	0.35×10^{-7} mole/ml at 618°C; 1.88×10^{-7} mole/ml at 872°C
Cl_2	NaCl	77		2.31×10^{-7} mole/ml at 840°C; 5.30×10^{-7} at 1031°C
Cl_2	KCl	77		13.3×10^{-7} mole/ml at 823°C; 25.5×10^{-7} at 1026°C
Cl_2	CsCl (RbCl calc.)	77		CsCl: 34.8×10^{-7} mole/ml at 661°C, 47.6×10^{-7} at 922°C; RbCl calc.: 19.5×10^{-7} mole/ml at 750°C, 31×10^{-7} mole/ml at 1000°C

TABLE I (continued)

Gas[a]	Melt	Ref.	Experimental technique	Solubility data
Cl_2	NaCl	62, 78		2.21×10^{-7} mole/ml at 847°C; 7.40×10^{-7} mole/ml at 1025°C
Cl_2	KCl	62, 78		10.40×10^{-7} mole/ml at 848°C; 20.32×10^{-7} at 1038°C
Cl_2	KCl–NaCl (0.50)	62, 78, 79	Stripping method and diffusion studies	6.10^{-7} mole/ml or 3.65×10^{-7} mole/ml at 750°C; 14.10 $\times 10^{-7}$ mole/ml at 1028°C
Cl_2	KCl–$MgCl_2$ (0.50)	62, 78, 79		2.18×10^{-7} mole/ml at 565°C; 10.20×10^{-7} at 1027°C
Cl_2	NaCl–$MgCl_2$ (0.50)	62, 78		0.48×10^{-7} mole/ml at 569°C; 5.82×10^{-7} mole/ml at 1022°C
Cl_2	$MgCl_2$	62, 78		5.72×10^{-7} mole/ml at 785°C; 8.33×10^{-7} at 953°C
Cl_2	$PbCl_2$–KCl (0.23–0.70)	28	Stripping method; ^{36}Cl isotope tracer exchange	0.23 KCl: 1.41×10^{-6} mole/ml atm at 454°C, 5.40×10^{-6} at 635°C; 0.70 KCl: 1.77×10^{-6} at 588°C, 1.40×10^{-6} at 702°C
Cl_2	LiCl	80	Stripping method; pressure drop in a known volume above the melt	$1.5 \pm 0.5 \times 10^{-6}$ mole/ml at 650°C
Cl_2	LiCl–KCl (eut)	81		4.19×10^{-6} mole/mole at 400°C; 7.14×10^{-6} at 550°C
Cl_2	$PbCl_2$	81	Stripping method; dead volume avoided; chronopotentiometric measurements	2.95×10^{-6} mole/mole at 513°C; 6.47×10^{-6} at 639°C
Cl_2	AgCl	81		6.95×10^{-6} mole/mole at 518°C
Br_2	$PbBr_2$	81		92.1×10^{-6} mole/mole at 430°C; 83.2×10^{-6} at 519°C
Br_2	LiBr	81		61.0×10^{-6} mole/mole at 617°C; 37.2×10^{-6} at 709°C
Br_2	KBr–LiBr (eut)	27	High-temperature spectroscopy	10^{-4}–10^{-3} mole/mole melt at 400°C
I_2	KI–LI (eut)	27		10^{-4}–10^{-3} mole/mole melt at 400°C
I_2, Cl_2	KCl–LiCl (eut)	27		I_2 and Cl_2: 10^{-4}–10^{-3} mole/mole melt at 400°C
I_2	($LiNO_3$ (0.43)–KNO_3–KI	39	Spectroscopic analysis of quenched and dissolved samples at room temp.	1.46×10^{-3} mole/kg atm at 450°K

Solubilities of Reactive Gases in Molten Salts

Gas	Salt	Ref.	Method	Data
Br_2	AgBr	29	Thermobalance technique and high-temperature spectroscopy	36.6×10^{-4} mole/mole at 760°K; 14.6×10^{-4} mole/mole at 950°K
HCl	LiCl–KCl (eut)	82	As in Ref. 81	4.11×10^{-5} mole/mole at 490°C; 6.31×10^{-5} mole/mole at 675°C
HCl	NaCl	85	Stripping method	3.8×10^{-5} mole/mole or 10×10^{-7} mole/ml atm at 930°C
HCl	KCl	85	Stripping method	12.2×10^{-5} mole/mole or 24×10^{-7} mole/ml atm at 900°C
HCl	RbCl	85	Stripping method	23.5×10^{-5} mole/mole or 42×10^{-7} mole/ml atm at 830°C
HCl	NaCl–KCl (0.50)	85	Stripping method	9.0×10^{-5} mole/mole or 21×10^{-7} mole/ml atm at 750°C
HCl	NaCl, KCl, RbCl, CsCl	166	Stripping method	NaCl: 36×10^{-6} mole/mole at 878°C, 48×10^{-6} at 1010°C; KCl: 114.8×10^{-6} at 900°C, 144×10^{-6} at 1000°C; RbCl: 101×10^{-6} at 718°C, 206×10^{-6} at 942°C; CsCl: 92×10^{-6} at 664°C, 287×10^{-6} at 885°C
HCl	LiCl (0.50–0.686)–KCl	65	Volumetric technique	Overall ratio of evolved $H_2O/HCl = 4.5$; see $H_2O^{(65)}$
HCl	$SnCl_4$	86	$H^{36}Cl$ tracer technique	6.6×10^{-5} mole/mole mm Hg at 0°C; 5.2×10^{-5} mole/mole mm pressure at 27°C
HCl	KCl	87	Stripping method	0.0105 wt% at 840°C
HCl	NaCl	87	Stripping method	0.0199 wt% at 840°C
HCl	KCl–NaCl	87	Stripping method	0.0158 wt% at 840°C
HCl	$KCl–MgCl_2$	87	Stripping method	0.0775 wt% at 840°C
HCl	$MgCl_2$	87	Stripping method	0.0995 wt% at 840°C
HF	CsF	88	Cooling and warming curves without vapor pressure measurements	Reaction solubility; formation of four compounds
HF	NaF (0.45)–ZrF_4	63	Stripping method applied to a large volume of molten salt	0.78×10^{-5} mole/ml atm at 600°C; 0.51×10^{-5} mole/ml atm at 800°C
HF	NaF(0.195)–ZrF_4	63	Stripping method applied to a large volume of molten salt	12.80×10^{-5} mole/ml at 600°C; 4.43×10^{-5} mole/ml atm at 800°C

TABLE I (*continued*)

Gas[a]	Melt	Ref.	Experimental technique	Solubility data
HF	LiF(0.66)–BeF$_2$	83	Stripping method	5.62×10^{-4} mole HF/mole melt at 1.65 atm and 500°C; 2.48×10^{-4} mole/mole melt at 1.61 atm, 700°C
DF	LiF(0.66)–BeF$_2$	83	Stripping method	4.76×10^{-4} mole DF/mole melt at 1.62 atm and 500°C; 2.15×10^{-4} mole/mole melt at 1.70 atm, 700°C
BF$_3$	LiF–BeF$_2$–ZrF$_4$–ThF$_4$–UF$_4$, reactor fuels	84	Stripping method	0.276 mole/liter atm at 500°C; 0.0346 mole/liter atm at 700°C
H$_2$O	NaNO$_3$, KNO$_3$, KNO$_3$–Ba(NO$_3$)$_2$ (0.94)	71	Cryoscopy	At temp. near m. p., NaNO$_3$: 14.1×10^{-4} mole H$_2$O/mole melt; KNO$_3$: 3.9×10^{-4}; KNO$_3$–Ba(NO$_3$)$_2$: 6.0×10^{-4}
H$_2$O	CsNO$_3$, CsNO$_3$–Ba(NO$_3$)$_2$ (0.433), LiNO$_3$	71	Cryoscopy	At temp. near m. p., CsNO$_3$: 9.2×10^{-4} mole H$_2$O/mole melt; CsNO$_3$–Ba(NO$_3$)$_2$: 6.3×10^{-4}; LiNO$_3$: 13.0×10^{-4} calc. from Ref. 93
H$_2$O	LiNO$_3$	89	Thermobalance technique	265°C: 0.232×10^{-3} mole H$_2$O/mole Torr; 280°C: 0.165×10^{-3}
H$_2$O	LiNO$_3$–KNO$_3$	89	Thermobalance technique	25% KNO$_3$: 230°C: 0.278×10^{-3} mole/mole Torr; 265°C: 0.137×10^{-3}; 75% KNO$_3$: 230°C: 0.099×10^{-3}; 265°C: 0.054×10^{-3}
H$_2$O	LiNO$_3$–NaNO$_3$	89	Thermobalance technique	25% NaNO$_3$: 230°C: 0.354×10^{-3} mole H$_2$O/mole Torr; 265°C: 0.165×10^{-3}; 75% NaNO$_3$: 265°C 0.084×10^{-3}; 280°C: 0.064×10^{-3}
H$_2$O	NaCl	90	Equilibrium study	At 750°C, the K_P for the reaction NaCl + H$_2$O = NaOH + HCl is 1.6×10^{-7}
H$_2$O + HNO$_3$	NH$_4$NO$_3$	91	Cryoscopy	Reaction solubility; freezing pt. depression $\sim$8 deg per 3 mol% (H$_2$O + HNO$_3$)

Solubilities of Reactive Gases in Molten Salts

Gas	Salt	Ref	Method	Notes
H_2O	LiCl (0.50–0.686)–KCl	65	Volumetric technique	0.6 mol% LiCl: 600 mmole H_2O/mole LiCl at 17 mm pressure H_2O and 390°C; 200 mmole at 17 mm and 480°C
H_2O	LiCl–KCl (eut)	92	Polarography and chemical techniques	OH^- content equivalent to 1.0 ml N/400 "acid" per 100 g of melted, HCl treated, vacuum-dried salt.
H_2O	$LiClO_4$	93	Volumetric technique	240°C: $M = 54.9$; 290°C: $M = 20.2$. where M = (moles H_2O/mole salt × mm Hg pressure of H_2O vapor) × 10^6
H_2O	$LiNO_3$–$NaNO_3$–KNO_3	93	Volumetric technique	LiCl(0.3)–NaCl(0.23): $M = 23.8$ at 145°C; $M = 3.2$ at 205°C; LiCl(0.125)–NaCl(0.464): $M = 1.02$ at 240°C; LiCl(0.869)–NaCl(0.069): $M = 49.7$ at 240°C
H_2O	LiCl–NaCl–KCl(0.5), LiCl–KCl (eut)	94	Polarography; standard addition method	Ordinary dry salt flushed with pure Ar: 3×10^{-4} mol% H_2O at 740°C
H_2O	LiCl–KCl	95	Polarography	1.90 g moisture collected from 195.4 g of salt during 94 hr
H_2O	Fused carnalite	96	Polarography	735°C: Limiting current 90 mA at 0.03% moisture, 30 mA at 0.01% moisture, respectively, before and after 20 min HCl bubbling
H_2O	LiF–NaF–KF (eut)	97, 98	Electrochemical studies	OH^- content 1/100 N described as "not unusual."
H_2O	$LiClO_3$	99	Cryoscopy, density, viscosity, surface tension, electrical conductance and their temperature dependence	Solutions up to 0.251 mole H_2O/mole salt studied at temperatures 132.5–159°C
Methanol propyl-alcohol	$LiClO_3$	100	Specific conductance	Solutions up to 0.03 mole methanol/mole salt, studied at 131.6–138.6°C; propyl alcohol solution up to 0.026 mole/mole, studied at 131.8°C
Ethylene glycol, methanol, H_2O	KSCN–NaSCN (3:1)	101	Optical observation cryoscopy	"Very soluble"

TABLE I (continued)

Gas[a]	Melt	Ref.	Experimental technique	Solubility data
SO_2	V_2O_5–$K_2S_2O_7$	102	Kinetic study	Reaction solubility
SO_2	V_2O_5	103	Equilibrium constant measurements by flow method through a contact plug; X-ray of solid phases	Reaction solubility
S_2	LiCl–KCl, LiBr–KBr, KSCN	64	High-temperature spectroscopy	No numerical data available; sulfur dissolves as diatomic molecule
CO_2	Na_3AlF_4–Al_2O_3	24, 60	Chilling method: evolution of dissolved CO_2 upon quenching and subsequent absorption	10 mol% Al_2O_3: about 10^{-6} mole/ml atm; 20%: about 5×10^{-6} mole/ml atm) at 1030°C
CO_2	KF, NaF, $NaNO_3$	104	Chilling method	KF: $\log K$ [mole$\times$ml$^{-1}\times$atm] $= -5.64$ at 950°C; NaF: -5.73 at 1000°C; $NaNO_3$: -5.9 at 322°C; -5.7 at 400°C; possible COF_2 formation
CO_2	RbF, CsF	105	Chilling method	CO_2 solubility much higher than in other alkali fluorides; COF_2 formation
CO_2	Sodium borate melts	25	Quenching technique; evolution of absorbed CO_2 during vacuum fusion	54 wt% Na_2O melt: CO_3^{2-}: 1 wt% at 1100°C; CO_3^{2-}: 3.2 wt% at 950°C; 38.8 wt% Na_2O melt: CO_3^{2-}: 0.0032 wt% at 1100°C, 0.017 wt% at 900°C
CO_2	Soda–silica melts	26		56 wt% Na_2O melt; CO_3^{2-}: 5 wt% Na_2O melt: CO_3^{2-}: 0.001 wt% at 1100°C; CO_3^{2-}: 0.0032 wt% at 900°C

Solubilities of Reactive Gases in Molten Salts

Gas	Salt	Ref.	Method	Results
CO_2	$Na_2O(0.14)$-$CaO(0.29)$-SiO_2	106	$^{14}CO_2$ tracer technique	78×10^{-3} mg CO_2/g glass at 1108°C; 42×10^{-3} mg CO_2/g glass at 1315°C
CO_2	CaO	107	Pressure–temperature studies; thermogravimetry and quenching techniques	Reaction solubility; CaO–$CaCO_3$ eutectic point at 1242°C, 39.5 atm, and 88.3 mol% $CaCO_3$
CO_2	SrO	108, 109		Reaction solubility; SrO–$SrCO_3$ eutectic point at 1270°C, 1.12 atm, and 79.3 mol% $SrCO_3$
CO_2	BaO	110		Reaction solubility; BaO–$BaCO_3$ eutectic point at 1060°C, 0.00661 atm, and 64 mol% $BaCO_3$
CO_2	$NaNO_3$	74	Modified stripping (elution) method	1.8×10^{-7} mole CO_2/ml melt at 1 atm and 315°C; mole/ml melt at 1 atm and 370°C
NH_3	$LiNO_3$–$NaNO_3$–KNO_3 (eut)	111	Volumetric technique	160°C: 2.1×10^{-5} mole NH_3/mole salt mm Hg ammonia pressure; 250°C: 0.7×10^{-5}
NH_3	$LiClO_4$–$KClO_4$ (eut)	111		234°C: 128×10^{-5} mole NH_3/mole salt mm Hg ammonia pressure; 311°C: 23×10^{-5}
NH_3	$LiNO_3$, $LiNO_3$–KNO_3 (0.25–0.57)	111		$LiNO_3$: 260°C: 12.5×10^{-5} mole/mole mm; 315°C: 3.1×10^{-5}; $LiNO_3$–KNO_3(0.57): 160°C: 4.1×10^{-5}; 250°C: 1.05×10^{-5}
$TiCl_4$	CsCl	42	Equilibrium vapor pressure measurements	Reaction solubility
$TiCl_4$	KCl–NaCl	40	Thermobalance technique	Reaction solubility
$TiCl_4$	KCl–LiCl	41	Equilibrium vapor pressure measurements	Reaction solubility
$ZrCl_4$	KCl–K_2ZrCl_6	30	Combination of equilibrium vapor pressure measurements, cryoscopy and cooling curves	Reaction solubility
$ZrCl_4$	NaCl–Na_2ZrCl_6	30		Reaction solubility

TABLE I (continued)

Gas[a]	Melt	Ref.	Experimental technique	Solubility data
$ZrCl_4$	$LiCl-Li_2ZrCl_6$	43	Combination of equilibrium vapor pressure measurements, cryoscopy, and cooling curves	Reaction solubility
$HfCl_4$	$LiCl-Li_2HfCl_6$	43		Reaction solubility
$ZrCl_4$	$NaCl-Na_2ZrCl_6$	44, 124	Combination of equilibrium vapor pressure measurements, cryoscopy, and cooling curves	Reaction solubility
$HfCl_4$	$NaCl-Na_2HfCl_6$	44, 124		Reaction solubility
$ZrCl_4$	NaCl, KCl, CsCl	45, 46	Combination of transpiration method and thermal analysis	Reaction solubility
$HfCl_4$	NaCl, KCl, CsCl	45, 46		Reaction solubility
$ZrCl_4$	NaCl, KCl	47	Combination of dew point (vapor pressure) measurements and cooling curves	Reaction solubility
$HfCl_4$	NaCl	48	Optical crystal-melt equilibrium observations and cooling curves	Reaction solubility
$ZrCl_4$	NaCl, KCl	49	Cooling curves	Reaction solubility
$ZrCl_4$	$NaCl-Na_2ZrCl_6$	50	Transpiration method	Reaction solubility
$NbCl_5$, $ZrCl_4$	$NbCl_5-ZrCl_4-NaCl$, $NbCl_5-NaCl$	112	Cooling curves and vapor pressure measurements	Reaction solubility

Solubilities of Reactive Gases in Molten Salts

Gas	Solvent	Ref	Method	Reaction solubility
I_2	Nb_nI_m	113	Thermal and differential thermal analysis; X-ray diffraction study	
$VOCl_3$	KCl, NaCl, NaCl–KCl (1:1)	116	⎫	Weight %: KCl: 4.5 at 800°C, 25.3 at 900°C; NaCl: 1.9 at 820°C, 5.45 at 900°C; NaCl–KCl (1:1) 5.23 at 700°C, 10.8 at 900°C
$TiCl_4$	KCl, NaCl, NaCl–KCl (1:1)	116	⎬ Equilibrium vapor pressure measurements	Weight %: KCl: 3.87 at 850°C, 2.94 at 900°C; NaCl 1.20 at 820°C, 0.38 at 900°C; NaCl–KCl (1:1) 2.84 at 700°C, 2.18 at 900°C
$SiCl_4$	KCl, NaCl, NaCl–KCl (1:1)	116	⎭	Weight %: KCl: 0.95 at 850°C, 0.67 at 900°C; 0.37 at 820°C, 0.16 at 900°C; NaCl–KCl (1:1) 0.56 at 700°C, 0.51 at 900°C
$TiCl_4$	KCl	117		$\log N_{KCl} = \log P - 2.86 + (2000/T)$ at 1096–1135°K
$TiCl_4$	NaCl, NaCl–KCl, KCl, RbCl, CsCl	118		$\log N = \log P - 3.24 + (5920/T) + (0.52 - 5250/T)/r$; at 700–800°C ($r$ radius of the solvent cation)
$TiCl_4$	LiCl–KCl, RbCl	119		$\log N_{RbCl} = \log P - 2.92 - (2430/T)$; $\log N_{LiCl-KCl} = \log P - 2.29 - (230/T)$; for 823–1072°K and 0.101–0.821 atm
$TiCl_4$	CsCl	120	Equilibrium at fixed vapor pressure; subsequent analysis of the melt for Ti	$\log N_{CsCl} = \log P - 2.95 + (2770/T)$ at 660–990°C and 60–540 mm Hg pressure
$TiCl_4$	NaCl, NaCl–KCl (1:1)	121		$\log N_{NaCl} = \log P - 2.73 + (580/T)$ at 1089–1224°K and 0.169–0.537 atm $\log N_{NaCl-KCl(1:1)} = \log P - 2.79 + (1300/T)$ at 961–1214°K and 0.256–0.699 atm
$TiCl_4$	$MgCl_2$	122		$\log N_{MgCl_2} = \log P - 0.50 - (1890/T)$ at 720–931°C and 119–716 mm Hg

[a] At 1 atm except when stated otherwise.

investigated the solubility of CO_2 in cryolite–alumina melts and assumed that all CO_2 escaped from the melt during quenching. Wartenberg[61] analyzed the excess halogen trapped in a quenched sample.

A dew point method was used by Howell et al.[47] to measure the vapor pressures of reactive gases in equilibrium with a melt.

Cryoscopic measurements have been used with many systems exhibiting extensive solubility and provide valuable information when equilibrium vapor pressure data are also available. Thermal analysis from cooling and heating curves or the visual observation for solubility do not supply all necessary data and require additional vapor pressure measurements. The vapor pressure measurements[30,40,44] for reactive gases listed in Table 1 were taken with the use of a quartz-spoon Bourdon-type gauge or an isoteniscope in which a molten tin manometer served as a device for equilibrating the $ZrCl_4$ or $HfCl_4$ vapor pressures with measured argon counterpressures.

Figure 2 shows the apparatus used by one of the authors[30] to measure the vapor pressures of $TiCl_4$ and $ZrCl_4$ in equilibrium with binary melts containing $ACl-A_2TiCl_6$ and $ACl-A_2ZrCl_6$, respectively, where A was an alkali metal. In this method solubilities were measured indirectly from the melting temperature of a preweighed mixture in equilibrium with its decomposition pressure.

The decomposition pressures of $ZrCl_4$ have also been measured by the transpiration method.[50] The transported zirconium chloride condensate was dissolved and analyzed spectrographically for the Zr content.

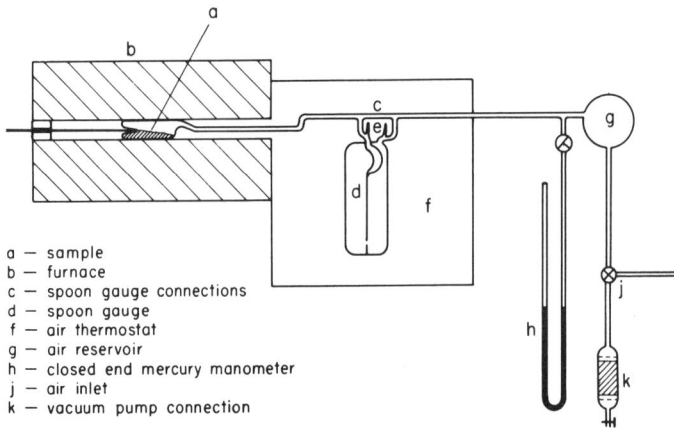

Fig. 2. Quartz-spoon-gauge type vapor pressure measuring apparatus with counterpressure arrangement.[30]

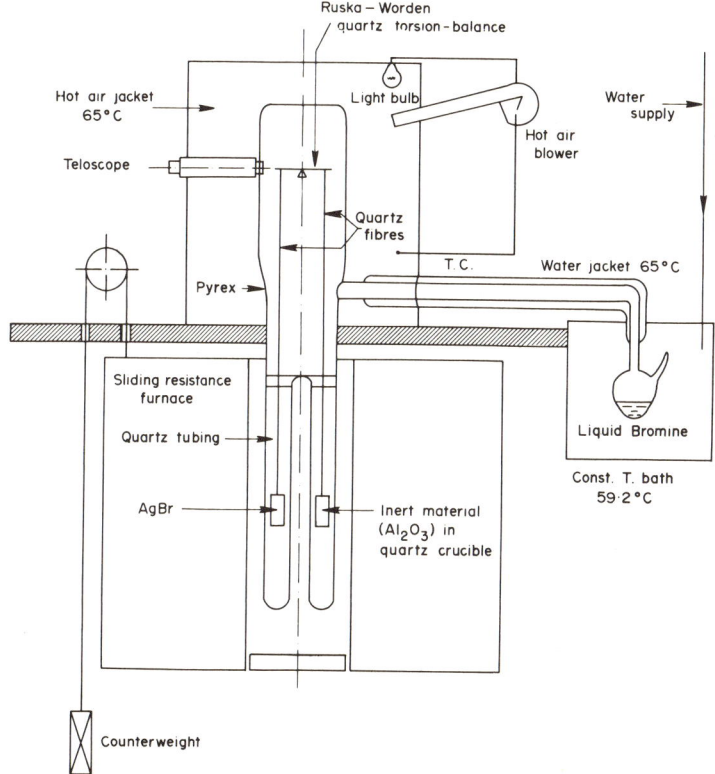

Fig. 3. Quartz microbalance with counterweight beam arrangement for the direct determination of solubility of Br_2 in molten AgBr.[29]

Stripping or elution methods, also widely used, are based on the saturation of the melt with the gas. The melt is stripped by an inert gas carrier and the evolved gas mixture is measured and analyzed.[62,63] In this method it is assumed that the inert gas removes completely the dissolved reactive gas by displacing the equilibrium representing the solubility reaction. However, this method, although applicable to nonreactive systems, does not appear to be reliable with reactive systems.

Gravimetric methods [24,29,40] are also useful with reactive gases where the solubilities are much higher than with the inert gases and can be detected. To avoid significant buoyancy corrections, beam-type balances should be used rather than spring balances. However, weight losses due to high-temperature vaporization are difficult to be accounted for in these methods.

Figure 3 shows a quartz microbalance which was adapted by the authors[29] for the measurement of the solubility of gaseous bromine in molten silver bromide.

Spectroscopic measurements of visible and UV absorption spectra[27,64] may also be combined with other independent solubility measurements[29] and can provide further solubility data from transmittance or absorbance curves. For this purpose a modified visible–light UV spectrophotometer in which the sample is situated between the light source and the monochromator (to eliminate the high-temperature background noise) may be used, or a more sophisticated instrument fitted with a chopper-type light beam. In a recent publication[28] an elegant radioactive tracer exchange method between Cl$^-$ ions in the melt and Cl$_2$ gas containing the isotope ^{36}Cl was also used successfully.

The gas-volumetric technique was combined with conductivity and viscosity measurements by Copeland and co-workers.[14–21] Burkhard and Corbett[65] also used the volumetric method to determine the solubility of HCl and H$_2$O in LiCl–KCl melts. This technique was also adopted by Bratland et al.[23] in their work on the solubilities of CO$_2$ shown in Table I. The technique appears to be sensitive and should give accurate results provided that the temperature gradients along the gas-volumetric column are kept constant.

3. SOLUBILITY DATA

A summary of solubility data for inert gases in fused salts is given in Table II, where the units are those used by the various authors. Solubilities in these systems which are characterized by low reactivity should be described as "physical."

Solubilities of reactive gases in fused salts are also included in Table I, and are expressed in the units chosen by the various authors.

The solubility of F$_2$ gas in chloride melts is high because of the displacement and complexing reaction[75]

$$F_2(\text{gas}) + 2Cl^-(\text{in melt}) \rightleftarrows Cl_2(\text{gas}) + 2[F^-(\text{in melt})] \qquad (47)$$

where F$^-$ is usually present as a fluoride compound. The solubility of Cl$_2$ in alkali chlorides appears to follow the distinctive trend that it increases with increasing size fo the alkali metal cation. The solubilities of Cl$_2$ in LiCl measured by Wartenberg,[61] which appear to follow the opposite trend, were obtained at a time when the preparation of anhydrous LiCl was probably difficult, and therefore the results should be considered to be of doubtful accuracy. The solubility of chlorine in alkali chlorides is generally higher than that in alkaline earth chlorides. Bromine and iodine solubilities indicate the same trends as with chlorine; again the solubility data for Br_2 in the hygroscopic $CaBr_2$ given by Wartenberg[61] should be considered to be less accurate. However, Wartenberg's work on many other systems conducted about 45 years ago is considered to be accurate.

The solubility of CO_2 in alkali fluoride melts[114,115] may be attributed to a reaction, as the product COF_2 was found to be present in the gas phase.

Considering the solubilities of the halogen gases with respect to a solvent salt containing the common halogen anion, it appears that in general solubilities increase in the order

$$I_2 > Br_2 > Cl_2$$

For chlorine gas in alkali chloride melts solubilities usually increase with increasing size of the alkali metal cation present. For example, in pure NaCl and KCl at about 1030°C the solubility of Cl_2 is 7.4×10^{-7} and 2.0×10^{-6} mole/ml of melt, respectively.

In mixed solvents the salt component having the larger ionic cation appears to predominate. Such behavior is followed by solutions of $ZrCl_4$ vapor in the mixed salt $\frac{1}{4}$ KCl — NaCl.[30] The decomposition pressures of $ZrCl_4$ vapor in the ternary $KCl-NaCl-ZrCl_4$ system are of the same order of magnitude as the corresponding pressures in equilibrium with the binary $KCl-ZrCl_4$ system containing the same fraction of $ZrCl_4$. However, the $NaCl-ZrCl_4$ solutions exhibited decomposition pressures which were larger by one or two orders of magnitude. Thus, at a constant applied pressure of $ZrCl_4$ the solubilities of $ZrCl_4$ in the mixed solvent $\frac{1}{4}$ KCl–NaCl are similar to those obtained in the binary system $KCl-ZrCl_4$.

It is evident that these trends are the result of competing interactions in which small cations tend to interact strongly with the halogen anions and therefore the latter do not possess the ability to react further with the XCl_4 species. Also, considering the relative reactivity of Br^- and I^- anions with respect to gaseous Br_2 and I_2, iodine is expected to form more stable triplet anions than bromine and therefore be more soluble. Solubility of Cl_2 is lowered by the presence of the F^- anion.[165]

Heats and entropies for solubility reported by various authors are presented in Table III. The tabulation includes the original data given by the authors as $\Delta H_P{}^\circ$, ΔH_T, or $\overline{\Delta H_X}$ and also includes a column of calculated enthalpies of vaporization which are given for the purpose of obtaining

TABLE II. Solubility of Inert Gases in Molten Salts

Gas	Melt	Ref.	Solubility data
He, Ne, Ar	LiF–NaF–KF	3	600°C: He, 11.3×10^{-8} mole/ml atm; Ne, 4.36×10^{-8}; Ar, 0.9×10^{-8}; 800°C: He, 23.0×10^{-8}; Ne, 11.18×10^{-8}; Ar, 3.40×10^{-8}
He, Ne, Ar, Xe	NaF–ZrF$_4$	66	600°C: He, 21.6×10^{-8} mole/ml atm; Ne, 11.3×10^{-8}; Ar, 5.06×10^{-8}; Xe 1.94×10^{-8}; 800°C: He, 42×10^{-8}; Ne, 24.7×10^{-8}; Ar, 12×10^{-8}; Xe, 6.32×10^{-8}
He, Xe	NaF–ZrF$_4$ UF$_4$	66	600°C: He, 20×10^{-8} mole/ml atm; Xe, 2.0×10^{-8}; 800°C: He, 41×10^{-8}; Xe, 6.5×10^{-8}
He, Ne, Ar, Xe	LiF–BeF$_2$	67	600°C: He, 11.55×10^{-8} mole/ml atm; Ne, 4.626×10^{-8}; Ar, 0.98×10^{-8}; Xe, 0.233×10^{-8}; 800°C: He, 19.48×10^{-8}; Ne, 9.01×10^{-8}; Ar, 2.66×10^{-8}; Xe, 0.863×10^{-8}
Xe	KNO$_3$–NaNO$_3$[a] (eut)	Unpublished work[a]	Unpublished preliminary work
He	Li silicates, binary alkali silicates, Na/Ca silicates,	114	20% mol. LiO$_2$, $K_1 = 0.0252$; 24.8% LiO$_2$, $K_1 = 0.0200$; at 1400°C; K_1 is Ostwald solubility coefficient
He, H$_2$O	Na/Ca/Al silicates	115	He: $K_1 = 0.012$ at 1200°C; $K_1 = 0.035$ at 1480°C; H$_2$O: $K_1 = 7$ at 1250°C; $K_1 = 45$ at 1750°C; K_1 is Ostwald solubility coefficient
H$_2$	CaO–Al$_2$O$_3$–SiO$_2$	68	0.35 mg/g slag 1400°C; 0.58 mg/g slag 1800°C; increasing considerably with CaO concentration
H$_2$[b]	NaOH, KOH	69	Less than 60 mg H$_2$ per 100 g of hydroxide at 410–500°C and pressures up to 800 psia; much higher solubility if Ni container subjected to corrosion attack
CO$_2$[c]	NaCl, KCl	22	NaCl: 4.2×10^{-6} mole/ml atm at 812.2°C; 5.71×10^{-6} at 948.3°C; KCl: 2.73×10^{-6} at 791°C; 8.90×10^{-6} at 969.9°C

Solubilities of Reactive Gases in Molten Salts

Gas	Salt	Ref	Data
CO_2	NaCl, KCl, KBr, KI	23, 24	NaCl: 5.20×10^{-7} mole/ml atm at 529°C; 7.58×10^{-7} at 997°C; KCl: 6.45×10^{-7} at 799°C; 8.95×10^{-7} at 1002°C; KBr 8.46×10^{-7} at 775°C; 10.89×10^{-7} at 927°C; KI: 19.2×10^{-7} at 700°C
CO_2	$LiCO_3$–Na_2CO_3–K_2CO_3	70	4 g CO_2/liter melt at 560°C
N_2, O_2	$NaNO_3$, KNO_3, $CsNO_3$	71	Below 10^{-4} mole/mole salt near m.p.
O_2	Li_2CO_3–Na_2CO_3, Li_2CO_3–Na_2CO_3–K_2CO_3	72	1.2×10^{-7} mole/ml atm at 600°C, 5×10^{-7} mole/ml atm at 800°C, 1.2×10^{-7} mole/ml atm at 600°C, 1×10^{-6} mole/ml atm at 800°C
Ar, He	$NaNO_3$, $AgNO_3$	14–18	Ar in $NaNO_3$ at 369°C: 1.10×10^{-4} mole/ml at 70 atm; 7.78×10^{-4} at 451 atm; He in $NaNO_3$ at 369°C: 2.65×10^{-4} mole/ml at 110 atm; 7.17×10^{-4} at 321 atm Ar in $AgNO_3$ at 250°C: 1.04×10^{-4} mole/ml at 47 atm; 12.6×10^{-4} at 375 atm; He in $AgNO_3$ at 250°C: 2.20×10^{-4} mole/ml at 51 atm; 9.37×10^{-4} at 247 atm
N_2	$NaNO_3$	19	N_2 in $NaNO_3$ 15.2×10^{-7} mole/ml atm at 727°K; 20.2×10^{-7} mole/ml atm at 628.5°K
He	$NaNO_3$, $LiNO_3$	73	He in $NaNO_3$ at 332°C: 1.86×10^{-7} mole/ml bar; 441°C: 2.80×10^{-7}; in $LiNO_3$ at 270°C: 1.51×10^{-7}
Ar	$NaNO_3$, $LiNO_3$	73	Ar in $NaNO_3$ at 331°C: 0.64×10^{-7} mole/ml bar, 440°C: 1.04×10^{-7}; in $LiNO_3$ at 273°C: 0.91×10^{-7}, 307°C: 1.09×10^{-7}
Ar	$RbNO_3$, $AgNO_3$	73	Ar in $RbNO_3$ at 331°C: 1.30×10^{-7} mole/ml bar, 440°C: 240×10^{-7}; in $AgNO_3$ at 234°C; 0.19×10^{-7}
N_2	$NaNO_3$	73	0.5×10^{-7} mole N_2/ml melt at 1 atm and 331°C, 0.84×10^{-7} mole/ml at 1 atm and 449°C
N_2	$LiNO_3$	73	0.73×10^{-7} mole N_2/ml melt at 1 atm and 277°C
N_2	$NaNO_3$	74	1.9×10^{-7} mole N_2/ml melt at 1 atm and 315°C; 2.2×10^{-7} mole/ml melt at 1 atm, 365°C
He	$NaNO_3$	74	315°C: 0.85×10^{-7} mole He/ml melt at 1 atm; 400°C: 1.25×10^{-7}
Ar	$NaNO_3$	74	315°C: 0.55×10^{-7} mole Ar/ml melt at 1 atm; 400°C: 0.85×10^{-7}

[a] Newton, cited in Ref. 66. [b] Inert if no corrosion products are present. [c] Values too high due to container corrosion.
[d] Additional data on the Ar, N_2, O_2, and CH_4 solubility in KNO_3–$NaNO_3$ (eut) by F. Paniccia and P. G. Zambonin: *J. Chem. Soc., Faraday Trans. I*, 2083 (1972) reported after completion of present manuscript.

TABLE III. Solution Properties of Selected Reactive Gases

Gas	Melt	Ref.	t, °C	ΔH°_P, cal/mole	ΔH_T^e, cal/mole	$\Delta H^\circ_V{}^a$, cal/mole	$\overline{\Delta H_Y}$, cal/mole	ΔS°_P, eu
Cl_2	$PbCl_2$-KCl (0.23)	28	454–635	7,930	—	(490)	8,420[b]	—
Cl_2	$PbCl_2$-KCl (0.48)	28	443–692	−995	—	(530)	−465[b]	—
Cl_2	$PbCl_2$-KCl (0.7)	28	588–702	−3,680	—	(510)	−3,170[b]	—
Cl_2	LiCl-KCl (eut)	81	400–550	3,700	—	(890)	4,590[b]	—
Cl_2	$PbCl_2$	81	513–639	9,400	—	(<100)	9,500[b]	—
Br_2	$PbBr_2$	81	703	−1,300	—	3,300	2,000[b]	—
			792	—	—	2,500		
Br_2	LiBr	81	890	−9,500	—	1,600	−7,900[b]	—
			982	—	—	700	—	
Br_2	AgBr	29	517	—	8,270	2,510	−5,760	—
			677	—	—	1,065	−7,210	
HCl	LiCl-KCl (eut)	82	490–675	3,500	—	—	—	—
HCl	KCl	87	840	19,100	—	} 3,860 at 188°K (b.p.)	—	—
HCl	KCl-NaCl	87	840	18,750	—		—	—
HCl	KCl-$MgCl_2$	87	840	2,170	—		—	—
HCl	$MgCl_2$	87	840	2,010	—		—	—

Solubilities of Reactive Gases in Molten Salts

Gas	Salt	Ref	Temp. range (°K)					
HF	NaF(0.45)–ZrF$_4$	63	550–800	−3,850	—	⎫	—	−14.0
HF	NaF(0.805)–ZrF$_4$	63	550–800	−9,700	—	⎬ 7,460 at 306.3°K (b.p.)	—	−15.2
HF	LiF(0.67)–BeF$_2$	83	500–700	−5,980	—	⎬	—	−7.07
DF	LiF(0.67)–BeF$_2$	83	500–700	−6,430	—	⎭	—	−7.92
BF$_3$	LiF(0.65)–BeF$_2$(0.28)–ZrF$_4$(0.05)–ThF$_4$(0.01)–UF$_4$(0.01)	84	500–700	−15,100	—	1,940–1,890	−13,160[b]	—
							—	−22.2
H$_2$O	LiNO$_3$	89	265–280	−13,580[c]	—	8,230–8,090	−4,270[b]	−12.7
H$_2$O	NaNO$_3$ (extrap.)	89	230–280	−10,900[c]	—	8,560–8,090	−1,340[b]	−11.0
H$_2$O	KNO$_3$ (extrap.)	89	230–265	−9,100[c]	—	8,560–8,230	460[b]	−9.2
CO$_2$	NaCl	23	829–997	5,900	—	(4,080)–(4,020)	9,980[b]	−0.6
CO$_2$	Cryolite–Al$_2$O$_3$ (0.18)	24	977–1100	−8,700[d]	—	(4,020)–(4,000)	−4,680[b]	—
CO$_2$	Sodium borate melts	25	950–1200	−19,000[d]	—	(4,030)–(4,000)	−15,970[b]	—
CO$_2$	Soda–silica melts	26	900–1200	−54,000[d]	—	(4,050)–(4,000)	−50,000[b]	−14.6
CO$_2$	NaNO$_3$	74	314–366	−2,680	—	(4,780)–(4,680)	2,100[b]	−21.7
NH$_3$	LiNO$_3$	111	260–315	−16,500	—	(2,980)–(2,420)	−13,520[b]	−9.5
NH$_3$	LiNO$_3$–KNO$_3$ (0.57)	111	160–250	−7,000	—	(3,720)–(2,950)	−3,280[b]	−8.0
NH$_3$	LiNO$_3$–KNO$_3$–NaNO$_3$ (eut)	111	160–250	−5,800	—	(3,720)–(2,950)	−2,080[b]	−14.4
NH$_3$	LiClO$_4$–KClO$_4$ (eut)	111	234–311	−13,800	—	(3,080)–(2,450)	−10,720[b]	

[a] Parentheses imply that the heat capacities for the liquefied gases are only approximate. Where heat capacity data were not available the enthalpy change for evaporation is given at the boiling temperature only.
[b] Data recalculated using Eqs. (7), (18), and (23).
[c] Data recalculated from ΔH_C values as given by the author.
[d] Calculated from author's graphs.
[e] ΔH_T as calculated from experimental results represents enthalpy changes during removal of 1 g-mole of dissolved gas from the solution.

the enthalpies of mixing on a common basis through the use of Eqs. (7), (18), and (23). Enthalpies of solution given as ΔH_C are not commonly used and are only found in connection with solubility data for nonreactive gases. Enthalpies of vaporization given in Table III in parentheses should be considered as approximate. In any case accuracy limits cannot be given because of the approximate nature of the heat capacity data, particularly for the liquefied gases and the long extrapolations implied by the calculations.

Considering the data in Table III, it becomes evident that the enthalpies of mixing in binary solvent melts are dependent upon the overall melt composition and on the salt components present. There are indications, however, that one of the components present promotes solubility more than others.

For example, the enthalpy of mixing of gaseous Cl_2 in $PbCl_2$-KCl melts,[28] given as ΔH_P, is positive at 23 mole % KCl and becomes negative at compositions higher than 48 mole % KCl. This behavior is understood if the results are interpreted in terms of Cl_2 interactions. Pure $PbCl_2$[81] shows positive enthalpies of mixing, which indicate the lack of reactivity with Cl_2. Enthalpies are also positive in the LiCl-KCl system,[81] indicating the decomplexing of the possible Cl_3^- species by the Li^+ cations. Similar trends are observed for the solubility of Br_2 in metal bromides. The positive partial molar heat of mixing indicates that gaseous Br_2 does not interact with molten $PbBr_2$.

The solubility of H_2O in Li-bearing melts generally is exothermic because of the hydration reaction. Furthermore solubilities in mixed melts increase with increasing Li content. Enthalpies of solution for H_2O vapor in the pure melts of $LiNO_3$,[89] $NaNO_3$,[89] and KNO_3[89] are given by the authors as ΔH_C°. In Table III, these have been converted to ΔH_P° using Eq. (18). The enthalpies of solution for the solubility of H_2O vapor in the $LiNO_3$-$NaNO_3$-KNO_3 ternary melt[93] were calculated from van't Hoff plots in which the equilibrium constant was expressed in moles of dissolved H_2O vapor per millimeter pressure per mole of Li^+ in the melt. Enthalpies obtained in this manner cannot be described as enthalpies of solution as they include part of the partial molar enthalpy of mixing for the $LiNO_3$ component in the melt. Similar criticism is also applicable to the enthalpy data for the solubility of H_2O vapor in LiCl-KCl melts.[65]

The solubility of CO_2[25,26] in soda-silica and sodium borate melts has been investigated by Pearce.[25,26] Solubilities are interpreted by the reaction

$$CO_2 + O^{2-}\text{(in melt)} \rightleftharpoons CO_3^{2-}\text{(in melt)} \qquad (48)$$

for which the equilibrium constant is written as

$$K = a_{CO_3^{2-}}/P_{CO_2} a_{O^{2-}} \qquad (49)$$

On the assumptions that the $a_{CO_3^{2-}}$ is proportional to the concentration of CO_2 in the melt expressed as wt% and that the activity of $a_{O^{2-}}$ is constant in the temperature range considered, the enthalpy changes for the reaction were calculated by the author from plots of log (wt%) versus 1/T. Since $a_{O^{2-}}$ is taken to be constant, the enthalpies of reaction should be classified as $\Delta H_P°$ and are listed in Table III.

NH_3[111] appears to behave similarly to H_2O vapor in that it is reactive toward Li-bearing melts. The solubilities of gaseous HCl in chloride melts appear to be much higher than those for chlorine. The enthalpy of solution for HCl in KCl[87] is highly endothermic, indicating that the solution process should be accompanied by significant ordering effects which could be characteristic of the H^+ ions. The enthalpy of solution becomes less endothermic in the KCl–LiCl[82] and the KCl–NaCl[7] melts. Solutions of HCl in $MgCl_2$[82] are only slightly endothermic and do not appear to become more endothermic in the presence of KCl. It is possible that water of hydration present in $MgCl_2$ as an impurity could create exothermic effects that would contribute to a lowering of the overall enthalpy of solution.

The solubilities of HF in fluoride melts are lower than the solubilities of HCl in chloride melts at comparative temperatures, with the main difference that the solution processes of HF are definitely exothermic, indicating their reactive nature.

Regarding the behavior of the highly reactive gases $TiCl_4$, $ZrCl_4$, and $HfCl_4$ in various binary melts containing alkali chlorides, it has already been shown that the solubilities are understood in terms of equilibria of the type

$$XCl_4 + 2MCl \rightleftarrows M_2XCl_6 \qquad (50)$$

where M is an alkali metal cation and X = Ti, Zr, or Hf. The thermodynamic treatment presented earlier[57] predicts that these gases can be dissolved in any amounts providing that the applied pressure of the XCl_4 vapor is maintained at the required equilibrium level. The vapor pressure versus temperature curves for various solubility conditions indicate inflections at the corresponding eutectic and liquidus points of the binary systems $MCl–M_2XCl_6$, as predicted from theory.

The compounds M_2XCl_6 which have been prepared as pure solids are listed in Table IV.

TABLE IV. Pressure Relationship in Pure M_nXCl_6

Reaction	$\log_{10} P_{XCl_4}$, mm Hg	Temperature range, °C	Decomposition temperature, °C	Physical properties of pure M_nXCl_6 compound			Ref.
				Color	Density, g/ml at 25°C	Melting point, °C	
$K_2TiCl_{6(s)} \rightarrow 2KCl_{(s)} + TiCl_{4(v)}$	$-5930/T + 10.364$	350–550	520	Light yellow	2.40	695 ± 5	51
$Rb_2TiCl_{6(s)} \rightarrow 2RbCl_{(s)} + TiCl_{4(v)}$	$-7524/T + 11.069$	525–650	645	Light green	2.92	686 ± 5	52
$Cs_2TiCl_{6(s)} \rightarrow 2CsCl_{(s)} + TiCl_{4(v)}$	$-8358/T + 11.762$	600–645 }	665	Yellow green	3.28	691 ± 5	52
$Cs_2TiCl_{6(s)} \rightarrow 2CsCl_{(L)} + TiCl_{4(v)}$	$-9223/T + 12.719$	645–675					
$Li_2ZrCl_{6(s)} \rightarrow 2LiCl_{(s)} + TiCl_{4(v)}$	$-3970/T + 7.90$	350–496 }	501	White	2.50	535 ± 1	53
$Li_2ZrCl_{6(s)} \rightarrow 2LiCl_{(L)} + TiCl_{4(v)}$	$-6760/T + 11.6$	496–505					
$Na_2ZrCl_{6(s)} \rightarrow 2NaCl_{(L)} + ZrCl_{4(v)}$	$-10,394/T + 14.34$	548–640	634	Light gray	2.43	646 ± 1	54, 124
$K_2ZrCl_{6(s)} \rightarrow 2KCl_{(L)} + ZrCl_{4(v)}$	$-18,380/T + 19.83$	735–804 }	831	Light gray	2.44	799 ± 2	54
$K_2ZrCl_{6(L)} \rightarrow 2KCl_{(L)} + ZrCl_{4(v)}$	$-4689/T + 7.12$	804–831					
$Rb_2ZrCl_{6(s)} \rightarrow 2RbCl_{(L)} + ZrCl_{4(v)}$	$-11,900/T + 13.4$	760–812 }	904	White	2.94	812	53
$Rb_2ZrCl_{6(L)} \rightarrow 2RbCl_{(L)} + ZrCl_{4(v)}$	$-5350/T + 7.4$	812–903					
$Cs_2ZrCl_{6(s)} \rightarrow 2CsCl_{(L)} + ZrCl_{4(v)}$	$-17,200/T + 17.9$	777–817 }	1040	White	(3.36)	808.7	44, 57
$Cs_2ZrCl_{6(L)} \rightarrow 2CsCl_{(L)} + ZrCl_{4(v)}$	$-5030/T + 6.71$	817–1040					
$BaZrCl_{6(s)} \rightarrow BaCl_{2(s)} + ZrCl_{4(v)}$	$-4780/T + 9.43$	330–460	455	White	3.15	—	53
$SrZrCl_{6(s)} \rightarrow SrCl_{2(s)} + ZrCl_{4(v)}$	$-4350/T + 9.16$	315–412	410	White	3.05	—	53
$Li_2HfCl_{6(s)} \rightarrow 2LiCl_{(s)} + HfCl_{4(v)}$	$-3810/T + 7.57$	350–500 }	513	Off-white	—	557 ± 1	53
$Li_2HfCl_{6(s)} \rightarrow 2LiCl_{(L)} + HfCl_{4(v)}$	$-4970/T + 9.12$	500–514					
Na_2HfCl_6				Off-white	—	680 ± 1	44, 57, 124

The plot of the decomposition temperatures, e.g., the temperature at which the reaction pressure of XCl_4 vapor over a given pure M_2XCl_6 compound reaches 1 atm, as a function of the ionic radius of the alkali metal cations is given in Fig. 4. It is evident that large alkali cations stabilize the M_2XCl_6 compounds which then decompose at higher temperatures. For example, the Cs_2ZrCl_6 complex compound decomposes at about 1300°K, while the Li_2ZrCl_6 decomposes at about 750°K. The complex compounds $BaZrCl_6$ and $SrZrCl_6$ are even more unstable.

This increasing thermodynamic stability with increasing size of the alkali metal cation may be understood in terms of the chloride ion donor capacity of an alkali chloride. For a given XCl_4 vapor the formation of the corresponding XCl_6^{2-} complex species is the result of competing interactions of XCl_4 and of an alkali cation M^+ for chloride ions. The small divalent, Ba^{2+} and Sr^{2+} cations exert a stronger electrostatic attraction with the result that the Zr—Cl bond in $ZrCl_6^{2-}$ is rather weak and the complexes decompose at relatively low temperatures. With the Cs_2ZrCl_6 compound

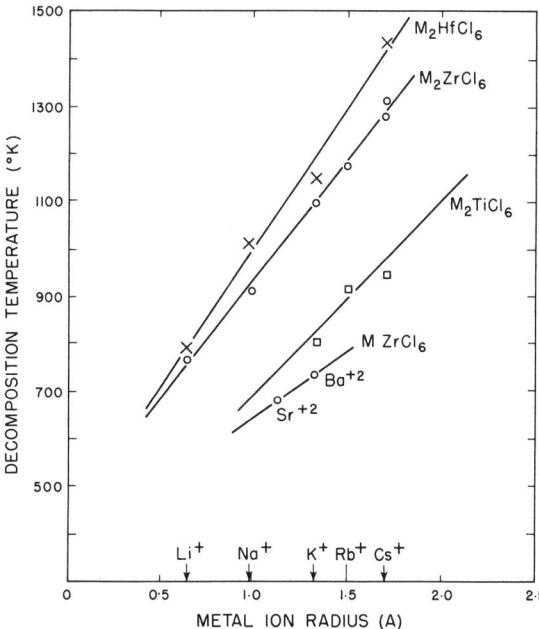

Fig. 4. Thermal stability of the hexachloro complex compounds of $TiCl_4$, $ZrCl_4$, and $HfCl_4$ as a function of the radii of the alkali metal cations.

the large Cs^+ cations interact rather weakly with the Cl^- anions and the Zr—Cl bonds in $ZrCl_6^{2-}$ are strengthened.

The reaction pressures for all pure hexachloro compounds investigated in this laboratory are also given in Table IV.

It is of interest to note that $TiCl_4$, $ZrCl_4$, and $HfCl_4$, which are covalently bonded and highly molecular species as vapors, appear to dissolve in the alkali halides in an ionic form.

The thermal stability of the hexachlorotitanate compounds is relatively low as compared to that of the corresponding hafnium and zirconium compounds, as the hexachlorotitanates exhibit high decomposition pressures even at low temperatures. Accordingly, the solubility of $TiCl_4$ vapor in alkali chloride melts under normal pressures is limited. For example, at 690 and 720°C the solubilities of $TiCl_4$ vapor in the equimolar melt of KCl–NaCl are only 14 and 5 wt% respectively,[40] when the applied pressure of the vapor was kept at 1 atm.

A general equation for the mole fraction of soluble titanium tetrachloride between 700 and 900°C in various alkali chloride melts is given by Smirnov et al.[118] as

$$\log N_{TiCl_4} = \log P - 3.24 + (5920/T) + [0.052 - (5250/T)/r_m^+] \quad (51)$$

where r_m^+ is the radius of the alkali metal cation. It can be seen that the solubility at constant temperature and pressure increases with increasing ionic radius of the alkali metal cation. For a given solvent salt the solubility at constant pressure decreases with increasing temperature.

The pressure of $TiCl_4$ over pure Rb_2TiCl_6 has been investigated.[52] For the system $CsCl–Cs_2TiCl_6$ the pressure curves for mixtures saturated in Cs_2TiCl_6 ($CsCl/TiCl_4 = 2$ and 3.85) indicate only liquidus melting. Previous assumptions[42] regarding the existence of the compound Cs_4TiCl_8 and of possible solid solubility in this system do not appear to be justified.

The pure compound Cs_2TiCl_6 has also been investigated at higher temperatures by Morozov and Toptygin.[123] However, their results differ significantly from the values obtained in this laboratory. The $CsCl–CsTiCl_6$ system is more stable than the $KCl–K_2TiCl_6$ system, and thus the reaction pressures at temperatures below the eutectic temperature have not been observed experimentally with accuracy. The increased thermal stability of the system is also reflected by the increased solubility of $TiCl_4$ vapor in molten CsCl. For example, pure molten Cs_2TiCl_6 may be prepared at about 660°C provided that the pressure of the $TiCl_4$ vapor is kept at about 1 atm.

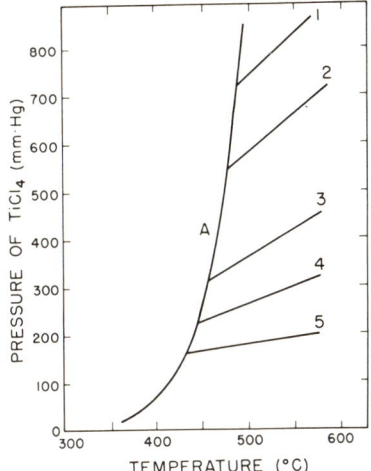

Fig. 5. Equilibrium decomposition pressures of K_2TiCl_6 in the KCl–LiCl eutectic melt at various compositions and temperatures. Curves 1–5, correspond to the following mole fractions of K_2TiCl_6: 12.7×10^{-3}, 9.1×10^{-3}, 6.3×10^{-3}, 5.7×10^{-3}, and 1.9×10^{-3}.

Solubilities of $TiCl_4$ in the eutectic melt of KCl–LiCl[41] were measured in the temperature range from 350 to 550°C. Part of the pressure–temperature diagram is given in Fig. 5. In this diagram the continuous part of the pressure curve represents solutions saturated with K_2TiCl_6, while the horizontal branches represent compositions of complete miscibility. The saturation solubility of $TiCl_4$ increases with temperature from 1.5 wt% at 420°C and a pressure of 160 mm Hg to about 5 wt% at 500°C and a pressure of 900 mm Hg $TiCl_4$. At constant pressure, however, the solubility decreases with increasing temperature. These results are in disagreement with the work by Smirnov et al.,[118] in which it was found that the isobaric solubility for this system increases with increasing temperature.

The pressure–temperature equilibrium diagram of the $LiCl–Li_2ZrCl_6$ system has been investigated over the entire concentration range[53] and the results are shown in Fig. 6. All compositions below the eutectic temperature are shown by a single vapor pressure curve, as predicted from theory. Eutectic melting is clearly indicated by a change in slope of the pressure curve in the diagram.

Because of the high decomposition pressures of solutions containing 80–100 mole % Li_2ZrCl_6, liquidus melting was not seen experimentally.

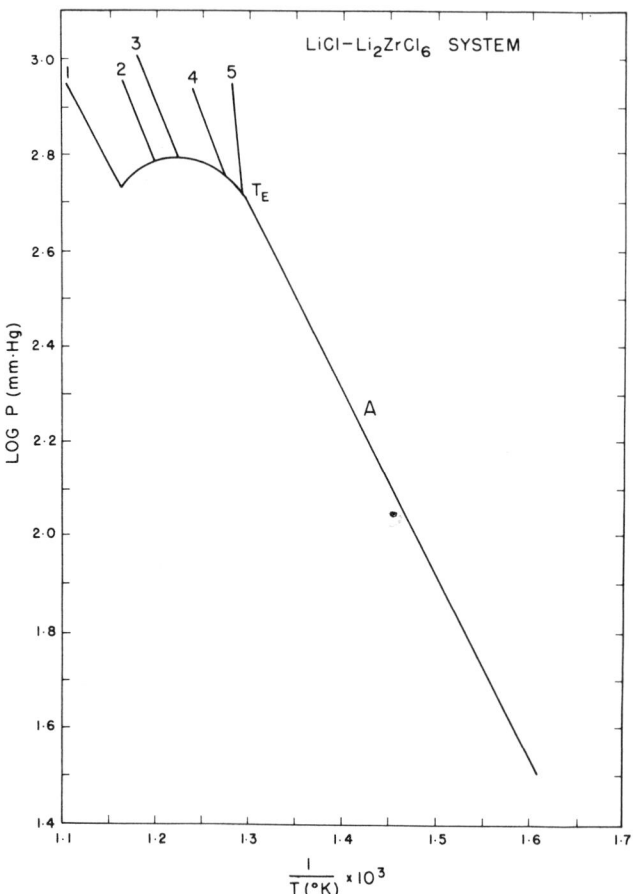

Fig. 6. Plots of log P_{ZrCl_4} versus $1/T$ in the system LiCl–Li$_2$ZrCl$_6$ for the following mole fractions of Li$_2$ZrCl$_6$: (1) 0.10. (2) 0.20, (3) 0.30, (4) 0.34, and (5) 0.80, 0.88, pure and Li$_2$ZrCl$_6$.

However, eutectic and liquidus melting was evident in solutions saturated with LiCl.

The pressure–temperature curves for the systems LiCl–Li$_2$HfCl$_6$, NaCl–Na$_2$ZrCl$_6$, and KCl–K$_2$ZrCl$_6$ are given in Fig. 7–9.

For the NaCl–Na$_2$ZrCl$_6$ system,[30] pressure data are only given at temperatures higher than the eutectic temperature. The curves clearly show the saturation tie-lines on the Na$_2$ZrCl$_6$-rich and NaCl-rich sides of the diagram, as well as liquidus melting.

The KCl–K$_2$ZrCl$_6$ system[30] was more stable and liquidus melting was observed only with respect to K_2ZrCl_6 saturation. The pressure curves for compositions varying between 10.9 and 58.8 mole % K_2ZrCl_6 correspond to the liquid solutions. Liquidus melting temperatures shown on the pressure diagrams are in agreement with the phase diagrams for these systems.[30]

Howell et al.[47] have also investigated reaction pressures in the systems $NaCl-ZrCl_4$ and $KCl-ZrCl_4$. Their investigation in the $MCl-M_2ZrCl_6$ subsystem was restricted to compositions rich in M_2ZrCl_6. In the $NaCl-ZrCl_4$ system compositions representing only the subsystem $Na_2ZrCl_6-ZrCl_4$

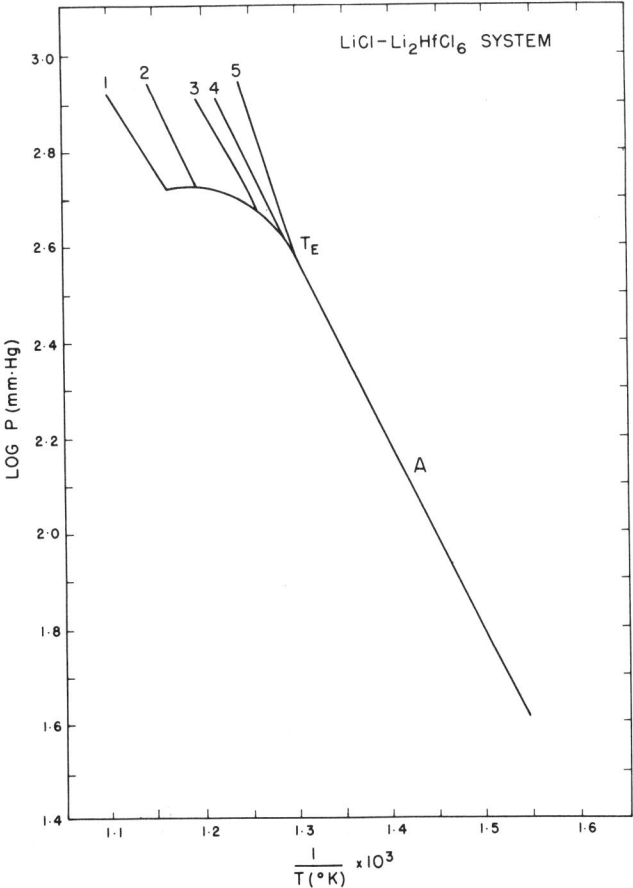

Fig. 7. Plots of log P_{HfCl_4} versus $1/T$ in the system $LiCl-Li_2HfCl_6$ for the following mole fractions of Li_2HfCl_6: (1) 0.10, (2) 0.20, (3) 0.30, (4) 0.40, (5) 0.60, 0.80, and pure Li_2HfCl_6.

were investigated. Solutions of the latter type are expected to have high reaction pressures and should represent the least useful range from the point of view of thermal stability. Thus at 400°C solutions containing between 50 and 75 mole % $ZrCl_4$ exhibited high vapor pressures of the order of 3 atm.

The reaction pressures of pure Na_2ZrCl_6, K_2ZrCl_6, and Cs_2ZrCl_6 were also studied by Morozov and Sun In-Chzhu.[46]

Considering these systems as potential electrolytes for the recovery of the metals Ti, Zr, or Hf by fused salt electrolysis, the equilibrium pressure measurements indicate that it is impossible to use directly "stable" solutions of $TiCl_4$ in alkali or alkaline earth metal chloride melts in open electrolytic cells operating under an inert gas atmosphere. The solutions would decompose thermally and the solubilities would appear to be extremely low. Closed electrolytic cells capable of maintaining the high $TiCl_4$ pressures required do not appear practical. Prereduction of the volatile $TiCl_4$ into nonvolatile lower chlorides appears to be a prerequisite for a successful electrolytic process for titanium.

The solutions of the hexachloro compounds of Zr and Hf in alkali chloride melts are more stable than the corresponding Ti systems and may

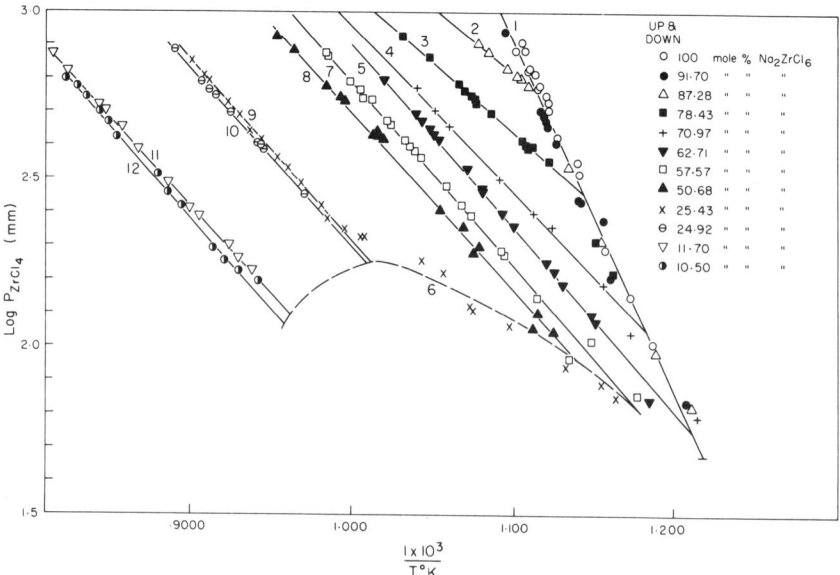

Fig. 8. Plots of log P_{ZrCl_4} versus $1/T$ in the system $NaCl$–Na_2ZrCl_6 at the indicated composition. Curves 1 and 6, respectively, represent saturation with Na_2ZrCl_6 and NaCl.

Solubilities of Reactive Gases in Molten Salts

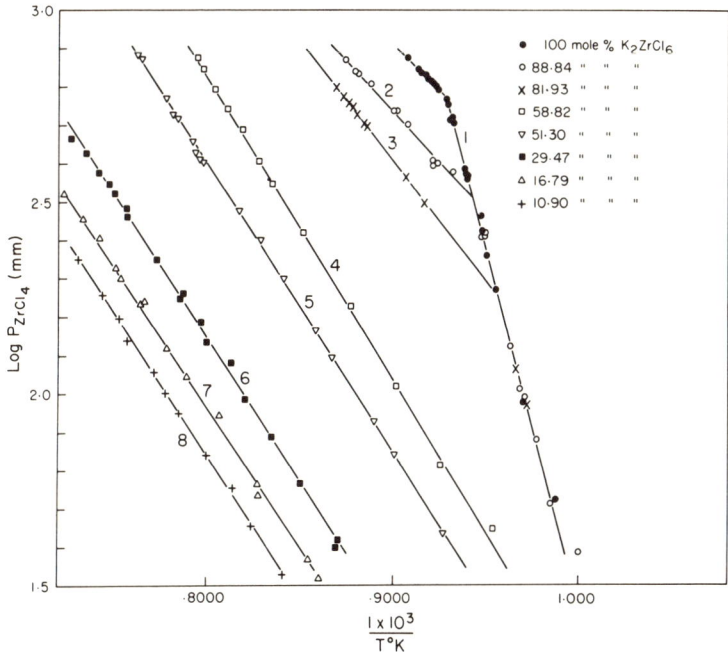

Fig. 9. Plots of log P_{ZrCl_4} versus $1/T$ in the system $KCl–K_2ZrCl_6$ at the indicated composition. Curves 2–8 represent the all-liquid solutions. Liquidus melting is only seen with solutions saturated with K_2ZrCl_6 (curve 1).

be readily prepared. Particularly stable are those formed between $ZrCl_4$ of $HfCl_4$ vapor and melts containing CsCl. For example, solutions containing 20 mole % Cs_2ZrCl_6 exhibit a decomposition pressure of $ZrCl_4$ vapor of only 3 mm Hg at the melting point of Cs_2ZrCl_6 (1090°K). Evidently at lower temperatures and for more dilute solutions the decomposition pressure should become negligibly small. Thus it may be concluded that electrolytes of this type could be of practical use in open cells kept under an inert atmosphere.

4. CHEMICAL SYNTHESES IN MOLTEN SALTS

Molten salt solutions provide a suitable environment for a number of chemical reactions involving inorganic and organic reactants.

According to Sundermeyer,[32] chemical processes in molten salts may be divided into the following categories.

(a) Reactions in which the melt acts a catalyst. For example,[125] in a melt containing KCl and $ZnCl_2$, methylchloride is formed according to

$$H_3C—O—CH_3 + 2HCl \rightarrow 2CH_3Cl + H_2O \qquad (52)$$

(b) Reactions involving the participation of the melt and the consumption of one or more of its components with no simple means for regenerating the latter. For example, the formation of ethylchloride[125] according to

$$3(H_5C_2—O—C_2H_5) + 2AlCl_3 \rightarrow Al_2O_3 + 6C_2H_5Cl \qquad (53)$$

taking place in a low-melting ternary melt containing $AlCl_3$–KCl–NaCl. (Ternary eutectic at 93°C.)

(c) Reactions in which the melt acts as a solvent for the reactants. For example, the synthesis of silicon hydride from silicon tetrachloride and lithium hydride according to[126]

$$4LiH + SiCl_4 \rightarrow SiH_4 + 4LiCl \qquad (54)$$

which takes place at 400°C in a melt containing LiCl and KCl. In this reaction one of the melt components is formed as the result of such reactions.

Sundermeyer,[32] who may be considered as one of the leading researchers in this field, has written a comprehensive review on fused salt reactions. A summary of selected syntheses involving only gaseous systems and molten salt is presented in Table V.

The main advantages for using molten salts as reaction media may be summarized as follows:

1. At high temperatures and under conditions of intimate contact in the homogeneous solution phase, or for heterogeneous systems in which the reactants are suspended in a molten salt, reactions usually proceed extremely rapidly.

2. For exothermic processes reaction rates may be controlled through dilution of the reacting gases with an inert carrier gas. Actual production rates are generally slow because of the small amounts present of the sparingly soluble gas. Thus local overheating is prevented and in addition the heat produced is dissipated by the high thermal capacity of the molten salt phase and its high thermal conductivity. Local overheating is the most common cause for explosions during such exothermic processes.

3. Interfacial type poisoning of the molten catalyst in catalytic type reactions is highly unlikely to occur.

4. For heterogeneous gas–liquid type reactions, kinetics are favored because of the continuous renewal of the molten interface. Stirring may be employed to enhance the rate for a given process.

5. Gaseous reaction products are easily removed from the melt because of their selective solubilities.

Very little quantitative information is available with regard to possible mechanisms. It appears, however, that the enhanced reactivity which occurs in the presence of a molten ionic phase may be attributed to a number of factors.

In catalytic gaseous reactions the molten salt solution should remain unchanged. The reaction is expected to take place either through the adsorption of the reactants at the liquid surface or by dissolution into the bulk of the melt, followed by the charge transfer processes. The gaseous products which normally have low solubilities in the melt are finally desorbed.

The solubility of an organic molecule in an ionic melt is expected to proceed by a mechanism which involves the polarization of the charge distribution within the molecule as the result of interactions with the strong electrostatic fields which characterize the ionic melts.

It is worthwhile to mention that the potential created by the electrostatic field of a single monovalent ion taken as a point charge at a distance equal to 10 Å from its geometric centre should be of the order of 14.5 million volts.[127] In catalytic reactions the smaller cations of high charge density appear to be the best catalysts. They function as electron acceptors and the reaction proceeds through the formation of intermediate unstable reaction products which are usually described as "activated complexes." For example, benzene may be chlorinated between 200 and 400°C in a melt containing $NaCl-AlCl_3-FeCl_3$ to yield monochlorobenzene.[32,128,129] The reaction is catalyzed by the Al^{3+} and Fe^{3+} cations which are described as the electron acceptors. By analogy with known mechanisms for similar reactions[130] in other media the following steps may be postulated:

$$(AlCl_3) + Cl_2 \rightarrow [AlCl_4]^- + Cl^+ \quad (55)$$
in solution ; gas in solution ; coordinated complex in solution

$$(C_6H_6) + Cl^+ \rightarrow [C_6H_6{-}Cl^+] \quad (56)$$
in solution ; activated complex in solution

$$[C_6H_6{-}Cl^+] + [AlCl_4]^- \rightarrow C_6H_5Cl\uparrow + (AlCl_3) + HCl\uparrow \quad (57)$$
(gas) ; (gas)

TABLE V. Selected Syntheses Involving Gases in Molten Salts

Gaseous reagent	Temperature range, °C	Melt	Main products	Ref.	Overall reaction	Type of reaction[a]
NH_3	850–950	$NaCl$–KCl–B_2O_3	BN	135, 32	$B_2O_3 + 2NH_3 \rightarrow 3BN + 3H_2O$	1
$COCl_2$	550–900	Alk.Cl–$AlCl_3$–Al_2O_3	$AlCl_3$, CO_2	136, 32	$Al_2O_3 + 3COCl_2 \rightarrow 2AlCl_3 + 3CO_2$	1
Cl_2	800	$NaCl$–$BaCl_2$–$CaCl_2$–B_2O_3–C	BCl_3, CO	137	$B_2O_3 + 3C + 3Cl_2 \rightarrow 2BCl_3 + 3CO$	2
Air; HCl	350–425	KCl–$CuCl$	—	138	$\tfrac{1}{2}O_2(\text{air}) + 2CuCl \rightarrow CuO \cdot CuCl_2 \; (+ \, 2HCl)$ $\rightarrow 2CuCl_2 + H_2O$	3, 4
	500–800		Cl_2, CuCl		$CuCl_2 \rightarrow CuCl + \tfrac{1}{2}Cl_2$	
Air	350–400	KCl–$CuCl$	Air without oxygen	139, 32	$\tfrac{1}{2}O_2(\text{air}) + 2CuCl \rightarrow CuO \cdot CuCl_2$	3
	450–525		O_2		$CuO \cdot CuCl_2 \rightarrow \tfrac{1}{2}O_2 + 2CuCl$	—
$CsCl_2$	500–600	LiF–NaF–KF	CSF_2	140	$CSCl_2 + 2F^- \rightarrow CSF_2 + 2Cl^-$	1
CO, HCl, O_2	500–525	$CuCl$–$CuCl_2$ (molten on carrier)	$COCl_2$	141	$CO + 2HCl + \tfrac{1}{2}O_2 \rightarrow COCl_2 + H_2O$	3
CO_2	530	Na_2CO_3–$Ca(CN)_2$	NaOCN	32	$Ca(CN)_2 + Na_2CO_3 + CO_2 \rightarrow NaOCN + CaCO_3$	1
C_2H_2, NH_3	400–550	Alk.Cl–$ZnCl_2$	CH_3CN	32, 142	$C_2H_2 + NH_3 \rightarrow xCH_3CN + xH_2 + (1-x)$ pyridine derivatives	3

Solubilities of Reactive Gases in Molten Salts

Gas	Temp (°C)	Solvent	Product	Refs	Reaction	
CH_4, Cl_2, O_2	400–550	Alk.Cl–CuCl$_2$ (or FeCl$_3$)	$CH_{4-x}Cl_x$	32, 143	$CH_4 + \frac{1}{2}xCl_2 + \frac{1}{4}xO_2 \rightarrow CH_{4-x}Cl_x + \frac{1}{2}xH_2O$	3
CH_4, HCl, O_2	400–550	Alk.Cl–CuCl$_2$ (or FeCl$_3$)	$CH_{4-x}Cl_x$	32, 143	$CH_4 + xHCl + \frac{1}{2}xO_2 \rightarrow CH_{4-x}Cl_x + xH_2O$	3
$S_2O_2Cl_2$; $SOCl_2$, $COCl_2$	280–550	LiF–NaF–KF or ZnCl$_2$–KCl–CaF$_2$	SO_2F_2; SOF_2; COF_2	144	$SO_2Cl_2(SOCl_2; COCl_2) + 2F^- \rightarrow SO_2F_2$ (SOF$_2$, COF$_2$) + 2Cl$^-$	1
CCl_4	280–550	LiF–NaF–KF or ZnCl$_2$–KCl–CaF$_2$	CF_xCl_{4-x}	144	$CCl_4 + xF^- \rightarrow CF_xCl_{4-x} + xCl^-$	1
PCl_3	500	LiF–NaF–KF	PF_3	140, 144	$PCl_3 + 3F^- \rightarrow PF_3 + 3Cl^-$	1
Cl_2	350–500	CaC$_2$–Alk.F	F_xCCl_{4-x}	32, 145	$CaC_2 + 2xF^- + 5Cl_2 \rightarrow 2F_xCCl_{4-x} + CaCl_2 + 2xCl^-$	1
$C_2H_2Cl_4$; O_2	300–500	CuCl$_2$–ZnCl$_2$	C_2Cl_4	32, 146	$C_2H_2Cl_4 + \frac{1}{2}O_2 \rightarrow C_2Cl_4 + H_2O$	3
C_2H_4Cl; Cl_2	400–480	NaCl–AlCl$_3$–FeCl$_3$	Dichloro-ethylenes, trichloro-ethylene	32, 147	$6C_2H_4Cl + 11Cl_2 \rightarrow 2CH_2:CCl_2 + 2CHCl:CHCl + 2CHCl:CCl_2 + 14HCl$	3
C_3H_8; HNO$_3$	372–444	Fused nitrates	Nitropropane	32, 148	$C_3H_8 + HNO_3 \rightarrow C_3H_7NO_2 + H_2O$	3
SO_2; SO_3; CO	430	Li$_2$CO$_3$–Na$_2$CO$_3$–K$_2$CO$_3$	CO_2	149	$SO_2(SO_3) + M_2CO_3 \rightarrow M_2SO_3(M_2SO_4) + CO_2$; $2CO + H_2 + M_2SO_3 \rightarrow M_2S + 2CO_2 + H_2O$	1
O_2	400	KCN–alkali nitrates	KOCN	32	$2KCN + O_2 \rightarrow 2KOCN$	1
Cl_2	400	KCN–alkali halides	$(CN)_2$	32	$2KCN + Cl_2 \rightarrow (CN)_2 + KCl$	1

TABLE V (continued)

Gaseous reagent	Temperature range, °C	Melt	Main products	Ref.	Overall reaction	Type of reaction[a]
BCl_3	400	LiCl-KCl-LiH	B_2H_6	150, 151	$2BCl_3 + 6LiH \rightarrow B_2H_6 + 6LiCl$; (via $LiBH_4$)	1
$SiCl_4$	400	LiCl-KCl-LiH	SiH_4	150, 151	$SiCl_4 + 4LiH \rightarrow SiH_4 + 4LiCl$	1
$SiCl_4$	400	LiCl-KCl-KCNS	$Si(CNS)_4$	152	$SiCl_4 + 4KCNS \rightarrow Si(CNS)_4 + 4KCl$	1
n-Butyl chloride	400	$ZnCl_2$-CuCl	Butene	129	$C_4H_9Cl + CuCl \rightarrow C_4H_8 + HCl$	3
C_2H_6, Cl_2	400	NaCl-$AlCl_3$-$CuCl_2$	Ethylchloride, Vinylchloride	153	$2C_2H_6 + 3Cl_2 \rightarrow C_2H_5Cl + C_2H_3Cl + 4HCl$	3
C_6H_6, Cl_2	200–400	NaCl-$AlCl_3$-$FeCl_3$	C_6H_5Cl	129	$C_6H_6 + Cl_2 \rightarrow C_6H_5Cl + 4HCl$	3
$COCl_2$	380–400	LiCl-KCl-K_2S	COS	134	$COCl_2 + K_2S \rightarrow COS + 2KCl$	1
SO_2	380	$K_2S_2O_7$–V_2O_5	SO_3	154	$V_2O_5 + SO_2 \rightarrow SO_3 + V_2O_4$	3
CH_3Cl (C_2H_5Cl)	>370	LiCl-KCl-KCN	CH_3CN (C_2H_5CN)	152	$CH_3Cl (C_2H_5Cl) + KCN \rightarrow CH_3CN (C_2H_5CN) + KCl$	1
Tertiary alcohols	350 or above	NaCl-$ZnCl_2$; or CuCl	Olefins	32, 129, 155	Example: $(CH_3)_3COH \rightarrow C_4H_8 + H_2O$	3
CH_3OH, HCl	350	KCl-$ZnCl_2$	CH_3Cl	33	$CH_3OH + HCl \rightarrow CH_3Cl + H_2O$	3, 5

Solubilities of Reactive Gases in Molten Salts

Gas(es)	Temp	Melt	Product	Reaction	Ref.	Type
Dimethyl ether, HCl	300–350	KCl–ZnCl$_2$	CH$_3$Cl	C$_2$H$_6$O + 2HCl → 2CH$_3$Cl + H$_2$O	33	3
1-1 dichloroethane, C$_2$H$_2$, HCl	330	CuCl–ZnCl$_2$ or KCl–ZnCl$_2$	Vinylchloride	C$_2$H$_4$Cl$_2$ → C$_2$H$_3$Cl + H$_2$O	156	3
CH$_4$, Cl$_2$	320	KHSO$_4$–ZnCl$_2$–Alk.Cl–Alk.Earth Cl$_2$–FeCl$_3$ (CuCl$_2$)	CH$_3$Cl	CH$_4$ + Cl$_2$ → xCH$_3$Cl + xHCl + (1 − x)(CH$_2$Cl$_2$ + CHCl$_3$)	32, 157	3
C$_2$Cl$_4$, H$_2$	300	CuCl–ZnCl$_2$	C$_2$HCl$_3$	C$_2$Cl$_4$ + H$_2$ → C$_2$HCl$_3$ + HCl	32, 158	3
O$_2$ or air	240–300	Alk.OH; MnO$_2$	K$_3$MnO$_4$	2MnO$_2$ + 6KOH + ½O$_2$ → 2K$_3$MnO$_4$ + 3H$_2$O	32, 159–161	1
O$_2$ or air	240–300	Alk.OH; K$_3$MnO$_4$	K$_2$MnO$_4$	2KMnO$_4$ + ½O$_2$ + H$_2$O → 2K$_2$MnO$_4$ + 2KOH	32, 159–161	1
C$_2$H$_2$, Cl$_2$, (CCl$_4$)	175–250	NaCl–AlCl$_3$–FeCl$_3$	C$_2$H$_4$Cl, C$_2$Cl$_4$, CCl$_4$	5C$_2$H$_2$ + CCl$_4$(dilutant) + 10Cl$_2$ → 2C$_2$H$_4$Cl + 2C$_2$Cl$_4$ + 2CCl$_4$ + 2HCl	32, 162	3
C$_6$H$_6$, C$_2$H$_5$Cl	150–170	Alk.Cl–AlCl$_3$	Ethylbenzene	C$_6$H$_6$ + C$_2$H$_5$Cl → xC$_6$H$_5$·C$_2$H$_5$ + xHCl + (1 − x) tarry products	129, 163	3
C$_2$H$_5$Cl	>140	NaCNS–KCNS–KCl	C$_2$H$_5$CNS	C$_2$H$_5$Cl + KCNS → C$_2$H$_5$CNS + KCl	152	1
SiH$_4$, SiCl$_2$	120	NaCl–AlCl$_3$; (164): AlCl$_3$ under pressure	SiH$_2$Cl$_2$	SiH$_4$ + SiCl$_2$ → 2SiH$_2$Cl$_2$	151, 164	3
C$_2$H$_4$, HCl	100	NaCl–KCl–AlCl$_3$	C$_2$H$_5$Cl	C$_2$H$_4$ + HCl → C$_2$H$_5$Cl	32	3

[a] (1) Thermodynamically controlled reactions in a homogenous molten phase. (2) Chemical reactions in heterogenous systems. (3) Catalytic reactions. (4) Two-step Deacon process. (5) No water-removing agents necessary.

A similar sequence of reactions may also be postulated for the $FeCl_3$ catalytic action. The NaCl component in this melt is not participating in the catalytic reaction. However, it should be realized that the catalytic activity of either $AlCl_3$ or $FeCl_3$ depends indirectly upon the type of alkali chloride present. Alkali chlorides with large cations like CsCl are expected to interact strongly with the $AlCl_3$ or $FeCl_3$ components forming relatively stable permanent complexes[131] within which the Al^{3+} and Fe^{3+} cations are shielded by their ligands. On the contrary, NaCl, because of the rather strong local interaction between the small Na^+ cation and the Cl^- anions, is a poor ligand donor and thus it enhances the catalytic activity of the melts. In the general Lewis concept of acids and bases a salt is an acid if it behaves like an electron acceptor and a base if it behaves as an electron donor. An acid–base reaction represents the sharing of an electron pair between the acidic and the basic components. In this sense all complex-ion-forming reactions are of the acid-base type and salts like $TiCl_3$, $ZnCl_2$, and $SnCl_2$ and most salts of the transition metals are expected to be good catalysts. It follows that an acidic molten salt component behaves as a better catalyst in the presence of a weak base. On the contrary, a strong base is expected to reduce the catalytic activity of a given acidic component. This effect of the second salt component on the catalytic behavior of the active element should be described as "poisoning."

It should be emphasized that the initial polarization of the organic molecule entering the melt phase, which is a requirement for solubility, should activate the molecule to some extent and facilitate further the charge transfer step of a possible catalytic reaction.

It has also been pointed out by Duke et al.[132,133] that anions can also cause catalytic reactions. For example, molten bromates are decomposed in the presence of bromide ions according to the reactions

$$BrO_3^- + Br^- \rightarrow BrO_2^- + BrO^- \quad \text{(intermediate unstable configurations)} \quad (58)$$

$$BrO_2^- \rightarrow Br^- + O_2(g) \quad (59)$$

$$3BrO^- \rightarrow BrO_3^- + 2Br^- \quad (60)$$

For this reaction the Br^- anions are the basic species and the BrO_3^- anions are the acidic species.

Because most of the industrially significant reactions are exothermic, it is important that they may be conducted at as low temperatures as possible. Accordingly, in selecting a suitable molten salt catalyst it is necessary to choose low-melting solutions which, however, do not contain salts that may permanently "poison" the catalyst. Furthermore, since the best salt catalysts are also semiionic in nature and are highly volatile it is essential to select low-melting solvents.

Regarding reactions involving at least one gaseous reactant, the removal of the gaseous products from the molten salt phase ensures their completion. For example, for the reaction

$$COCl_2(gas) + (K_2S)(in\ melt) \rightarrow COS(gas) + 2KCl(in\ melt) \quad (61)$$

which takes place in a K_2S–$LiCl$–KCl melt at temperatures between 380 and 400°C, the ratio P_{COS}/P_{COCl_2} is determined by the equilibrium constant for reaction (61) according to

$$P_{COS}/P_{COCl_2} = Ka_{K_2S}/a_{KCl}^2 \quad (62)$$

The reaction should proceed to completion provided the ratio P_{COS}/P_{COCl_2} in the reactor is maintained at all times at levels below the equilibrium value, a condition which is easily realized in flowing systems.

The curves representing the equilibrium P_{COS}/P_{COCl_2} ratios as functions of temperature and melt composition are expected to obey the same general relationships as for the reactive systems of the type XCl_4–MCl discussed earlier and presented in Fig. 1. Thus, for a given composition the P_{COS}/P_{COCl_2} curves are expected to be continuous for the all-liquid range in the K_2S–KCl–$LiCl$ ternary system and should change slopes at the characteristic liquidus and eutectic temperatures of the corresponding phase diagram. Several exchange reactions of this type which are thermodynamically controlled are given in Table V.

REFERENCES

1. H. Reiss, H. L. Frisch, E. Hefland, and J. L. Lebowitz, *J. Chem. Phys.* **32**:119 (1960).
2. H. H. Uhlig, *J. Phys. Chem.* **41**:1215 (1937).
3. M. Blander, W. R. Grimes, N. V. Smith, and G. M. Watson, *J. Phys. Chem.* **63**:1164 (1959).
4. W. Altar, *J. Chem. Phys.* **5**:577 (1937).
5. R. Fürth, *Proc. Cambr. Phil. Soc.* **37**:252, 276, 281 (1941).
6. H. Bloom and J. O'M. Bockris, in: *Modern Aspects of Electrochemistry*, (J. Bockris, ed.), Vol. 2, Butterworths, London (1959).
7. R. Lorenz and W. Herz, *Z. Anorg. Chem.* **145**:88 (1925).
8. A. Eucken and W. Dannöhl, *Z. Electrochem.* **40**:814 (1934).
9. J. W. Johnson and M. A. Bredig, in Bauer's and Porter's chapter, *Molten Salt Chemistry* (M. Blander, ed.), p. 621, Interscience, New York (1964).

10. H. A. Levy and M. D. Danforth, in: *Molten Salt Chemistry* (M. Blander, ed.), Interscience, New York (1964).
11. T. Førland, "On the properties of some mixture of fused salts," *Norwegian Techn. Acad. Sci., Series* 2, **1957** (4).
12. A. R. Ubbelohde, *Chem. Ind.* **1968**:313 (1968).
13. A. K. K. Lee and E. F. Johnson, *Ind. Eng. Chem. Fund.* **8**:726 (1969).
14. J. L. Copeland and W. C. Zybko, *J. Phys. Chem.* **69**:3631 (1965).
15. J. L. Copeland and W. C. Zybko, *J. Phys. Chem.* **86**:4734 (1964).
16. J. L. Copeland and W. C. Zybko, *J. Phys. Chem.* **70**:181 (1966).
17. J. L. Copeland and J. R. Christie, *J. Phys. Chem.* **73**:1205 (1969).
18. J. L. Copeland and S. Radak, *J. Phys. Chem.* **71**:4360 (1967).
19. J. L. Copeland and L. Seibles, *J. Phys. Chem.* **70**:1811 (1966).
20. J. L. Copeland and L. Seibles, *J. Phys. Chem.* **72**:603 (1968).
21. J. L. Copeland and S. Radak, *J. Phys. Chem.* **70**:3356 (1966).
22. K. Grjotheim, P. Heggelund, C. Krohn, and K. Motzfeldt, *Acta Chem. Scand.* **16**:689 (1962).
23. D. Bratland, K. Grjotheim, C. Krohn, and K. Motzfeldt, *Acta Chem. Scand.* **20**:1811 (1966).
24. D. Bratland, K. Grjotheim, C. Krohn, and K. Motzfeldt, *J. Metals* **19**(10):13 (1967).
25. M. L. Pearce, *J. Am. Ceram. Soc.* **48**:175 (1965).
26. M. L. Pearce, *J. Am. Ceram. Soc.* **47**:342 (1964).
27. J. Greenberg and B. Sundheim, *J. Chem. Phys.* **29**:1029 (1958).
28. M. Kowalski and G. W. Harrington, *Inorg. Nucl. Chem. Letters* **3**:121 (1967).
29. A. Block-Bolten and S. N. Flengas, *Can. J. Chem.* **49**:2266 (1971).
30. R. L. Lister and S. N. Flengas, *Can. J. Chem.* **43**:2947 (1965).
31. I.C.I. Belgian Patent 614467 (1962).
32. W. Sundermeyer, *Angew. Chem.* **77**:241 (1965).
33. W. Sundermeyer, *Chem. Ber.* **97**:1069 (1964).
34. M. Blander, in: *Molten Salt Chemistry* (M. Blander, ed.), Interscience, New York (1964).
35. R. Battino and H. L. Clever, *Chem. Rev.* **66**:395 (1966).
36. H. Bloom and J. W. Hastie, in: *Non-aqueous Solvent Systems* (T. C. Waddington, ed.), Academic Press, New York (1965).
37. G. J. Janz *Molten Salts Handbook*, Academic Press, New York (1965).
38. G. Zimmerman and F. C. Strong, *J. Am. Chem. Soc.* **79**:2063 (1957).
39. G. A. Sachetto, G. G. Bombi, and M. Fiorani, *J. Electroanalytical Chem. Interfacial Electrochem.* **20**:89 (1969).
40. S. N. Flengas, *Ann. N.Y. Acad. Sci.* **79**:853 (1960).
41. J. H. Mui and S. N. Flengas, *Can J. Chem.* **40**:997 (1962).
42. R. L. Lister and S. N. Flengas, *J. Electrochem. Soc.* **111**:343 (1964).
43. J. E. Dutrizac and S. N. Flengas, *Can. J. Chem.* **45**:2314 (1967).
44. P. Pint and S. N. Flengas, unpublished results.
45. I. S. Morozov and Sun-In-Chzhu, *Russ. J. Inorg. Chem.* **4**:307 (1959).
46. I. S. Morozov and Sun-In-Chzhu, *Russ. J. Inorg. Chem.* **4**:1176 (1959).
47. L. J. Howell, R. C. Sommer, and H. H. Kellogg, *Trans. AIME.* **209**:193 (1957).
48. A. S. Roy, L. J. Howell, and H. H. Kellogg, *Trans. AIME* **212**:817 (1958).
49. G. J. Barton, R. J. Sheil, and W. R. Grimes, Phase Diagrams of Nuclear Reactor Materials, ORNL-2548.

50. B. G. Korshunov and N. W. Gregory, *Inorg. Chem.* **3**:451 (1964).
51. S. N. Flengas and T. R. Ingraham, *Can. J. Chem.* **38**: 813 (1960).
52. R. L. Lister and S. N. Flengas, *Can J. Chem.* **41**:1548 (1963).
53. J. E. Dutrizac and S. N. Flengas, Advances in Extractive Metallurgy Symposium, April 1967, Inst. Mining and Metallurgy, paper 24.
54. R. L. Lister and S. N. Flengas, *Can. J. Chem.* **42**:1102 (1964).
55. A. A. Palko, A. D. Ryon, and D. W. Kuhn, *J. Phys. Chem.* **62**:319 (1958).
56. G. P. Luchinskii, *Russ. J. Phys. Chem.* **40**:318 (1966).
57. S. N. Flengas and P. Pint, *Can. Met. Quart.* **8**:167 (1969).
58. A. Block-Bolten and S. N. Flengas, *Can. J. Chem.* **49**: 3327 (1971).
59. J. A. Wasastjerna, *Soc. Sc. Fennica, Comm. Phys. Math.* **XIV**:3 (1948).
60. T. Førland, H. Storegraven, and S. Urnes, *Z. Anorg. Allg. Chem.* **279**:205 (1955).
61. H. von Wartenberg, *Z. Elektrochem.* **32**:330 (1926).
62. Yu. M. Ryabukhin, *Russ. J. Inorg. Chem.* **7**:565 (1962).
63. J. H. Shaffer, W. R. Grimes, and G. M. Watson, *J. Phys. Chem.* **63**:1999 (1959).
64. J. Greenberg, B. R. Sundheim, D. Gruen, *J. Chem. Phys.* **29**:461 (1958).
65. W. J. Burkhard and J. D. Corbett, *J. Am. Chem. Soc.* **79**:6361 (1957).
66. W. R. Grimes, N. V. Smith, and G. M. Watson, *J. Phys. Chem.* **62**:862 (1958).
67. G. M. Watson, R. B. Evans, W. R. Grimes, and N. V. Smith, *J. Chem. Eng. Data* **7**:285 (1962).
68. I. A. Novochatskii, O. A. Esin, and S. K. Chuchmarev, *Chem. Abstr.* **56**:9795 (1962).
69. E. A. Sullivan, S. Johnson, and M. D. Banus, *J. Am. Chem. Soc.* **77**:2023 (1955).
70. J. Dubois, *Ann. Chim. (Paris)* **10**:145 (1965).
71. J. P. Frame, E. Rhodes, and A. R. Ubbelohde, *Trans. Faraday Soc.* **57**:1075 (1961).
72. M. Schencke, G. H. J. Broers, and J. A. A. Keterlaar, *J. Electrochem. Soc.* **113**:404 (1966).
73. B. Cleaver and D. E. Mather, *Trans. Faraday Soc.* **66**:2469 (1970).
74. P. E. Field and W. J. Green, *J. Phys. Chem.* **75**:821 (1971).
75. W. Klemm and E. Hus, *Z. Anorg. Chem.* **258**:221 (1949).
76. D. R. Olander and J. L. Camahort, *AIChE J.* **12**:693 (1966).
77. Yu. M. Ryabukhin and N. G. Bukun, *Russ. J. Inorg. Chem.* (Engl. Transl.) **13**:597 (1968).
78. Yu. M. Ryabukhin, *Russ. J. Phys. Chem.* **39**:1563 (1965).
79. Yu. M. Ryabukhin, *Russ. J. Inorg. Chem.* **11**:1296 (1966).
80. S. B. Tricklebank, Abstr. No. 48, Electrochemical Soc. extended abstracts, Meeting Oct. 1969.
81. J. D. van Norman and R. J. Tivers, in: *Molten Salts* (Gleb Mamantov, ed.), Marcel Dekker, New York (1969).
82. J. D. van Norman and R. J. Tivers, *J. Electrochem. Soc.* **118**:258 (1971).
83. P. E. Field and J. H. Shaffer, *J. Phys. Chem.* **71**:3218 (1967).
84. J. H. Shaffer, W. R. Grimes, and G. M. Watson, *Nucl. Sci. Eng.* **12**:337 (1962).
85. V. N. Devyatkin and E. A. Ukshe, *Zh. Prikl. Khim* **38**:1612 (1965).
86. R. A. Howald and J. E. Willard, *J. Am. Chem. Soc.* **77**:2046 (1955).
87. T. L. Lukmanova and Ya. E. Vilnyanskii, *Chem. Abstr.* **64**:13812c.
88. R. V. Winsor and G. H. Cady, *J. Am. Chem. Soc.* **70**:1500 (1948).
89. G. Bertozzi, *Z. Naturforsch.* **A22**:1748 (1967).
90. S. Bretsznajder, *Roczniki Chemii* **10**:729 (1930).

91. A. G. Keenan, *J. Phys. Chem.* **61**:780 (1957).
92. J. B. Raynor, *Z. Elektrochemie* **67**:360 (1963).
93. F. R. Duke and A. S. Doan, *Iowa State College J. Sci.* **32**:451 (1958).
94. D. L. Maricle and D. N. Hume, *J. Electrochem. Soc.* **107**:354 (1960).
95. H. A. Laitinen, W. S. Ferguson, and R. A. Osteryoung, *J. Electrochem. Soc.* **104**:516 (1957).
96. S. I. Rempel, *Dokl. Akad. Nauk. SSSR* **74**:331 (1950).
97. S. Pizzini, R. Morlotti, and E. Roemer, *Nucl. Sci. Abstr.* **20**:43 (1966).
98. G. Mamantov, in: "Molten Salts," (G. Mamantov, ed.), p. 540, Marcel Dekker, New York (1969).
99. A. N. Campbell and D. F. Williams, *Can. J. Chem.* **42**:1778 (1964).
100. A. N. Campbell and D. F. Williams, *Can. J. Chem.* **42**:1984 (1964).
101. T. I. Crowell and P. Hillery, *J. Org. Chem.* **30**:1339 (1965).
102. A. R. Glueck and C. N. Kenney, *Chem. Eng. Sci.* **23**:1257 (1968).
103. H. Flood and O. J. Kleppa, *J. Am. Chem. Soc.* **69**:998 (1947).
104. D. Bratland and C. Krohn, *Acta Chem. Scand.* **23**:1839 (1969).
105. D. Bratland, K. Grjotheim, and C. Krohn, Unpublished results, private communication by H. Øye.
106. J. Mahieux, *Compt. Rend.* **240**:2521 (1955).
107. E. H. Baker, *J. Chem. Soc.* **1962**:464.
108. E. H. Baker, *J. Chem. Soc.* **1962**:2525.
109. E. H. Baker, *J. Chem. Soc.* **1963**:339.
110. E. H. Baker, *J. Chem. Soc.* **1963**:699.
111. S. Allulli, *J. Phys. Chem.* **73**:1084, (1969).
112. I. S. Morozov and B. G. Korshunov, *Russ. J. Inorg. Chem.* **1**:145 (1956).
113. P. W. Seabaugh and J. D. Corbelt, *Inorg. Chem.* **4**:176 (1965).
114. H. Scholze and H. O. Mulfinger, *Angew. Chem.* **74**:75 (1962).
115. H. Scholze, H. O. Mulfinger, and H. Franz, *Chem. Abstr.* **58**:326e.
116. R. Kh. Kumarev and S. A. Amirova, *Zh. Neorg. Khim.* **13**:2258 (1968).
117. M. V. Smirnov and V. S. Maksimov, *Elektrokhimiya* **1**:727 (1965).
118. M. V. Smirnov, V. S. Maksimov, and A. P. Khaimenov, *Zh. Neorg. Khim.* **11**:1765 (1966).
119. V. S. Maksimov and M. V. Smirnov, *Zh. Prikl. Khim.* **39**:931 (1966).
120. M. V. Smirnov and V. S. Maksimov, *Tr. Inst. Elektrokhim. Akad. Nauk SSSR Uralsk. Fil.* **8**:35 (1966).
121. V. S. Maksimov and M. V. Smirnov, *Tr. Inst. Elektrokhim. Akad. Nauk SSSR Uralsk. Fil.* **9**:41 (1966).
122. M. V. Smirnov and V. S. Maksimov, *Tr. Inst. Elektrokhim. Akad. Nauk SSSR Uralsk. Fil.* **10**:49 (1967).
123. I. S. Morozov and D. A. Toptygin, *Russ. J. Inorg. Chem.* **5**:42 (1960).
124. P. Pint and S. N. Flengas, *Can. J. Chem.* **49**:2285 (1971).
125. W. Sundermeyer, *Chem. Ber.* **97**:1069 (1964).
126. W. Sundermeyer, German Patent Appl. 1,080,077, 1957.
127. B. V. Nekrasov, *General Chemistry*, 9th ed, Chemizdat, Moscow, 1952.
128. J. H. Reilly, U. S. Patent 2140550 (1938).
129. O. Glemser and K. Kleine-Weischede, *Liebigs Ann. Chem.* **17**:659 (1962).
130. C. A. Vanderwerf, *Acids, Bases, ...*, Reinhold, New York (1961).
131. S. N. Flengas and A. S. Kucharski, *Can. J. Chem.* **49**: 3971 (1971).

132. F. R. Duke and E. Shute, *J. Phys. Chem.* **66**:2114 (1962).
133. F. R. Duke, *Reaction Kinetics in Fused Salts* (B. R. Sundheim, ed.), McGraw-Hill, New York (1964).
134. M. Fild, W. Sundermeyer, and O. Glemser, *Chem. Ber.* **97**:620 (1964).
135. L. M. Litz, German Patent Appl. 1096884 (1961).
136. J. Hille and W. Dürrwachter, *Angew. Chem.* **72**:850 (1960).
137. M. Baccaredda and F. G. Nencetti, *Chim. e. Ind.* **42**:1084 (1960).
138. E. Gorin, U. S. Patent 2,418,931 (1947).
139. C. M. Fontana, U. S. Patent 2,447,323 (1948).
140. W. Sundermeyer and W. Meise, *Z. Anorg. Allg. Chem.* **317**:334 (1962).
141. E. Gorin and C. B. Miles, U. S. Patent 2,444,289 (1948).
142. R. S. Hanner and S. Swann, *Ind. Eng. Chem.* **41**:325 (1949).
143. E. Gorin, U. S. Patent 2,498,546 (1950).
144. W. Sundermeyer, *Z. Anorg. Allg. Chem.* **314**:100 (1962).
145. H. L. Roberts, Brit. Patent 874099 (1961).
146. R. E. Feathers and R. H. Rogerson, U. S. Patent 2,914,575 (1956).
147. J. H. Reilly, U. S. Patents 2,140,548 and 2,140,549 (1938).
148. D. C. Coldiron, L. F. Albright, and L. G. Alexander, *Ind. Eng. Chem.* **50**:991 (1958).
149. L. F. Grantham, L. A. Heredy, D. E. McKenzie, R. D. Oldenkamp, and S. J. Yosim, Report on contract # PH 86-67-128, U. S. Public Health Service, Dept. of Health, Education, and Welfare.
150. W. Sundermeyer and O. Glemser, *Angew. Chem.* **70**:625 (1958).
151. W. Sundermeyer and O. Glemser, *Angew. Chem.* **70**:628 (1958).
152. W. Sundermeyer, *Z. Anorg. Allg. Chem.* **313**:290 (1961).
153. J. H. Reilly, U. S. Patent 2,140,547 (1938).
154. A. R. Glueck and C. N. Kenney, *Chem. Eng. Sci.* **23**:1257 (1968).
155. N. E. Norman and H. F. Johnstone, *Ind. Eng. Chem.* **43**:1553 (1951).
156. W. Sundermeyer, O. Glemser, and K. Kleine-Weischede, *Chem. Ber.* **95**:1829 (1962).
157. Holzverkohlungsindustrie A. G., German Patent 393550 (1924).
158. Uddeholms, A. B. Belgian Patent 616504 (1962).
159. H. Lux, R. Kuhn, and T. Niedermaier, *Z. Anorg. Allg. Chem.* **298**:285 (1959).
160. H. Lux, E. Renauer, and E. Betz, *Z. Anorg. Allg. Chem.* **310**:305 (1961).
161. H. Lux and T. Niedermaier, *Z. Anorg. Allg. Chem.* **285**:246 (1956).
162. J. H. Reilly, U. S. Patent 2,140,551 (1938).
163. W. Sundermeyer and O. Glemser, *Angew. Chem.* **70**: 629 (1958).
164. A. Stock and C. Somieski, *Ber. deutsch. Chem. Ges.* **52**:695 (1919).
165. N. A. *Krasilnikova*, M. V. Smirnov, and I. H. Ozeryanaya, *Tr. Inst. Elektrokhim. Akad. Nauk SSSR Uralsk. Fil.* **12**:50 (1969).
166. N. A. Krasilnikova, M. V. Smirnov, and I. H. Ozeryanaya, *Tr. Inst. Elektrokhim. Akad. Nauk SSSR Uralsk. Fil.* **14**:3 (1970).

Chapter 3

HIGH-TEMPERATURE COORDINATION CHEMISTRY OF GROUP VIII

Keith E. Johnson
Department of Chemistry
University of Saskatchewan, Regina Campus
Regina, Saskatchewan, Canada

and

John R. Dickinson
Department of Chemistry
University of Virginia
Charlottesville, Virginia

> "When I use a word," Humpty Dumpty said, in rather a scornful tone, "it means just what I choose it to mean—neither more nor less."
> "The question is," said Alice, "whether you can make words mean so many different things."
> "The question is," said Humpty Dumpty, "which is to be master—that's all."
>
> L. Carroll

1. INTRODUCTION

1.1. Scope

The bulk of our discussion is concerned with electronic spectroscopy but we shall indicate the various other methods available to help in the elucidation of the coordination of Group VIII metal ions at high temperatures. The two types of information which electronic spectroscopy can provide under favorable circumstances are the number and arrangement of ligands about the central metal and the distribution of electronic charge within the moiety, albeit the expression of the latter is usually restricted to a listing of the first few energy levels.

Since our experience and any modicum of expertise are restricted to molten salt systems, a word of explanation for Section 3 is called for. Many inorganic chemists appear to have become somewhat insular in their discussions of coordination and bonding, not drawing enough upon the experiences of their scientific colleagues who work with different-sized molecules in different phases or at other temperatures. It is hoped, therefore, to create some awareness of the gas-phase work in the minds of fellow molten salt advocates and vice versa. With a few notable exceptions (the writings of Herzberg, Jørgensen, Gray, and Gruen, for example) it appears that questions such as the *raison d'etre* of the formation by nickel of the series of entities NiCl, $NiCl_2$, $NiCl_3^-$, and $NiCl_4^{2-}$ in fluid phases have received little attention. We believe that the time is ripe for raising them even though no good answers are yet apparent.

Within Sections 3 and 4 we have introduced results from other techniques where they clarify the electronic spectroscopy questions under discussion but we have not, for example, gone into details about every nickel halide complex in solids or low-temperature solutions. Again, coordination in solids such as spinels at several hundred degrees Celsius is high-temperature coordination but on the whole the techniques involved and the issuant details of the interactions are sufficiently different for us simply to have to commend the topic to the editors of this series or of *Advances in High-Temperature Chemistry*.

Whereas 1500°C is considered by some people to be the lower limit for "high temperatures," we prefer the more flexible approach which treats high temperatures as those from which the majority of chemists shy away other than for wasteful sources of heat such as burners or for classical or pseudoclassical analytical operations. Ferrocene, nickel carbonyl, phthalocyanine vapors, anhydrous cobalt nitrate, and the like, but not nitrate melts, are thus excluded from our deliberations.

In Section 5 we have suggested directions in which we think some exciting results are awaiting revelation and in Section 2 and the appendices we have set out some of the theoretical background which has been or could be used in the interpretation of the electronic spectra of transition metal compounds.

1.2. Methods of Studying Coordination at High Temperatures

Of the several methods available for studying the structure of coordination compounds as room-temperature solids or solutions, very few have obvious simple application to the liquid and vapor phases at high temperatures. X-ray and neutron diffraction studies did reveal the average coordination numbers in single alkali halides, for example,[1] and we might anticipate some future success with electron diffraction of simple vapors.

The method of vibrational spectroscopy has found useful application to groups such as the NO_3^- ion[2] but studies of such entities as $NiCl_4^{2-}$ or $PtCl_4^{2-}$ by the laser Raman technique are still awaited. While NMR,[3] ESR,[4-6] and magnetic susceptibility studies[7,8] of molten salt systems have been made, only the last have been of much use in coordination assignment. The techniques used to measure stability constants of complexes, such as polarography,[9] potentiometry,[10,11] and chromatography,[12] are beginning to be applied to molten salt systems but the results to date require caution in interpretation.

Phase rule studies,[13,14] vapor pressure measurements,[15] and mass spectrometry[16] are useful in so far as they give indications of stoichiometry but we must not assume that the formation of the compound Cs_3NiCl_5,[17] for example, is evidence for 5-coordinate nickel.

To date the most useful method for studying coordination of transition metal ions in melts and a few vapors has been electronic spectroscopy and this will form the kernel of the succeeding sections.

1.3. General Survey of the Coordination Chemistry of Group VIII

A thorough review of the coordination chemistry of Group VIII would, by this time, fill a book of several volumes. We would draw the attention of our readers to three texts[18-20] and to three serial publications.[21-23] The diversity of complexes formed by Group VIII metals is apparent from the table of contents of any issue of *Inorganic Chemistry* or the *Journal of the Chemical Society A* or by scanning Sections 65–79 of a current issue of *Chemical Abstracts*.

In Table I we present examples of oxidation states and coordination of Group VIII metals which might have some bearing on their behavior at elevated temperatures if and when a wider variety of systems are studied.

TABLE I. Some Complexes of Group VIII Elements

Element	Oxidation state	Symmetry	Examples
Fe	0	D_{3h}	$Fe(CO)_5$
	I	C_{4v}	$Fe(NO)(H_2O)_5^{2+}$
	II	T_d	$FeCl_4^{2-}$
	II	O_h	$Fe(CN)_6^{4-}$
	III	T_d	$FeCl_4^{-}$
	III	O_h	$Fe(H_2O)_6^{3+}$
	VI	T_d	FeO_4^{2-}
Co	$-$I	T_d	$Co(CO)_4^{-}$
	II	T_d	$CoCl_4^{2-}$
	II	O_h	$Co(H_2O)_6^{2+}$
	II	D_{2d}	$Co(NO_3)_4^{2-}$
	II	C_{4v}	$Co(CN)_5^{3-}$
	III	O_h	CoF_6^{3-}
Ni	0	T_d	$Ni(CN)_4^{4-}$, $Ni(CO)_4$
	I	—	$Ni_2(CN)_6^{4-}$
	II	O_h	$CsNiCl_3$
	II	T_d	$NiCl_4^{2-}$
	II	D_{4h}	$Ni(CN)_4^{2-}$
	IV	O_h	NiF_6^{2-}
Ru	II	C_{4v}	$Ru(NO)Cl_5^{3-}$
	III	O_h	$RuCl_6^{3-}$
	IV	O_h	$RuCl_6^{2-}$
	VIII	T_d	RuO_4
Rh	I	D_{4h}	$Rh(SnCl_3)_4^{3-}$
	III	O_h	$RhCl_6^{3-}$
	IV	O_h	RhF_6^{2-}
Pd	II	D_{4h}	$PdCl_4^{2-}$, $Pd(CN)_4^{2-}$
	II	O_h	PdF_2
	IV	O_h	$PdCl_6^{2-}$
Os	II	O_h	$Os(CN)_6^{4-}$
	III	D_3	$Os(acac)_3$
	IV	O_h	$OsCl_6^{2-}$
	VI	O_h	OsF_6
	VIII	T_d	OsO_4
	VIII	C_{3v}	OsO_3N^{-}

TABLE I (continued)

Element	Oxidation state	Symmetry	Examples
Ir	III	O_h	$IrCl_6^{3-}$
	IV	O_h	$IrCl_6^{2-}$
	VI	O_h	IrF_6
Pt	II	D_{4h}	$PtCl_4^{2-}$, $Pt(CN)_4^{2-}$
	II	D_{3h}	$Pt(SnCl_3)_5^{3-}$
	II(?)	C_{4v}	$Pt(NO)Cl_5^{2-}$
	IV	O_h	PtF_6^{2-}
	VI	O_h	PtF_6

Oh! my name is John Wellington Wells,
I'm a dealer in magic and spells,
  In blessings and curses
  And ever-filled purses,
In prophecies, witches and knells.

If you want a proud foe to "make tracks"—
If you'd melt a rich uncle in wax—
  You've but to look in
  On our resident Djinn,
Number seventy, Simmery Axe!

The Sorcerer, W. S. Gilbert

2. ELECTRONIC SPECTROSCOPY AND BONDING

2.1. Atomic Spectra and Coupling Schemes

In this section we wish to discuss the relationships between the stationary electronic states of free atoms and ions. The quantum mechanical derivation of these relationships is to be found in almost any required depth in several texts so that we shall be content to classify the states and list expressions for their energies.

Electronic states are conveniently classified in terms of the electronic configuration, as described by the principal and orbital quantum numbers, and the interactions between orbital and spin momenta of the electrons outside closed shells. These interactions are generally denoted by reference to vector coupling schemes and we shall examine viable schemes for several systems, not being restricted to the customary configurations of Group VIII elements.

The most common coupling scheme is known as *LS*—or Russell-Saunders—coupling. In this it is considered that all the orbital momenta of the electrons, represented by vectors $\mathbf{l}_i$, combine to form a resultant $\mathbf{L}$ and all the spin momenta, represented by vectors $\mathbf{s}_i$, combine to form a resultant $\mathbf{S}$. The vectors $\mathbf{L}$ and $\mathbf{S}$ then combine to form the total angular momentum vector $\mathbf{J}$, taking all integral values from $|L+S|$ to $|L-S|$. The stationary states are then labeled in terms of the multiplicity, $2S+1$, the L value, as shown by a capital letter:

$$L = 0, 1, 2, 3, 4, 5, 6, \text{ etc.}$$

$$\text{S P D F G H I}$$

and the J value; thus $^4F_{9/2}$ means a state with $S = 3/2$, $L = 3$, $J = 9/2$. In the limit of zero interaction between spin and orbital momenta all states of a given multiplet will have the same energy, in other words, the multiplet will be $(2S+1)$-fold degenerate (provided $L \geq S$).[24] Even in the very light elements this situation is not quite realized and separation into components with different J values occurs. This separation is easily accounted for in the Hamiltonian of the system by introducing spin–orbital coupling as a small perturbation[24]:

$$H = \sum_{i=1}^{N} \S(r_i)(\mathbf{s}_i \cdot \mathbf{l}_i)$$

where r_i is the radial coordinate of the ith electron. The magnitude of the separation between adjacent J values of a multiplet is then

$$E_J - E_{J-1} = \pm \zeta_{nl} J/2S$$

where ζ_{nl} is the one-electron spin–orbital coupling constant.[24,25]

For $3p$ and $3d$ configurations it is often sufficient to neglect off-diagonal elements in calculations of the spin–orbital energy but with $4p$, $4d$, and $4f$ configurations this is not such a good approximation and for the $5p$ and $5d$ configurations the *LS* coupling scheme leaves much to be desired.

In *jj* coupling the individual $\mathbf{l}_i$ and $\mathbf{s}_i$ are coupled to form $\mathbf{j}_i$ and then the various $\mathbf{j}_i$ are coupled together to form states again characterized by J. Interactions such as $\mathbf{l}_i + \mathbf{l}_k$, $\mathbf{l}_i + \mathbf{s}_k$, and $\mathbf{s}_i + \mathbf{s}_k$ are neglected to a first approximation and L and S are no longer good quantum numbers. States

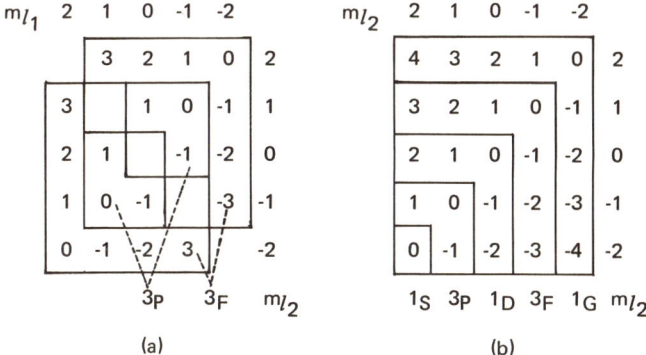

Fig. 1. Spectroscopic terms in LS coupling for a d^2 configuration.[25,27,28] Possible m_l values for d electrons range from 2 to -2. Resultant $M_L = \sum m_l$ values are arranged in blocks. Number of microstates $= {}^{10}C_2 = 45$. The microstates computed from the principal diagonal of the left-hand square are excluded and those in the upper right half are identical with those in the lower left. Allowed terms: ^1S, ^3P, ^1D, ^3F, ^1G. (a) $S_1 = S_2 = \pm \frac{1}{2}$, $m_{l_1} = m_{l_2}$, $M_s = \sum S = \pm 1$, triplet states ^3P and ^3F each require $M_s = 1, 0, -1$. (b) $S_1 = -S_2 = \frac{1}{2}$, $M_s = \sum S = 0$, singlets and remainder of triplets.

may be labeled by the j values: Thus the (3/2, 5/2) state of a d^2 configuration has $j_1 = 3/2$, $j_2 = 5/2$, and one $J = 4$.

For electrons having the same principal quantum number the Pauli exclusion principle operates to reduce the number of terms from that predicted by a *carte blanche* operation of the vector coupling models. In LS coupling the exclusion principle is best operated by introducing the magnetic quantum number m_l having integral values from $+l$ to $-l$ for an electron with orbital quantum number l [the $(2l+1)$-fold orbital degeneracy of a one-electron state is lifted by a magnetic field]. If we then specify that no two electrons can have all four of n, l, m_l, and s identical, we can derive the allowable terms as illustrated for the d^2 configuration in Fig. 1.[25,27,28] In jj coupling the exclusion principle may be operated in a similar manner although it is often expressed somewhat differently[24,25] for two electrons: If $j_1 \neq j_2$, then J may take all values from $|j_1 + j_2|$ to $|j_1 - j_2|$ but if $j_1 = j_2$ then odd values of J are excluded. For systems with several electrons the allowed J values, given by Condon and Shortley,[24] are reproduced in Table II. Tables IV and V list the LS terms of configurations with various numbers of equivalent p and d electrons and Table III gives the corresponding jj states.

TABLE II. Allowed J Values for Groups of $nl\,j$-Equivalent Electrons[24]

	J		J		J
$l^0_{1/2}, l^2_{1/2}$	0	$l^0_{3/2}, l^4_{3/2}$	0	$l^0_{5/2}, l^6_{5/2}$	0
$l_{1/2}$	1/2	$l^1_{3/2}, l^3_{3/2}$	3/2	$l^1_{5/2}, l^5_{5/2}$	5/2
		$l^2_{3/2}$	0, 2	$l^2_{5/2}, l^4_{5/2}$	0, 2, 4
				$l^3_{5/2}$	3/2, 5/2, 9/2

TABLE III. (j, j) States of Equivalent p^n and d^n Configurations

p^2, p^4	1/2, 1/2	3/2, 1/2	3/2, 3/2		
p^3	3/2, 1/2, 1/2	3/2, 3/2, 1/2	3/2, 3/2, 3/2		
d^2, d^8	3/2, 3/2	5/2, 3/2	5/2, 5/2		
d^3, d^7	3/2, 3/2, 3/2	5/2, 3/2, 3/2	5/2, 5/2, 3/2	5/2, 5/2, 5/2	
d^4, d^6	$(3/2)^4$	$(3/2)^3(5/2)$	$(3/2)^2(5/2)^2$	$(3/2)(5/2)^3$	$(5/2)^4$
d^5		$(3/2)^4(5/2)$	$(3/2)^3(5/2)^2$	$(3/2)^2(5/2)^3$	$(3/2)(5/2)^4$ $(5/2)^5$

TABLE IV. Interaction Energies for the Configurations p^2, p^3, and p^4 in LS Coupling[31]

Configuration	Term	Energy
p^2	^1S	$F_0 + 10F_2$
	^1D	$F_0 + F_2$
	^3P	$F_0 - 5F_2$
p^3	^2P	$3F_0$
	^2D	$3F_0 - 6F_2$
	^4S	$3F_0 - 15F_2$
p^4	^1S	$6F_0$
	^1D	$6F_0 - 9F_2$
	^3P	$6F_0 - 15F_2$

TABLE V. Relative Term Energies for the Configurations $d^{n(31)}$

d^2, d^8	Energy	d^4, d^6	Energy	d^5	Energy
^3F	0	^5D	0	^6S	0
^1D	$5B + 2C - 6\alpha$	^3P	$16B + 5.5C - 4\alpha - R_1$	^4G	$10B + 5C + 20\alpha$
^3P	$15B - 10\alpha$	^3H	$4B + 4C + 24\alpha$	^4P	$7B + 7C + 2\alpha$
^1G	$12B + 2C + 8\alpha$	^3F	$16B + 5.5C + 6\alpha - R_2$	^4D	$17B + 5C + 6\alpha$
^1S	$22B + 7C - 12\alpha$	^3G	$9B + 4C + 14\alpha$	^2I	$11B + 8C + 42\alpha$
		^1G	$16B + 7.5C + 14\alpha - R_3$	^2D	$32B + 11C + 6\alpha - R'$
d^3, d^7		^3D	$16B + 4C$	^2F	$10B + 10C + 12\alpha$
^4F	0	^1I	$6B + 6C + 36\alpha$	^4F	$22B + 7C + 12\alpha$
^4P	$15B - 10\alpha$	^1S	$31B + 10C - 6\alpha - R_4$	^2H	$13B + 10C + 30\alpha$
^2G	$4B + 3C + 8\alpha$	^1D	$30B + 7.5C - R_5$	^2G	$22B + 8C + 20\alpha$
^2P	$9B + 3C - 10\alpha$	^1F	$21B + 6C + 6\alpha$	$^2F'$	$26B + 8C + 12\alpha$
^2D	$20B + 5C - 6\alpha - R$	$^3P'$	$16B + 5.5C - 4\alpha + R_1$	^2S	$32B + 8C$
^2H	$9B + 3C + 10\alpha$	$^3F'$	$16B + 5.5C + 6\alpha + R_2$	$^2D'$	$31B + 10C + 6\alpha$
^2F	$24B + 3C$	$^1G'$	$16B + 7.5C + 14\alpha + R_3$	$^2G'$	$38B + 10C + 20\alpha$
$^2D'$	$20B + 5C - 6\alpha + R$	$^1D'$	$30B + 7.5C + R_5$	^2P	$55B + 10C + 2\alpha$
		$^1S'$	$31B + 10C - 6\alpha + R_4$	$^2D''$	$32B + 11C + 6\alpha + R'$

$R = (193B^2 + 8BC + 4C^2)^{1/2}$

$R_1 = \frac{1}{2}(912B^2 - 24BC + 9C^2)^{1/2}$
$R_2 = \frac{3}{2}(68B^2 + 4BC + C^2)^{1/2}$
$R_3 = \frac{1}{2}(708B^2 - 12BC + 9C^2)^{1/2}$
$R_4 = 2(193B^2 + 8BC + 4C^2)^{1/2}$
$R_5 = \frac{3}{2}(144B^2 + 8BC + C^2)^{1/2}$

$R' = 3(57B^2 + 2BC + C^2)^{1/2}$

The coupling in highly excited configurations and some involving f electrons may not be capable of description in LS or jj terms. Other coupling schemes are then introduced, e.g., $J_1 j$ for $4f^4(^5I)6s$ of Nd(II),[29] $J_1 l$ or pair coupling for $4f5g$ of Ce(III),[29] and $p^5 nf$ of the noble gases,[30,31] $J_1 J_2$ for $4f^{13}6s6p$ of Yb(II),[29] and $J_1 L_2$ for $4f^{13}5d6s$ of Yb(II).[29] Pair coupling warrants consideration in connection with later remarks on diatomic molecules. In this scheme the orbital momentum of the outer electron is coupled to the total momentum of the inner electron(s) J_1 to give a resultant angular momentum K which is then coupled to the spin of the outer electron to give a total angular momentum $J = K \pm 1/2$.[31]

The terms in LS coupling and relative term energies of configurations of interest in following sections are given in Tables IV[31] and V.[31-33] The F's and G's are so-called Slater parameters[24,34,35] defined in terms of the Coulomb and exchange integrals, respectively (see Appendix A),[24,36] and B and C are Racah parameters[37] related to F_2 and F_4:

$$B = F_2 - 5F_4, \qquad C = 35F_4$$

The energy of terms of configurations $d^k s$ is given by[38]

$$E(d^k s) = E(d^k) + kF_0(ds) - \tfrac{1}{2}G_2[k + \tfrac{1}{2}(M^2 - M_0^2 - 3)]$$

where $M = 2S + 1$ and $M_0 = 2S_0 + 1$, S and S_0 being the total spin of the terms of $d^k s$ and d^k, respectively.

Fortunately, we can treat the interactions in the heavier elements by a coupling scheme intermediate between LS and jj coupling since the latter scheme is an extreme which is difficult to manipulate and is only approximated by a few real systems anyway.

For intermediate coupling situations the matrix elements of the electrostatic and spin–orbit interactions are calculated in LS coupling (in theory they could be done in jj coupling) and the energy levels are given as the roots of algebraic equations of a degree which is equal to the number of levels having the same J value.

For ls configurations the energy levels are[24,31]

$$^3L_{l+1} = F_0 - G_1 + \tfrac{1}{2}l\zeta_{nl}$$
$$^1L_l, {}^3L_l = F_0 - \tfrac{1}{4}\zeta_{nl} \pm [(G_1 + \tfrac{1}{4}\zeta_{nl})^2 + \tfrac{1}{4}l(l+1)\zeta_{nl}^2]^{1/2}$$
$$^3L_{l-1} = F_0 - G_1 - \tfrac{1}{2}(l+1)\zeta_{nl}$$

For p^2 we obtain the energy levels[24,31]

$$^1D_2, {}^3P_2 = F_0 - 2F_2 + \tfrac{1}{4}\zeta_p \pm [(3F_2)^2 - \tfrac{3}{2}F_2\zeta_p + (\tfrac{3}{4}\zeta_p)^2]^{1/2}$$

$$^3P_1 = F_0 - 5F_2 - \tfrac{1}{2}\zeta_p$$

$$^1S_0, {}^3P_0 = F_0 + \tfrac{5}{2}F_2 - \tfrac{1}{2}\zeta_p \pm [(\tfrac{15}{2}F_2)^2 + \tfrac{15}{2}F_2\zeta_p + (\tfrac{3}{2}\zeta_p)^2]^{1/2}$$

which transform to the levels for p^4 by replacing F_0 by $6F_0 - 10F_2$ and ζ_p by $-\zeta_p$. For p^3 we obtain the energy levels[24,31]

$$^2P_{1/2} = 3F_0, \quad ^2D_{5/2} = 3F_0 - 6F_2, \quad ^4S_{3/2}, {}^2P_{3/2}, {}^2D_{3/2} = 3F_0 + x$$

where x represents the roots of the cubic equation

$$x^3 + 21F_2 x^2 + [90F_2^2 - (9/4)\zeta_p^2]x - (99/4)F_2\zeta_p^2 = 0$$

According to whether the spin–orbital coupling is much larger or much smaller than the appropriate Slater parameter (G_1 or F_2), these formulas reduce to the SL and jj extremes.*

The transformation form LS through intermediate to jj coupling may be represented diagramatically for sl, p^2, p^3, and p^4 by means of the following substitutions:

for sl: $\quad \psi = \chi/(1+\chi), \quad \chi = (2l+1)\zeta_l/4G_1$

for p^2, p^3, and p^4: $\quad \psi = \chi/(1+\chi), \quad \chi = \zeta_p/5F_2$

Figures 2–4 reproduce such diagrams for p^2, p^3, and p^4.[31] These diagrams are not immediately useful in the interpretation of spectra and we have had cause to calculate the energy differences for p^2, p^3, and p^4 for a variety of F_2 and ζ_p values. One curious result is that $^3P_2 - {}^3P_0 \simeq \tfrac{3}{2}\zeta_p$ for p^2. For d^n systems intermediate coupling calculations are much more complex, though not beyond current computing technology. It is of interest to draw a correlation diagram for d^2 in LS and jj coupling (Table VI). We shall see later how each array of energy levels is modified by the electrostatic field of neighboring atoms or ions in a chemical compound.

To complete our sketch of atomic spectral phenomena the effects of electric and magnetic fields and the selection rules for electronic transition will be noted.

---

* Note the absence of first-order ζ_p terms for p^3 in LS coupling.

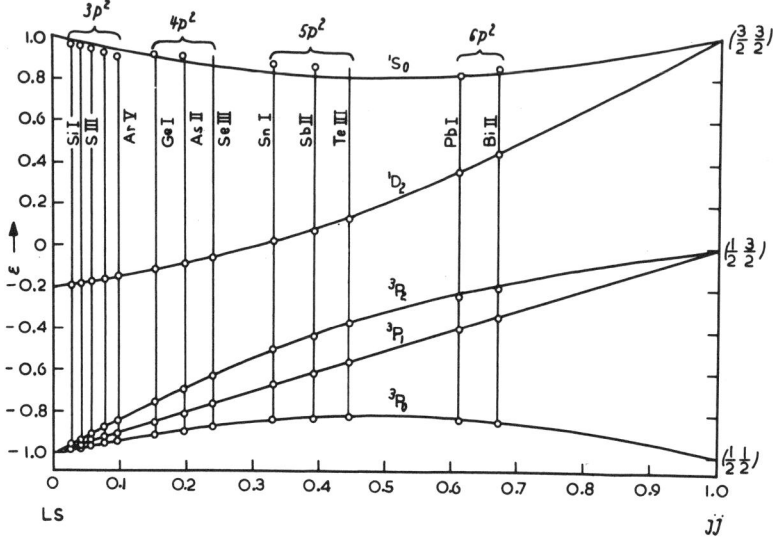

Fig. 2. Intermediate-coupling diagram for the configuration p^2.[31]

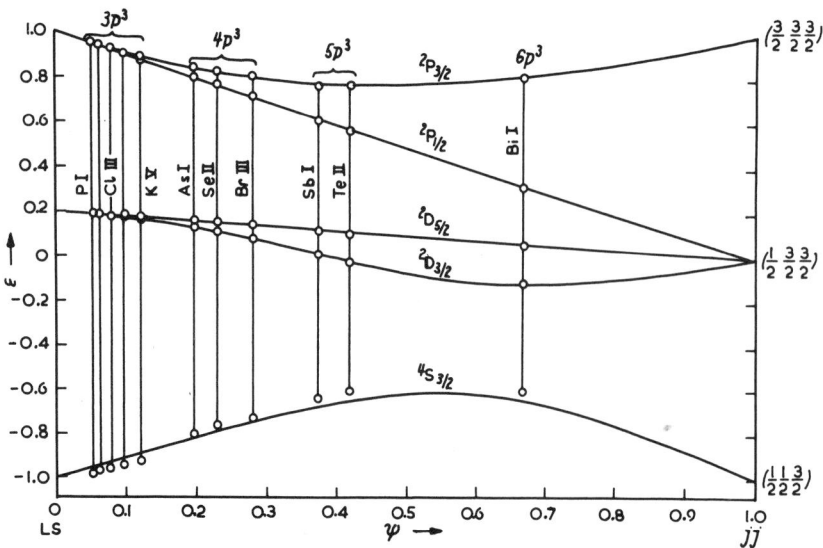

Fig. 3. Intermediate-coupling diagram for the configuration p^3.[31]

High-Temperature Coordination Chemistry of Group VIII

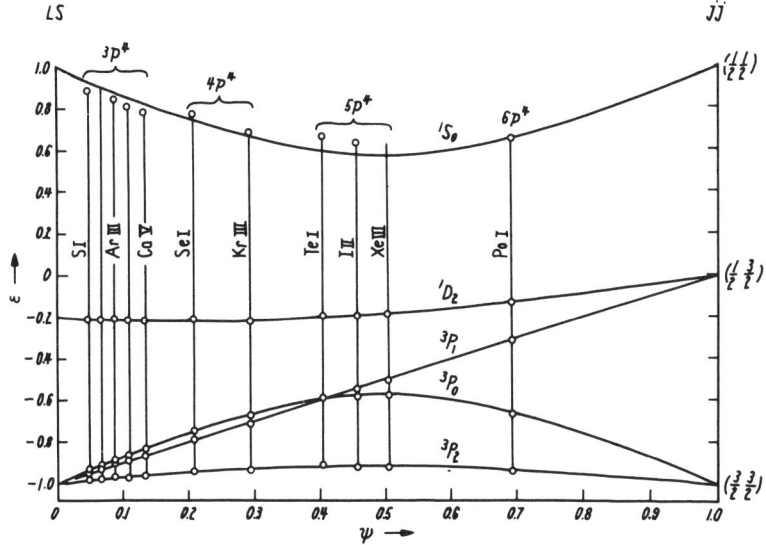

Fig. 4. Intermediate-coupling diagram for the configuration p^4.[31]

TABLE VI. d^2 Configuration in *LS* and *jj* Coupling

LS term	Energy	Energy	cf.	*jj* term
$^1S_0$	$22B + 7C - 12\alpha$	$5\zeta + 427B + (311C/5)$	$^1S_0$	5/2, 5/2
$^1G_4$	$12B + 2C + 8\alpha$	$5\zeta + 175B + (122C/5)$	$^3P_2$	5/2, 5/2
$^3P_2$	$15B - 10\alpha + 2\zeta$	$5\zeta + 7B + (4C/5)$	$^1G_4$	5/2, 5/2
$^3P_1$	$15B - 10\alpha + (3/2)\zeta$	$(5/2)\zeta + 343B + 49C$	$^3P_1$	5/2, 3/2
$^3P_0$	$15B - 10\alpha$	$(5/2)\zeta + 217B + 31C$	$^3F_4$	5/2, 3/2
$^1D_2$	$5B + 2C - 6\alpha$	$(5/2)\zeta + 119B + 17C$	$^1D_2$	5/2, 3/2
$^3F_4$	2ζ	$(5/2)\zeta - 7B - C$	$^3F_3$	5/2, 3/2
$^3F_3$	$(3/2)\zeta$	$392B + 56C$	$^3P_0$	3/2, 3/2
$^3F_2$	0	0	$^3F_2$	3/2, 3/2

2.1.1. Effects of Electric and Magnetic Fields

This subject is considered in Appendix B.

2.1.2. Selection Rules

We will not derive but merely state the selection rules for atomic spectral transitions.[25] For a one-electron transition the l value of that electron changes by unity. In LS coupling[25] we have electric dipole transitions if

$$\Delta S = 0$$
$$\Delta L = 0, \pm 1 \quad \text{but} \quad L = 0 \nleftrightarrow L = 0$$
$$\Delta J = 0, \pm 1 \quad\quad J = 0 \nleftrightarrow J = 0$$

and the rule of parity: $g \leftrightarrow u$ but not $g \leftrightarrow g$ or $u \leftrightarrow u$, the parity being even (*gerade*) or odd (*ungerade*) according as $\sum l_i$ is even or odd. For jj coupling[25] the selection rules for electric dipole transitions are

$$\Delta j_1 = 0, \quad\quad \Delta j_2 = 0, \pm 1$$
$$\Delta j = 0, \pm 1 \quad \text{but} \quad J = 0 \nleftrightarrow J = 0$$

If the initial and final states of transitions are written in each coupling scheme, it will be apparent that transitions allowed under LS coupling are still allowed under jj coupling but that the latter also permits further transitions otherwise spin forbidden. This accounts for the frequent occurrence of intense intercombination lines in the spectra of heavy elements.[26]

For LS coupling the selection rules for magnetic dipole transitions are the same as for electric dipole transitions with one exception: no parity change is allowed. Conservation of parity also holds for electric quadrupole transitions but we have

$$\Delta J = 0, \pm 1, \pm 2 \quad (J_1 + J_2 \geq 2)$$
$$\Delta L = 0, \pm 1, \pm 2, \quad \Delta S = 0$$

2.2. Spectra of and Bonding in Diatomic Molecules

2.2.1. Spectra

When we come to consider the electronic states of molecules we must realise that the spectra are now complicated by vibration or relative motion of the nuclei and rotation or orientation of the molecule in space. Fortunately, provided vibration and rotation are not too strong, the total eigenfunction can be factorized into electronic, vibrational, and rotational components. In other words, the electronic energy can be discussed realistically with reference to a fixed nuclear frame. This is a qualitative description of the Born–Oppenheimer approximation.[40]

In a diatomic molecule the electrons move in a (usually strong) cylindrical electric field.[41] The component of orbital angular momentum along the internuclear axis, described by a quantum number Λ, thus defines the electronic state. Where **L** is the orbital angular momentum vector of all the electrons, Λ takes the values $\Lambda = |M_L| = 0, 1, 2, \ldots, L$. This is an internal Stark effect (cf. Appendix B). As these electronic states are usually well separated, those of given Λ values are labeled $\Sigma, \Pi, \Delta, \Phi$, ... as $\Lambda = 0, 1, 2, 3, \ldots$ analogously to atomic states of given L values. The resultant spin angular momentum vector **S** for molecules has values of $\Sigma h/2\pi$ in the direction of the internuclear axis, where Σ takes integral values from S to $-S$. The quantum number Σ is not defined for Σ states. The addition of Λ and Σ gives the total electronic angular momentum about the internuclear axis Ω, analogous to J for atoms. Ω has $2S+1$ values for each Λ provided $\Lambda \neq 0$. This situation resembles that of an atomic term with $L = 0$ and $S \neq 0$ in a magnetic field. The splitting of a multiplet is proportional to Λ so that $T_e = T_0 + A \cdot \Lambda \cdot \Sigma$ gives the energy of a term component with respect to the baricenter T_0. Here A is the spin–orbital coupling constant, which can be positive or negative, giving rise to multiplets with the smallest and highest Ω values, respectively, as the lowest-energy component. One observes terms labeled as $^3\Delta_r$ or $^3\Delta_i$ indicating a regular $(A + ve)$ or inverted $(A - ve)$ multiplet.[41]

To complete the classification of molecular electronic states, the symmetry properties of the eigenfunctions must be specified. For Σ states ($\Lambda = 0$) one can distinguish between two eigenfunctions[26,42] according to whether the sign is unchanged (Σ^+ state) or reversed (Σ^- state) upon reflection at any plane passing through the line joining the nuclei. For states with $\Omega = 0$ ($\Lambda \neq 0$) one can distinguish between two components (0^+ and 0^-). Similarly, for a molecule formed from two nuclei with the same charge, one can differentiate between a state for which the sign of the electronic eigenfunction is unchanged ($g = $ gerade $\equiv$ even) or reversed ($u = $ ungerade $\equiv$ odd) upon reflection at the center of symmetry.[41]

The coupling of the angular momenta in the molecule may proceed in a variety of ways known as Hund's coupling cases (a) through (e).[43] These are summarized in Table VII. Mulliken's reviews[42,44] are recommended for a full account.

TABLE VII. Hund's Coupling Cases[a]

Case	Coupling of L to axis	Coupling of L to S	Important quantum numbers	Restrictions
(a)	Large	Moderate	Λ, S, Ω	$\Lambda > 0, S > 0$
(b)	Large	Small	Λ, S, K	
(c)	Moderate	Large	Ω	$L > 0, S > 0$
(d)	Moderate	Small	R, K	
(e)	Small	Moderate	L, S, R	$L > 0, S > 0$

[a] After Ref. 44. K is total angular momentum quantum number apart from spin, R is molecular rotational quantum number.

In addition to the selection rule for J, the total angular momentum, as in atoms, and the rule of parity, we now have[41] a conservation of symmetry rule for homonuclear diatomics, a rule for Σ states:

$$\Sigma^+ \leftrightarrow \Sigma^+, \quad \Sigma^- \leftrightarrow \Sigma^-, \quad \Sigma^+ \not\leftrightarrow \Sigma^-$$

a more general rule for Λ and Σ:

$$\Delta\Lambda = 0, \pm 1; \quad \Delta\Sigma = 0$$

for Ω even if Λ and Σ are no longer defined: $\Delta\Omega = 0, \pm 1$; and finally a rule for K where it applies:

$$\Delta K = 0, \pm 1 \quad \text{but not} \quad \Delta K = 0 \text{ for } \Sigma \leftrightarrow \Sigma \text{ transitions}$$

It is possible to correlate the molecular electronic states with the states of the separated atoms and independently, with the united atom having the same number of nucleons and electrons as the molecule. This correlation proceeds according to the Wigner–Witmer rules.[41,45,46,48] and will be outlined in Appendix C.

If we consider the effect of a very strong electric field on the united atom, then L loses its meaning and the l of each electron is a good quantum number with an associated m_l. The M_L, Λ, and S still retain meaning but we can add a quantum number $\lambda = |m_l|$ for each electron and describe electrons in the molecule as $\sigma, \pi, \delta, \phi, \ldots$ electrons for $\lambda = 0, 1, 2, 3, \ldots$[46,49,50]

The Pauli principle must still operate for molecules and takes the form that n, l, and m_l cannot be alike for two electrons unless their spins are antiparallel. However, n and l may not always be defined. Equivalent electrons become defined as those with identical λ and closed shells have *two* electrons for $\lambda = 0$ (σ^2) but *four* electrons for $\lambda \neq 0$. The molecular electronic configurations can then be built up and the possible spectroscopic terms derived by a vector coupling procedure. The small size of closed shells makes this a much simpler exercise than for d^5 or f^7 atomic configurations. Table VIII presents the terms for a selection of configurations. The order of filling orbitals is, however, complicated by the greater number of possibilities and one is restricted to applying some sort of *Aufbauprinzip* within sets of like molecules only.[41]

TABLE VIII. Spectroscopic Terms of Electronic Configurations for Diatomic Molecules[a]

Configuration	Terms
σ	$^2\Sigma^+$
π	$^2\Pi_r$
$\sigma\sigma$	$^1\Sigma^+$, $^3\Sigma^+$
$\sigma\pi$	$^1\Pi$, $^3\Pi_r$
$\sigma\delta$	$^1\Delta$, $^3\Delta_r$
$\pi\pi$	$^1\Sigma^+$, $^3\Sigma^+$, $^1\Sigma^-$, $^3\Sigma^-$, $^1\Delta$, $^3\Delta_r$
$\pi\delta$	$^1\Pi$, $^3\Pi$, $^1\Phi$, $^3\Phi_r$
$\delta\delta$	$^1\Sigma^+$, $^3\Sigma^+$, $^1\Sigma^-$, $^3\Sigma^-$, $^1\Gamma$, $^3\Gamma_r$
σ^2	$^1\Sigma^+$
π^2	$^1\Sigma^+$, $^3\Sigma^-$, $^1\Delta$
π^3	$^2\Pi_i$
π^4	$^1\Sigma^+$
δ^2	$^1\Sigma^+$, $^3\Sigma^-$, $^1\Gamma$
δ^3	$^2\Delta_i$
δ^4	$^1\Sigma^+$
$\delta^2\pi$	$^2\Pi(2)$, $^2\Phi$, ^2H, $^4\Pi$
$\sigma\delta^2\pi$	$^1\Pi(2)$, $^3\Pi(3)$, $^1\Phi$, $^3\Phi$, ^1H, ^3H, $^5\Pi$
$\delta^3\pi^2\sigma$	$^1\Sigma^+$, $^1\Sigma^-$, $^1\Delta(2)$, $^1\Gamma$, $^3\Sigma^+$, $^3\Sigma^-$, $^3\Delta(3)$, $^3\Gamma$, $^5\Delta$
$\sigma\pi^2$	$^2\Sigma^+$, $^2\Sigma^-$, $^2\Delta$, $^4\Sigma^-$
$\sigma\sigma\pi^3$	$^2\Pi(2)$, $^4\Pi$
$\delta^2\pi^2$	$^1\Sigma^+(2)$, $^1\Pi$, $^1\Delta$, $^1\Gamma$, ^1K, $^3\Sigma^+$, $^3\Sigma^-(2)$, $^3\Delta$, $^3\Gamma$, $^5\Sigma^+$

[a] See Ref. 47. A following number in parentheses indicates the number of terms of that particular type.

2.2.2. Bonding

Having considered, albeit briefly, the electronic states produced from separated or united atom correlations and from electron configurations, we must now consider which represent stable states of the molecule. In the first instance, stability refers to physical stability, i.e., a molecule is stable if its ground state is lower in energy than those of its constituent atoms. Chemical stability is a much more relative term.

There are two basic approximate approaches to the bonding in molecules—the valence bond (VB) and molecular orbital (MO) theories. Both are applicable to a study of ground states, though in their more usual approximate forms tend to answer different problems. Only the MO theory is useful for a study of excited states. Ligand-field theory is a simpler application of MO theory to systems consisting of a central atom with a partly filled shell surrounded by closed-shell atoms, ions, or dipoles.

The valence-bond theory of Heitler and London,[51] Slater,[52,53] and Pauling[54] developed from a theory of the hydrogen molecule. Briefly, the starting point is two separate H atoms and their bonding is related to the *exchange degeneracy* at large internuclear distances. This is split when the atoms are brought together, giving a symmetric state ($^1\Sigma_g^+$) with a minimum in its potential energy function and an antisymmetric state ($^3\Sigma_u^+$) with no minimum. Bond formation then occurs when the two electrons have antiparallel spins. Generalizing for any diatomic molecule, where S_1 and S_2 are the spins of the separated atoms x and y, we may write the energies of the resulting states (versus $E = 0$ at $R_{xy} = \infty$):

$$E = J_{xy} + (q - n_x n_y) K_{xy}$$

where $J_{xy} = \sum J_{ij}$ is the Coulomb energy summed over all pairs of electrons, K_{xy} is the exchange integral formed for one of the q created electron pairs, and n_x and n_y are the numbers of electrons of x and y remaining unpaired.[55] The Heitler–London theory assumes that the atoms are in S states and that other states are far removed in energy. Slater and Pauling neglected the separation of states within a configuration and started from the atomic electron configurations. The success of hybridization and resonance in explaining directed valence in non-transition metal compounds followed.

The molecular orbital theory of Hund,[56,57] Mulliken,[45] Herzberg,[55] and Lennard-Jones[58–60] starts from the motion of the electron in the field of two separated H nuclei (a and b), i.e., the ion H_2^+.[55] The eigenfunctions of the two lowest states are

$$\psi(1\sigma_g) = \{1/[2(1+S)]^{1/2}\}[\phi(1sa) + \phi(1sb)]$$
$$\psi(1\sigma_u) = \{1/[2(1-S)]^{1/2}\}[\phi(1sa) - \phi(1sb)]$$

where $S = \int \phi(1sa)\phi(1sb)\, d\tau$, and their eigenvalues are

$$E(1\sigma_g) = (H_{aa} + H_{ab})/(1+S) \quad \text{and} \quad E(1\sigma_u) = (H_{aa} - H_{ab})/(1-S)$$

with $H_{aa} = \int \phi a W \phi a\, d\tau$ and $H_{ab} = \int \phi a W \phi b\, d\tau$, where W is the potential energy of the system referred to $H + H^+$ at large separation R. The bonding energy is roughly proportional to the overlap of the constituent atomic wave functions. In larger molecules each electron is considered to be bonding or antibonding, i.e., to decrease or increase the total energy.[55] The order of molecular orbitals normally encountered in light atom diatomic molecules is $\sigma_g < \sigma_u$ and $\pi_u < \pi_g$.[57]

The questions now are what is the form of the wave functions and what values does the energy take? The answers, in favorable cases, are the solutions of the Hartree–Fock equations.[36,61–65]

The potential energy term $V(r_i)$ in the Schrödinger equation

$$\left[-\sum_i - (h^2/2m)\nabla_i^2 + V(r_i)\right]\psi(r_i) = E\psi(r_i)$$

will have the same symmetry as the molecule ($D_{\infty h}$ or $C_{\infty v}$ for diatomic molecules) and the molecular wave functions may be classified according to the irreducible representations to which they belong as well as to their S and M_s values.[35] $\psi(r_i)$ is written as a linear combination of determinantal functions θ, each an antisymmetrized product of spin orbitals:

$$\theta = [Q/2n!]^{1/2} u_1(1)\alpha(1) u_1(2)\beta(2) \ldots u_n(2n-1)\alpha(2n-1) u_n(2n)\beta(2n)$$

where the u_i are the molecular orbitals. For a closed-shell electronic structure we deal with one θ but otherwise we require the linear combination.[35]

The energy for a closed shell is given by $E = 2\sum_i H_i + \sum_{i>j}(2J_{ij} - K_{ij})$, where H_i is an orbital energy and J_{ij} and K_{ij} are Coulomb and exchange integrals with[64]

$$0 \leq K_{ij} \leq J_{ij} \leq \tfrac{1}{2}(J_{ii} + J_{jj}) \quad \text{and} \quad K_{ii} = J_{ii}$$

For minimum energy

$$F_{\text{op}} u_i = \left[\sum_i f_i + \sum_i (2J_i - K_i)\right]_{\text{op}} u_i = \varepsilon_i u_i$$

where the operator is the Hartree–Fock Hamiltonian operator with

$$J_i(1)u_k(1) = u_k(1) \int u_i{}^*(2)(1/r_{12})u_i(2)\,dv_2$$
$$K_i(1)u_k(1) = u_i(1) \int u_i{}^*(2)(1/r_{12})u_k(2)\,dv_2$$
(35)

Any MO u is written as a linear combination of basis functions $u = \sum_k c_k u_k{}^\circ$ and the variation principle applied to evaluate the coefficients. The whole procedure is best carried out in matrix form when the Hartree–Fock equations become

$$\mathbf{F}\mathbf{C}_i = \sum_k \mathbf{S}\mathbf{C}_k E_{ki}$$

with $\mathbf{S}$ as an overlap integral matrix. The work of Roothaan[64,65] has made this possible for diatomic molecules.

The basis functions used are not (now) simple Slater atomic functions but include further functions to represent the radial and angular distortion of atomic orbitals upon the combination of atoms (contra Moffitt's atoms-in-molecules approach[66]). One problem usually remains, however. An implicit assumption in the Hartree–Fock scheme is that each electron undergoes uncorrelated motion in the field of the nuclei and the other electrons, whereas, in fact, the electrons have to avoid each other. Thus, the calculated kinetic energy is very low and the potential energy is rather high, it being implied that two electrons of opposite spin occupy the same orbital in space. The difference between the exact eigenvalue of the Hamiltonian and its expectation value in the Hartree–Fock approximation for a state is known as the correlation energy for that state.[67] Several ways of handling correlation energies, which can reach 50% of the binding energy, exist, including the setting up of different orbitals for different spins.[68] For alkali halides the molecules and *ions* have equivalent correlation energies.[69]

Calculations on diatomic molecules of transition elements have begun but so far are restricted to the first few elements. A ligand-field theory for these molecules[70] based on the anticipated transition metal ion configuration (e.g., $3d^2$ for Ti^{2+} in TiO) was not successful and the configurations including one or two $4s$ electrons of the transition metal ion must be considered.[71] Cheetham and Barrow[72] list the ground states predicted for d^n, $d^{n-1}s$, and $d^{n-2}s^2$ configurations of diatomics containing a transition metal atom. We shall return to this topic later.

2.3. Bonding in and Spectra of Polyatomic Species

Just as for diatomic molecules, it is possible to find the electronic states of polyatomic molecules by the separated-atom[73] or united-atom approaches. Several examples are given by Herzberg.[55] It is more useful, however, to build up the electronic configuration of a molecule by assigning electrons to atomic orbitals having defined behavior under the symmetry operations of the point group of the nuclear framework. We can use our knowledge of the relative energies of the orbitals in the atoms to derive an order of energies for the molecular orbitals, which are, *usually*, linear combinations of atomic orbitals of like symmetry species. Figures 5–7[55] illustrate the results of this approach for linear, tetrahedral, and octahedral molecules. The molecular electronic states resulting from the addition of electrons to orbitals are, like the nature, degeneracy, and number of the orbitals, necessarily a function of the molecular symmetry. The states resulting from electrons in different orbitals follow by multiplying the species of the different orbitals according to the group multiplication table. If we use a system in which the spin and orbital wave functions are not

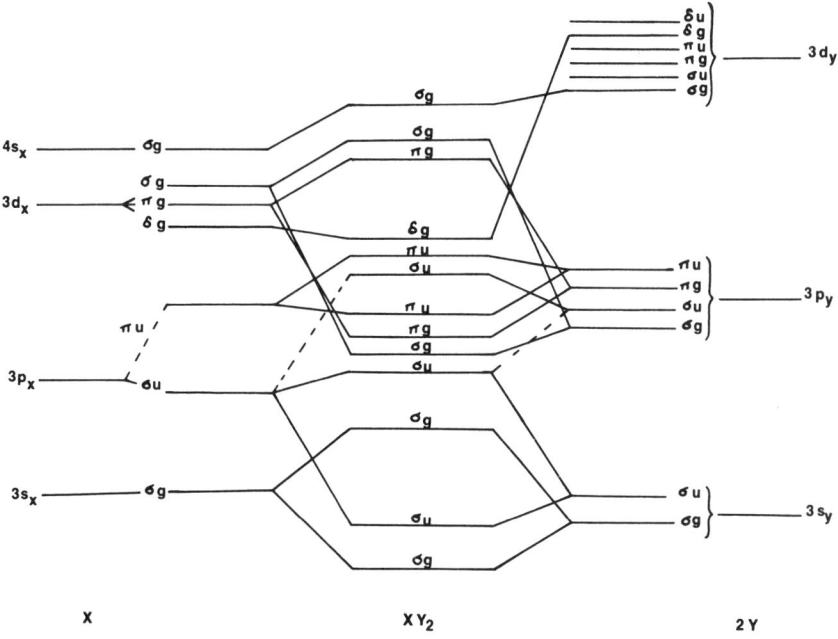

Fig. 5. Correlation of orbitals of linear XY_2 molecules to those of the separated atoms, including d orbitals.[55] (Reproduced with permission of Litton Educational Publishing.)

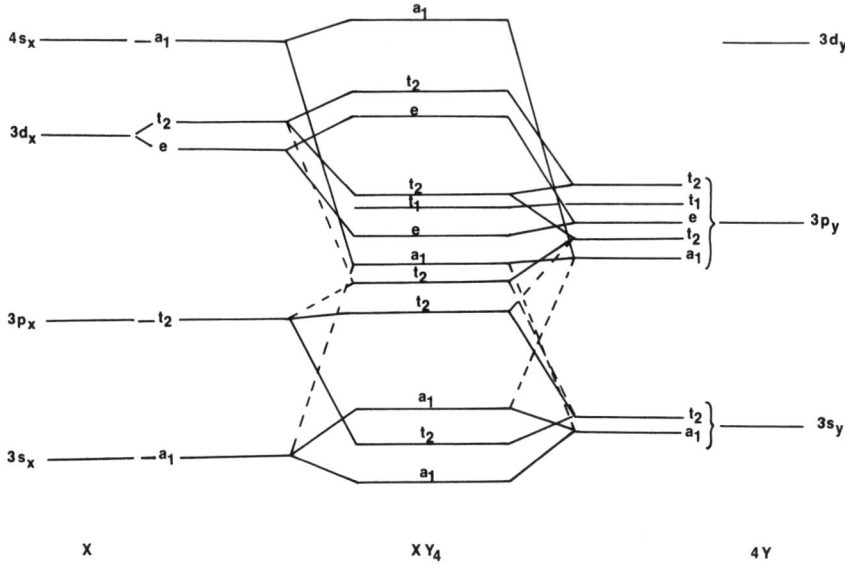

Fig. 6. Correlation of orbitals of tetrahedral XY_4 molecules to those of the separated atoms, including d orbitals.[55] (Reproduced with permission of Litton Educational Publishing.)

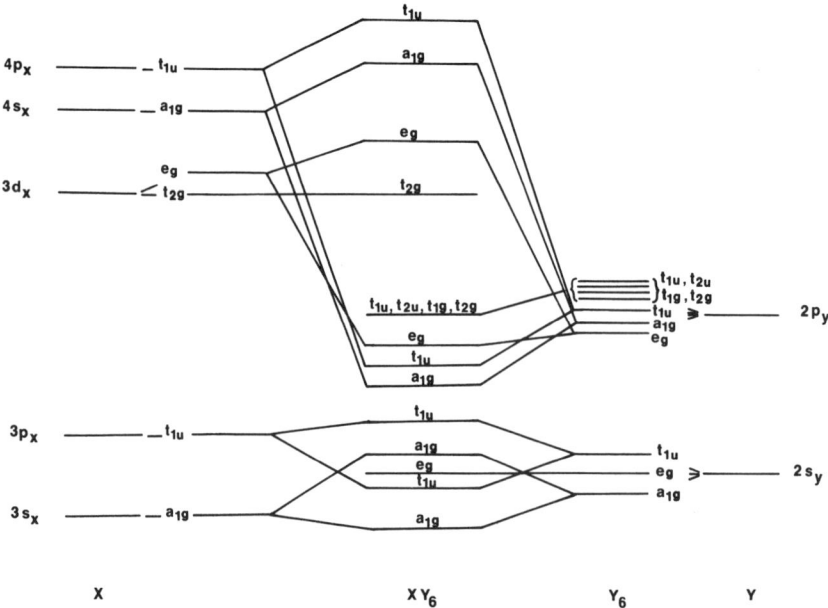

Fig. 7. Correlation of orbitals between the separated atoms and the molecule for octahedral XY_6.[55] (Reproduced with permission of Litton Educational Publishing.)

TABLE IX. Molecular States of Electronic Configurations for a Selection of Point Groups[55]

Point group	Config.	States	Point group	Config.	States
O, T_d	e	^2E	$C_{\infty v}$	π	$^2\Pi$
	e^2	$^1A_1, {}^1E, {}^3A_2$		π^2	$^1\Sigma^+, {}^1\Delta, {}^3\Sigma^-$
	e^3	^2E		π^3	$^2\Pi$
	e^4	$^1A_1$		δ	$^2\Delta$
	t_1	$^2T_1$		δ^2	$^1\Sigma^+, {}^1\Gamma, {}^3\Sigma^-$
	t_1^2	$^1A_1, {}^1E, {}^1T_2, {}^3T_1$		δ^3	$^2\Delta$
	t_1^3	$^2E, {}^2T_1, {}^2T_2, {}^4A_1$	D_4	e	^2E
	t_1^4	$^1A_1, {}^1E, {}^1T_2, {}^3T_1$		e^2	$^1A_1(2), {}^1B_1, {}^3B_1$
	t_1^5	$^2T_1$		e^3	^2E
	t_1^6	$^1A_1$		e^4	$^1A_1$
	t_2	$^2T_2$	C_{3v}	a_2	$^2A_2$
	t_2^2	$^1A_1, {}^1E, {}^1T_2, {}^3T_1$		a_2^2	$^1A_1$
	t_2^3	$^2E, {}^2T_1, {}^2T_2, {}^4A_2$		e	^2E
	t_2^4	$^1A_1, {}^1E, {}^1T_2, {}^3T_1$		e^2	$^1A_1, {}^1E, {}^3A_2$
	t_2^5	$^2T_2$		e^3	^2E
	t_2^6	$^1A_1$		e^4	$^1A_1$

independent, then the process of evaluating molecular states must be carried out in the spinor group or double group of the molecule[74] which contains further elements and double-valued representations[75] to allow for a (fictitious) change upon rotation through 2π. Table IX[55] gives the molecular states of various electronic configurations of a selection of point groups, Table X[74] lists all the possible states of d^n configurations in cubic symmetry and Table XI shows into which species any species of the octahedral group O_h decomposes upon lowering the symmetry.[75] In Table XII are the species to which the various atomic and spin–orbital functions reduce in a selection of symmetries.

The latter results incorporate the very simple ideas of crystal field theory in which the electronic states of a central ion are split by the electric field of surrounding ions or dipoles. Although commonly applied to d^n and f^n configurations, this theory is merely a special case of molecular orbital theory applicable in principle to any system. Of course, the surrounding ligands are not point charges and do contribute to the molecular orbitals and the bonding in such a way that the electronic parameters of the free gaseous central ion are larger than the corresponding parameters of the ion in combination—Jorgensen's nephelauxetic effect in what is termed ligand-field theory.[76,77]

TABLE X. Allowed States for d^n Configurations ($t_2^m e^{n-m}$) in a Strong Octahedral Field

Configuration	State
e^1, e^3	^2E
e^2	$^1A_1, {}^3A_2, {}^1E$
t_2^1, t_2^5	$^1T_2$
t_2^2, t_2^4	$^1A_1, {}^1E, {}^3T_1, {}^1T_2$
t_2^3	$^4A_2, {}^2E, {}^2T_1, {}^2T_2$
$t_2^1e^1, t_2^5e^1, t_2^1e^3, t_2^5e^3$	$^1T_1, {}^3T_1, {}^1T_2, {}^3T_2$
$t_2^2e^1, t_2^4e^1, t_2^2e^3, t_2^4e^3$	$^2A_1, {}^2A_2, {}^2E(2), {}^2T_1(2), {}^4T_1, {}^2T_2(2), {}^4T_2$
$t_2^3e^1, t_2^3e^3$	$^1A_1, {}^3A_1, {}^1A_2, {}^3A_2, {}^1E, {}^3E(2), {}^5E, {}^1T_1(2), {}^3T_1(2), {}^1T_2(2), {}^3T_2(2)$
$t_2^1e^2, t_2^5e^2$	$^2T_1(2), {}^4T_1, {}^2T_2(2)$
$t_2^2e^2, t_2^4e^2$	$^1A_1(2), {}^1A_2, {}^3A_2, {}^1E(3), {}^3E, {}^1T_1, {}^3T_1(3), {}^1T_2(3), {}^3T_2(2), {}^5T_2$
$t_2^3e^2$	$^2A_1(2), {}^4A_1, {}^6A_1, {}^1A_2, {}^4A_2, {}^2E(3), {}^4E(2), {}^2T_1(4), {}^4T_1, {}^2T_2(4), {}^4T_2$

TABLE XI. Decomposition of Species upon Symmetry Lowering[75]

O_h	T_d	D_{4h}	D_{2d}	C_{2v}	D_3	C_{2h}
A_{1g}	A_1	A_{1g}	A_1	A_1	A_1	A_g
A_{2g}	A_2	B_{1g}	B_1	A_2	A_2	B_g
E_g	E	$A_{1g} + B_{1g}$	$A_1 + B_1$	$A_1 + A_2$	E	$A_g + B_g$
T_{1g}	T_1	$A_{2g} + E_g$	$A_2 + E$	$A_2 + B_1 + B_2$	$A_2 + E$	$A_g + 2B_g$
T_{2g}	T_2	$B_{2g} + E_g$	$B_2 + E$	$A_1 + B_1 + B_2$	$A_1 + E$	$2A_g + B_g$
A_{1u}	A_2	A_{1u}	B_1	A_2	A_1	A_u
A_{2u}	A_1	B_{1u}	A_1	A_1	A_2	B_u
E_u	E	$A_{1u} + B_{1u}$	$A_1 + B_1$	$A_1 + A_2$	E	$A_u + B_u$
T_{1u}	T_2	$A_{2u} + E_u$	$B_2 + E$	$A_1 + B_1 + B_2$	$A_2 + E$	$A_u + 2B_u$
T_{2u}	T_1	$B_{2u} + E_u$	$A_2 + E$	$A_2 + B_1 + B_2$	$A_1 + E$	$2A_u + B_u$

TABLE XII. Reduction to Irreducible Representations of the Icosahedral (I), Cubic (O), and Linear ($C_{\infty v}$) Groups with given J[74]

J	I	O	$C_{\infty v}$
0	A	A_1	Σ
1	T_1	T_1	$\Sigma + \Pi$
2	H	$E + T_2$	$\Sigma + \Pi + \Delta$
3	$T_2 + G$	$A_2 + T_1 + T_2$	$\Sigma + \Pi + \Delta + \Phi$
4	$G + H$	$A_1 + E + T_1 + T_2$	$\Sigma + \Pi + \Delta + \Phi + \Gamma$
5	$T_1 + T_2 + H$	$E + 2T_1 + T_2$	$\Sigma + \Pi + \Delta + \Phi + \Gamma + H$
6	$A + T_1 + G + H$	$A_1 + A_2 + E + T_1 + 2T_2$	etc.
7	$T_1 + T_2 + G + H$	$A_2 + E + 2T_1 + 2T_2$	
8	$T_2 + G + 2H$	$A_1 + 2E + 2T_1 + 2T_2$	
$\tfrac{1}{2}$	Γ_6	Γ_6	$E_{1/2}$
$1\tfrac{1}{2}$	Γ_8	Γ_8	$E_{1/2} + E_{3/2}$
$2\tfrac{1}{2}$	Γ_9	$\Gamma_7 + \Gamma_8$	$E_{1/2} + E_{3/2} + E_{5/2}$
$3\tfrac{1}{2}$	$\Gamma_7 + \Gamma_9$	$\Gamma_6 + \Gamma_7 + \Gamma_8$	$E_{1/2} + E_{3/2} + E_{5/2} + E_{7/2}$
$4\tfrac{1}{2}$	$\Gamma_8 + \Gamma_9$	$\Gamma_6 + 2\Gamma_8$	etc.
$5\tfrac{1}{2}$	$\Gamma_6 + \Gamma_8 + \Gamma_9$	$\Gamma_6 + \Gamma_7 + 2\Gamma_8$	
$6\tfrac{1}{2}$	$\Gamma_6 + \Gamma_7 + \Gamma_8 + \Gamma_9$	$\Gamma_6 + 2\Gamma_7 + 2\Gamma_8$	
$7\tfrac{1}{2}$	$\Gamma_8 + 2\Gamma_9$	$\Gamma_6 + \Gamma_7 + 3\Gamma_8$	

2.3.1. Cubic Coordination

There are several ways of calculating and expressing the term energies of a polyatomic system based on a central transition metal. Depending on the magnitude of the central ion versus ligand interaction, we start from the free central ion states and consider the effect of a small perturbation on them or we start from orbitals already possessing the appropriate molecular symmetry. For states represented by pure LS coupling we obtain, for the weak-field coupling scheme, the energies listed in Table V together with a set of crystal field terms which are a function of L only unless species of the same type arise from more than one L value.[36] The energies of the d^n configuration in O_h symmetry under the strong-field coupling scheme are presented in Table XIII.[76–78] The two sequences (weak-field and strong-field) for d^2 in O_h symmetry are correlated in

TABLE XIII. Energy Levels for d^n—Systems in Strong Octahedral Fields[a]

A. Ground state is one of maximum multiplicity[b]

d^2	t_2^2	$^3T_1$	0
		^1E	$6B + 2C - 6(B^2/\varDelta) + x$
		$^1T_2$	$6B + 2C - 12(B^2/\varDelta) + x$
		$^1A_1$	$15B + 5C - 108(B^2/\varDelta) + x$
	t_2e	$^3T_2$	$\varDelta - 3B + x$
		$^3T_1$	$\varDelta + 9B + 2x$
		$^1T_2$	$\varDelta + 5B + 2C + \cdots$
		$^1T_1$	$\varDelta + 9B + 2C + \cdots$
	e^2	$^3A_2$	$2\varDelta - 3B + x$
		^1E	$2\varDelta + 5B + 2C + \cdots$
		$^1A_1$	$2\varDelta + 13B + 4C + \cdots$
d^3	t_2^3	$^4A_2$	0
		^2E	$9B + 3C - 90(B^2/\varDelta)$
		$^2T_1$	$9B + 3C - 24(B^2/\varDelta)$
		$^2T_2$	$15B + 5C - 176(B^2/\varDelta)$
	t_2^2e	$^4T_2$	$\varDelta$
		$^4T_1$	$\varDelta + 12B - y$
	t_2e^2	$^4T_1$	$2\varDelta + 3B + y$
d^4	t_2^3e	^5E	0
	$t_2^2e^2$	$^5T_2$	$\varDelta$
	t_2^4	$^3T_1$	$-\varDelta + 6B + 5C - 64(B^2/\varDelta)^{(c)}$
		^1E	$-\varDelta + 12B + 7C - 82(B^2/\varDelta)$
		$^1T_2$	$-\varDelta + 12B + 7C - 208(B^2/\varDelta)$
		$^1A_1$	$-\varDelta + 21B + 10C - 436(B^2/\varDelta)$
d^5	$t_2^3e^2$	$^6A_1$	0
		$^4A_1$	$10B + 5C$
		^4E	$10B + 5C$
		$^4T_2$	$13B + 5C + \cdots$
		^4E	$17B + 5C$
		$^4T_1$	$19B + 7C - \cdots$
		$^4A_2$	$22B + 7C$
	t_2^4e	$^4T_1$	$-\varDelta + 10B + 6C - 26(B^2/\varDelta)$
		$^4T_2$	$-\varDelta + 18B + 6C - 38(B^2/\varDelta)$
	t_2^5	$^2T_2$	$-2\varDelta + 15B + 10C - 140(B^2/\varDelta)^{(c)}$

TABLE XIII (continued)

d^6	$t_2^4 e^2$	$^5T_2$	0
	$t_2^3 e^3$	^5E	Δ
	$t_2^5 e$	$^3T_1$	$-\Delta + 5B + 5C - 70(B^2/\Delta)$
		$^3T_2$	$-\Delta + 13B + 5C - 106(B^2/\Delta)$
		$^1T_1$	$-\Delta + 5B + 7C - 34(B^2/\Delta)$
		$^1T_2$	$-\Delta + 21B + 7C - 118(B^2/\Delta)$
	t_2^6	$^1A_1$	$-2\Delta + 5B + 8C - 120(B^2/\Delta)^{(c)}$
d^7	$t_2^5 e^2$	$^4T_1$	0
	$t_2^4 e^3$	$^4T_2$	$\Delta - 3B + x$
		$^4T_1$	$\Delta + 9B + 2x$
	$t_2^3 e^4$	$^4A_2$	$2\Delta - 3B + x$
	$t_2^6 e$	^2E	$-\Delta + 4B + 4C - 60(B^2/\Delta) + x^{(c)}$
d^8	$t_2^6 e^2$	$^3A_2$	0
		^1E	$8B + 2C - 6(B^2/\Delta)$
		$^1A_1$	$16B + 4C - 108(B^2/\Delta)$
	$t_2^5 e^3$	$^3T_2$	Δ
		$^3T_1$	$\Delta + 12B - y$
		$^1T_2$	$\Delta + 8B + 2C - 12(B^2/\Delta)$
		$^1T_1$	$\Delta + 12B + 2C$
	$t_2^4 e^4$	$^3T_1$	$2\Delta + 3B + y$
		^1E	$2\Delta + 9B - 2C + \cdots$
		$^1T_2$	$2\Delta + 9B + 2C + \cdots$
		$^1A_1$	$2\Delta + 18B + 5C + \cdots$

B. Ground state has low spin[d]

d^4	t_2^4	$^3T_1$	0
		^1E	$6B + 2C - 18(B^2/\Delta)$
		$^1T_2$	$6B + 2C - 144(B^2/\Delta)$
		$^1A_1$	$15B + 5C - 372(B^2/\Delta)$
	$t_2^3 e$	^5E	$\Delta - 6B - 5C + 64(B^2/\Delta)^{(c)}$
	$t_2^2 e^2$	$^5T_2$	$2\Delta - 6B - 5C + 64(B^2/\Delta)$

TABLE XIII (continued)

d^5	t_2^5	$^2T_2$	0
	$t_2^4 e$	$^4T_1$	$\Delta - 5B - 4C + 114(B^2/\Delta)$
		$^4T_2$	$\Delta + 3B - 4C + 102(B^2/\Delta)$
	$t_2^3 e^2$	$^6A_1$	$2\Delta - 15B - 10C + 140(B^2/\Delta)^{(c)}$
		$^4A_1$	$2\Delta - 5B - 5C + 140(B^2/\Delta)$
		^4E	$2\Delta - 5B - 5C + 140(B^2/\Delta)$
		$^4T_2$	$2\Delta - 2B - 5C + \cdots$
		^4E	$2\Delta + 2B - 5C - 140(B^2/\Delta)$
		$^4T_1$	$2\Delta + 4B - 3C - \cdots$
		$^4A_2$	$2\Delta + 7B - 3C + 140(B^2/\Delta)$
d^6	t_2^6	$^1A_1$	0
	$t_2^5 e$	$^3T_1$	$\Delta - 3C + 50(B^2/\Delta)$
		$^3T_2$	$\Delta + 8B - 3C + 14(B^2/\Delta)$
		$^1T_1$	$\Delta - C + 86(B^2/\Delta)$
		$^1T_2$	$\Delta + 16B - C + 2(B^2/\Delta)$
	$t_2^4 e^2$	$^5T_2$	$2\Delta - 5B - 8C + 120(B^2/\Delta)^{(c)}$
	$t_2^3 e^3$	^5E	$3\Delta - 5B - 8C + 120(B^2/\Delta)$
d^7	$t_2^6 e$	^2E	0
	$t_2^5 e^2$	$^4T_1$	$\Delta - 4B - 4C + 60B^2/\Delta - x^{(c)}$
	$t_2^4 e^3$	$^4T_2$	$2\Delta - 7B - 4C + 60(B^2/\Delta)$
		$^4T_1$	$2\Delta + 5B - 4C + 60(B^2/\Delta) + x$
	$t_2^3 e^4$	$^4A_2$	$3\Delta - 7B - 4C + 60(B^2/\Delta)$

[a] $x = 36B^2/(\Delta + 9B + x)$, $y = 36B^2/(\Delta - 9B + y)$.

[b] Ground-state energies are for: d^2 ($^3T_1$), $-0.8\Delta + 3B - x$; d^3 ($^4A_2$), -1.2Δ; d^4 (^5E), -0.6Δ; d^5 ($^6A_1$), 0; d^6 ($^5T_2$), -0.4Δ; d^7 ($^4T_1$), $-0.8\Delta + 3B - x$; d^8 ($^3A_2$), -1.2Δ. Excited-state energies referred to ground-state energy as zero.

[c] When this value is negative, this state becomes the ground state.

[d] Ground-state energies are for: d^4 ($^3T_1$), $-1.6\Delta + 6B + 5C - 64(B^2/\Delta)$; d^5 ($^2T_2$), $-2\Delta + 15B + 10C - 140(B^2/\Delta)$; d^6 ($^1A_1$), $-2.4\Delta + 5B + 8C - 120(B^2/\Delta)$; d^7 (^2E), $-1.8\Delta + 7B + 4C - 60(B^2/\Delta)$. Excited-state energies referred to ground-state energy as zero.

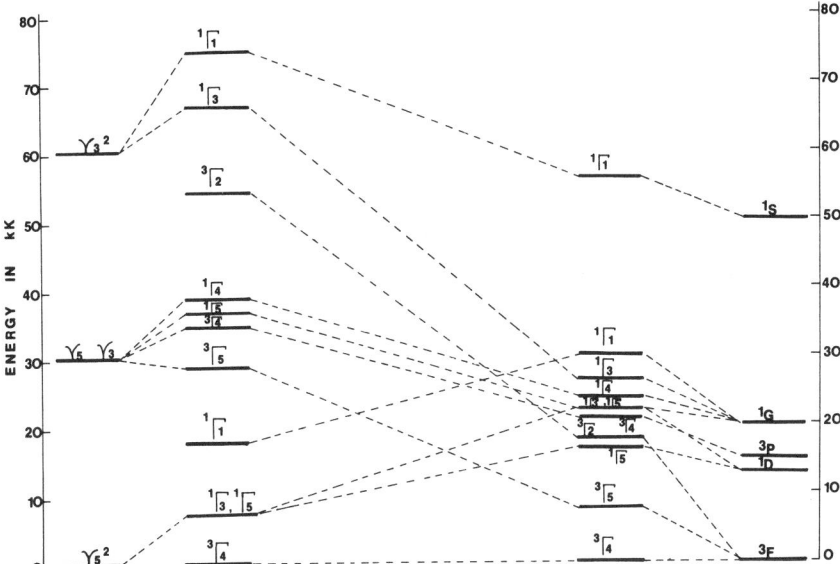

Fig. 8. Correlation of energies of d^2 configuration in strong (left) and weak (right) octahedral fields. The extreme Δ values used in the diagram are 30 kK and 10 kK with interelectronic repulsion parameters in the central arrays of $B = 0.5$ kK and $C = 2$ kK (left) and $B = 11$ kK and $C = 4$ kK (right).

Fig. 8, which includes the very approximate energies evaluated by a neglect of off-diagonal matrix elements. For an introduction to the calculations using determinantal functions see Stevenson[36] and for further details and results see Griffith.[74] Since all the essential data are tabulated,[79,80] it is probably more sensible to use Racah's methods[37,82-84,87] to evaluate the energies in any symmetry in a weak-field coupling case for all configurations and in a strong-field coupling case for the more complex configurations.[36]

However, for many cases of interest in high-temperature chemistry the available electronic spectrum consists of only a few of the expected bands and this spectrum is used to determine the symmetry. Several simplifications are therefore in order in treating the raw data. Unfortunately, consistency of simplification is not apparent in the work described below.* Most of the variants will be discussed under the individual elements.

---

* The authors themselves are open to this criticism.

2.3.2. Cubic Coordination with Spin–Orbital Coupling

In dealing with $3d^n$ configurations it is common to omit spin–orbital coupling from the calculations of energies, since its effect is usually to superimpose a splitting on a given band or intensify a spin-forbidden transition[76] (but see Section 3.2). With $4d^n$ and $5d^n$ configurations spin–orbital coupling is much larger[25,26] and its influence is shown by the greatly enhanced intensity of spin-forbidden transitions and even by the frequencies of the transitions observed (cf. the intercombination lines observed for heavy elements).[26]

Following Moffitt et al.[85] and Liehr,[86] one can write the electronic configurations $\gamma_8{}^a\gamma_7{}^b\gamma_8^{*c}$ ($a = 0, 1, 2, 3, 4$; $b = 0, 1, 2$; $c = 0, 1, 2, 3, 4$) for d^n systems in a strong octahedral field with spin–orbital coupling and then include interelectronic repulsion as a perturbation on the $\gamma_8{}^a\gamma_7{}^b\gamma_8^{*c}$ levels. For the single-electron system the energies are given by

$$E(\Gamma_7) = -0.4\Delta, \qquad E(\Gamma_8) = -\tfrac{1}{2}\zeta - 0.4\Delta - \tfrac{1}{2}\sqrt{6}\,\zeta \cot \tfrac{1}{2}\eta$$

where $-\sqrt{6}\,\zeta \cot \eta = \tfrac{1}{2}\zeta + \Delta$.[86]

Writing $\Gamma_7 - \Gamma_8 = \theta$ and $\Gamma_8^* - \Gamma_8 = \phi$, we obtain the energies of the spin–orbital terms as presented in Table XIV.[87] This table also includes the terms produced by adding the interelectronic repulsion but the complete energy calculations have yet to be performed.

In the meantime, certain approximations are in order. First, substitution of reasonable values of ζ_{5d} and Δ_{5d} (~3 and 30 kK, respectively) in the expressions for θ and ϕ indicates that θ is primarily a function of ζ and ϕ is primarily a function of Δ. Now the optical transitions of ions which we might hopefully expect to observe will include ones centered at θ and 2θ in the first half of a transition series ($n < 5$) but ones centered at $\phi - \theta$ and ϕ in the second half ($n > 5$), because of the electron–hole equivalence. Thus many of the transitions of $5d^n$ ions, where $n < 5$, can be treated as a function of spin–orbital and interelectronic interactions and many of the transitions of d^n, where $n > 5$, can be treated as a function of crystal field and interelectronic interactions.

In this connection use has been made on several occasions of the fact that the energies of the intra-t_{2g}^n configuration transitions (in the octahedral group O_h) for intermediate coupling correspond to the transitions of atomic p^{6-n}, because of the isomorphism of the two sets[85,88,89].

TABLE XIV. Strong-Field Configurations of d^n in Octahedral Symmetry Including Relativistic Effects

System	Configuration	Energy[a]	States produced by interelectronic repulsion
d^2	γ_8^2	0	$\Gamma_1 + \Gamma_3 + \Gamma_5$
	$\gamma_8\gamma_7$	θ	$\Gamma_3 + \Gamma_4 + \Gamma_5$
	γ_7^2	2θ	Γ_1
	$\gamma_8\gamma_8^*$	ϕ	$\Gamma_1 + \Gamma_2 + \Gamma_3 + 2\Gamma_4 + 2\Gamma_5$
	$\gamma_7\gamma_8^*$	$\theta + \phi$	$\Gamma_3 + \Gamma_4 + \Gamma_5$
	γ_8^{*2}	2ϕ	$\Gamma_1 + \Gamma_3 + \Gamma_5$
d^3	γ_8^3	0	Γ_8
	$\gamma_8^2\gamma_7$	θ	$\Gamma_6 + \Gamma_7 + 2\Gamma_8$
	$\gamma_8\gamma_7^2$	2θ	Γ_8
	$\gamma_8^2\gamma_8^*$	ϕ	$2\Gamma_6 + 2\Gamma_7 + 4\Gamma_8$
	$\gamma_8\gamma_7\gamma_8^*$	$\theta + \phi$	$3\Gamma_6 + 3\Gamma_7 + 5\Gamma_8$
	$\gamma_7^2\gamma_8^*$	$2\theta + \phi$	Γ_8
	$\gamma_8\gamma_8^{*2}$	2ϕ	$2\Gamma_6 + 2\Gamma_7 + 4\Gamma_8$
	$\gamma_7\gamma_8^{*2}$	$\theta + 2\phi$	$\Gamma_6 + \Gamma_7 + 2\Gamma_8$
	γ_8^{*3}	3ϕ	Γ_8
d^4	γ_8^4	0	Γ_1
	$\gamma_8^3\gamma_7$	θ	$\Gamma_3 + \Gamma_4 + \Gamma_5$
	$\gamma_8^2\gamma_7^2$	2θ	$\Gamma_1 + \Gamma_3 + \Gamma_5$
	$\gamma_8^3\gamma_8^*$	ϕ	$\Gamma_1 + \Gamma_2 + \Gamma_3 + 2\Gamma_4 + 2\Gamma_5$
	$\gamma_8^2\gamma_7\gamma_8^*$	$\theta + \phi$	$2\Gamma_1 + 2\Gamma_2 + 4\Gamma_3 + 6\Gamma_4 + 6\Gamma_5$
	$\gamma_8\gamma_7^2\gamma_8^*$	$2\theta + \phi$	$\Gamma_1 + \Gamma_2 + \Gamma_3 + 2\Gamma_4 + 2\Gamma_5$
	$\gamma_8^2\gamma_8^{*2}$	2ϕ	$3\Gamma_1 + \Gamma_2 + 4\Gamma_3 + 3\Gamma_4 + 5\Gamma_5$
	$\gamma_8\gamma_7\gamma_8^{*2}$	$\theta + 2\phi$	$2\Gamma_1 + 2\Gamma_2 + 4\Gamma_3 + 6\Gamma_4 + 6\Gamma_5$
	$\gamma_7^2\gamma_8^{*2}$	$2\theta + 2\phi$	$\Gamma_1 + \Gamma_3 + \Gamma_5$
	$\gamma_8\gamma_8^{*3}$	3ϕ	$\Gamma_1 + \Gamma_2 + \Gamma_3 + 2\Gamma_4 + 2\Gamma_5$
	$\gamma_7\gamma_8^{*3}$	$\theta + 3\phi$	$\Gamma_3 + \Gamma_4 + \Gamma_5$
	γ_8^{*4}	4ϕ	Γ_1

TABLE XIV (*continued*)

System	Configuration	Energy^a	States produced by interelectronic repulsion
d^5	$\gamma_8^4\gamma_7$	0	Γ_7
	$\gamma_8^3\gamma_7^2$	θ	Γ_8
	$\gamma_8^4\gamma_8^*$	$\phi - \theta$	Γ_8
	$\gamma_8^3\gamma_7\gamma_8^*$	ϕ	$3\Gamma_6 + 3\Gamma_7 + 5\Gamma_8$
	$\gamma_8^2\gamma_7^2\gamma_8^*$	$\theta + \phi$	$2\Gamma_6 + 2\Gamma_7 + 4\Gamma_8$
	$\gamma_8^3\gamma_8^{*2}$	$2\phi - \theta$	$2\Gamma_6 + 2\Gamma_7 + 4\Gamma_8$
	$\gamma_8^2\gamma_7\gamma_8^{*2}$	2ϕ	$6\Gamma_6 + 6\Gamma_7 + 12\Gamma_8$
	$\gamma_8\gamma_7^2\gamma_8^{*2}$	$\theta + 2\phi$	$2\Gamma_6 + 2\Gamma_7 + 4\Gamma_8$
	$\gamma_8^2\gamma_8^{*3}$	$3\phi - \theta$	$2\Gamma_6 + 2\Gamma_7 + 4\Gamma_8$
	$\gamma_8\gamma_7\gamma_8^{*3}$	3ϕ	$3\Gamma_6 + 3\Gamma_7 + 5\Gamma_8$
	$\gamma_7^2\gamma_8^{*3}$	$\theta + 3\phi$	Γ_8
	$\gamma_8\gamma_8^{*4}$	$4\phi - \theta$	Γ_8
	$\gamma_7\gamma_8^{*4}$	4ϕ	Γ_7

^{a} θ is the difference between the γ_7 and γ_8 orbitals and ϕ that between the γ_8^* and γ_8 orbitals.

2.3.3. Quadratic Coordination

Detailed ligand-field treatments of square-planar metal complexes have been provided by Martin and co-workers[90,91] and Gray and co-workers,[92,93] in particular. Most of the discussion has centered on d^8 complexes of halides and cyanides but other systems are reviewed by Gray.[94] The situation is not as simple as for cubic systems because more than one ordering of the subshells is feasible. Until recently both the energy level schemes $b_{1g}(x^2 - y^2) > b_{2g}(xy) > a_{1g}(z^2) \geq e_g(xz, yz)^{(92)}$ and $b_{1g}(x^2 - y^2) > b_{2g}(xy) > e_g(xz, yz) > a_{1g}(z^2)$ were seriously entertained[90,91,95] for $PtCl_4^{2-}$ with several variants on the order of states produced by the addition of interelectronic repulsion.

In the treatment of Martin *et. al.*[91] spectra are calculated from six parameters: three crystal field parameters, the Slater parameters F_2 and F_4, and the spin–orbit coupling constant. Some account of the relative intensities of the optical transitions is provided by the spin–orbit mixing

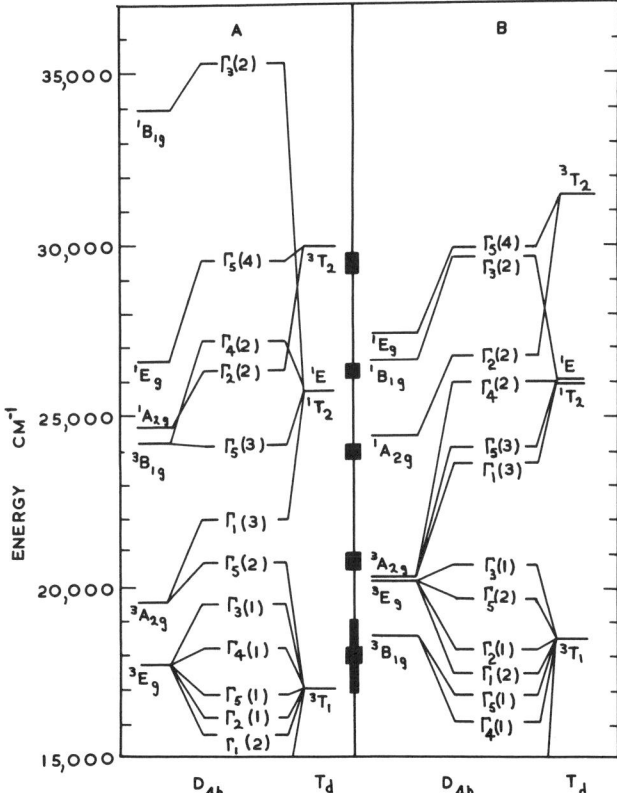

Fig. 9. Calculated energy levels for $PtCl_4^{2-}$.[91] (A) $\alpha = 1500$ cm^{-1}, $F_2 = 1000$ cm^{-1}, and $F_4 = 65$ cm^{-1}. (B) $\alpha = 1700$ cm^{-1}, $F_2 = 820$ cm^{-1}, and $F_4 = 54$ cm^{-1}. The observed bands are depicted in the center. (Reprint by permission of the American Chemical Society.)

of triplet and singlet states. Gray and Ballhausen,[92] however, assign the spectra by neglecting spin–orbit coupling and assuming $F_2 = 10F_4$ or some close relation as opposed to the trial ratio of $F_2/F_4 \simeq 15$ used by Martin et al.[91] Figure 9 depicts the energy level scheme of Martin et al. for d^8 with a correlation of the states to those of T_d symmetry. Table XV gives the first-order energy expressions for d–d transitions according to Gray and Ballhausen[92] as corrected for reassignment of transition by Mason and Gray.[93] The only consequence of a different order of subshells is to change the crystal field energy, the interelectronic repulsion (Slater–Condon) energy being appropriate to the transitions as listed.

TABLE XV. Energies of States of nd^8 Quadratic Complexes[a]

Orbital transition	States	Energy
$b_{1g}(x^2 - y^2) \leftarrow b_{2g}(xy)$	$^3A_{2g} \leftarrow {}^1A_{1g}$	$\Delta_1 - 105F_4 = \Delta_1 - 3C$
	$^1A_{2g} \leftarrow$	$\Delta_1 - 35F_4 = \Delta_1 - C$
$b_{1g}(x^2 - y^2) \leftarrow e_g(xz, yz)$	$^3E_g \leftarrow$	$\Delta_1 + \Delta_2 - 9F_2 - 60F_4 = \Delta_1 + \Delta_2 - 9B - 3C$
	$^1E_g \leftarrow$	$\Delta_1 + \Delta_2 - 3F_2 - 20F_4 = \Delta_1 + \Delta_2 - 3B - C$
$b_{1g}(x^2 - y^2) \leftarrow a_{1g}(z^2)$	$^3B_{1g} \leftarrow$	$\Delta_1 + \Delta_2 + \Delta_3 - 12F_2 - 45F_4 = \Delta_1 + \Delta_2 + \Delta_3 - 12B - 3C$
	$^1B_{1g} \leftarrow$	$\Delta_1 + \Delta_2 + \Delta_3 - 4F_2 - 15F_4 = \Delta_1 + \Delta_2 + \Delta_3 - 4B - C$

[a] Based on the material of Refs. 92 and 93.

2.3.4. Linear Coordination

For linear molecules or groups the splitting of orbitals or states is a very straightforward function of the appropriate quantum number, e.g.,[55]

$$J = 0 \quad \text{yields} \quad \Sigma_g^+$$
$$J = 1 \quad \text{yields} \quad \Sigma_g^- + \Pi_g$$
$$J = 2 \quad \text{yields} \quad \Sigma_g^+ + \Pi_g + \Delta_g$$
$$\vdots$$

whereas

$$J = 1/2 \quad \text{yields} \quad E_{\frac{1}{2}g}$$
$$J = 3/2 \quad \text{yields} \quad E_{\frac{1}{2}g} + E_{\frac{3}{2}g}$$
$$J = 5/2 \quad \text{yields} \quad E_{\frac{1}{2}g} + E_{\frac{3}{2}g} + E_{\frac{5}{2}g}$$
$$\vdots$$

For a d^1 system the energies of the states in a weak symmetric axial field ($D_{\infty h}$), neglecting spin–orbital effects, are given by[96,97]

$$E_{\sigma+} = 4A_2 + 12A_4, \quad E_\pi = 2A_2 - 8A_4, \quad E_\delta = -4A_2 + 2A_4$$

where A_2 and A_4 are the required crystal field parameters.

For $d^{2,7}$ with A_2 and A_4 positive and $d^{3,8}$ with A_2 and A_4 negative the energies of the six states of maximum multiplicity (three for $d^{2,8}$, four for $d^{3,7}$) are given by [97]

$$E_\Delta = 14A_4 + E_F, \quad E_\Phi = -2A_2 - 6A_4 + E_F$$

$$\begin{vmatrix} -E_\Pi + E_F + 1.2A_2 - 2A_4 & -(8/5)6^{1/2}A_2 - (2)6^{1/2}A_4 \\ -(3/5)6^{1/2}A_2 - (2)6^{1/2}A_4 & -E_\Pi + E_P + 2.8A_2 \end{vmatrix} = 0$$

$$\begin{vmatrix} -E_\Sigma + E_F + 1.6A_2 - 12A_4 & -4.8A_2 + 8A_4 \\ -4.8A_2 + 8A_4 & -E_\Sigma + E_P - 5.6A_2 \end{vmatrix} = 0$$

where E_F and E_P are the origins of the F and P states ($E_F - E_P = 15B$).

An energy level diagram including weak spin–orbital effects for weak and strong fields has been obtained for a d^8 system. In principle, energy level diagrams for linear systems containing the $5d$ ions can be derived starting from the spin–orbital configurations $e_{1/2}^z e_{3/2}^y e_{1/2}^{*x} e_{3/2}^{*w} e_{5/2}^v$.

2.3.5. General Molecular Orbital Considerations

Again without going into fine details, we would like to remind the reader of some more general molecular orbital calculations for transition metal ions by Gray and co-workers[99-103] following from the original Wolfsberg–Helmholz[104] work on tetrahedral oxyanions such as MnO_4^-.

The molecular orbitals are assumed to be linear combinations of atomic symmetry basis orbitals and the coefficients in the linear relations, together with the corresponding orbital energies, are derived from a set of secular determinants, one for each symmetry species, having the form $|F_{qs} - E_i G_{qs}| = 0$. Here the group overlap integral G_{qs}, defined by

$$G_{qs} = \int \chi_q(1)\chi_s(1)d\tau(1)$$

is calculated exactly by a judicious choice of orbital functions for the basis orbitals and from a knowledge of the geometry, including bond distances. The diagonal Hamiltonian matrix elements F_{qq} are taken to be the negative of the valence orbital ionization potential (VOIP)[105] and the off-diagonal elements are given by

$$F_{qs} = \tfrac{1}{2}KG_{qs}(\text{VOIP}_q + \text{VOIP}_s)$$

The evaluation of the VOIP's for several elements is detailed by Basch *et al.*[105] For metals the nd, $(n+1)s$, and $(n+1)p$ orbitals are considered as valence orbitals, and for ligands it is the $n's$ and $n'p$. The factor K is an empirical constant which takes a different value for σ and π interactions. Taking K^π as 2.10, it was possible to fit the first ligand field transition of a series of 16 octahedral complexes with K^σ as a function of atomic number:

$$K^\sigma(n) = (0.027n + 1.546) \pm 0.02$$

starting at $n = 1$ for Ti. A similar situation arose for several tetrahedral complexes but for a group of square planar complexes both K^σ and K^π varied from one complex to another. In Fig. 10–12[99] we reproduce the molecular orbital energy level schemes for these three geometries and in Table XVI[103,106] summarize some results of the calculations for Group VIII complexes.

2.3.6. Intensities

Absorption bands of compounds of transition metal ions exhibit a wide variety of intensities. Although, in many cases, the maximum value

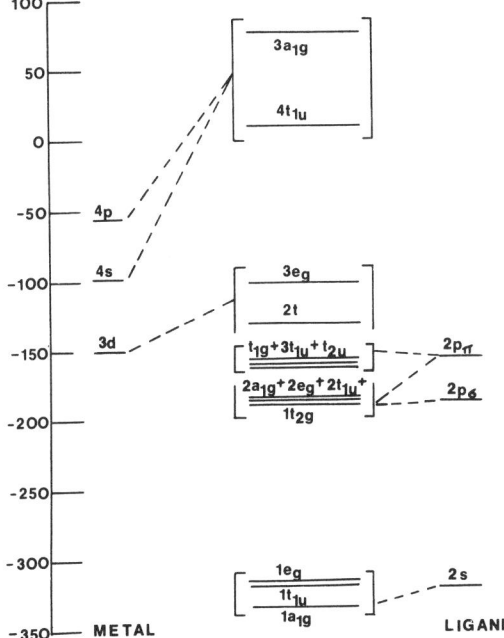

Fig. 10. Molecular orbital energy levels for octahedral complexes containing monatomic ligands. The scheme specifically represents the results of a calculation of FeF_6^{3-}. Energy is in kK.[99] (Reproduced by permission of W. H. Freeman and Company.)

only of the molar absorptivity (extinction coefficient) ε_{max} is quoted, it is more precise to quote the oscillator strength f of a transition[76]:

$$f = 4.32 \times 10^{-9} \int \varepsilon \, dv = 9.20 \times 10^{-9} \varepsilon_{max} \delta \qquad (76)$$

(assuming a Gaussian distribution in energy). Here v is the wave number in cm^{-1} and δ half the band-width at half the maximum absorption, $\varepsilon_{max}/2$. Observed bands may be broken down into three groups according to their f values:

$f = 10^{-7}$–10^{-5} spin-forbidden, parity forbidden

$f = 10^{-5}$–10^{-3} spin-allowed, parity forbidden

$f \sim 10^{-1}$ spin-allowed, parity allowed

The bandwidth is accounted for in terms of the Franck–Condon principle, in that the electronic transition takes place without a change in internuclear distance, whereas the potential energy functions of the two electronic levels may be minimal at different distances; an envelope of several vibrational levels associated with each electronic level is involved. The parity selection rule governing a transition between states a and b by a change in a moment R is[76]

$$\Gamma_a \times \Gamma_R \times \Gamma_b = \Gamma_1$$

where Γ_a and Γ_b are representations of the states in the given point group, Γ_1 is the totally symmetric representation, and Γ_R is the representation of the operator of the moment (electric dipole, magnetic dipole, electric quadruple). With the exception of Os(IV), we shall be concerned with electric dipole transitions only.

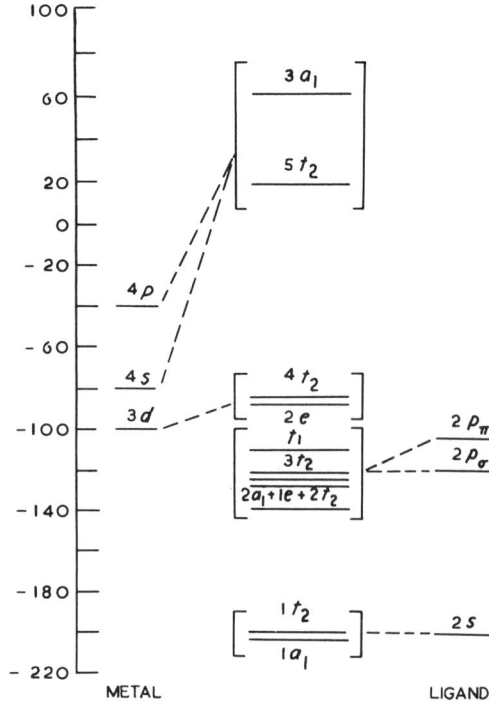

Fig. 11. Molecular orbital energy levels for tetrahedral complexes containing monatomic ligands. The levels were calculated for $FeCl_4^{2-}$. Energy is in kK.[99] (Reproduced by permission of W. H. Freeman and Company.)

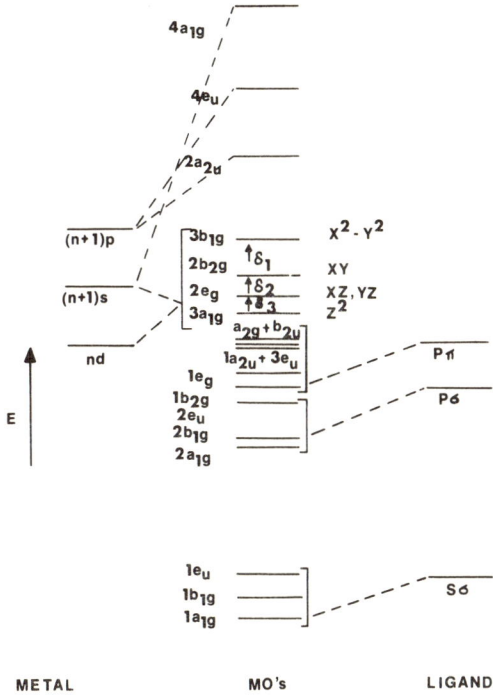

Fig. 12. Molecular orbital energy levels for square-planar halide complexes such as $PtCl_4^{2-}$.[99] (Reproduced by permission of W. H. Freeman and Company.)

TABLE XVI. Calculated MO Parameters for Transition Metal Complexes

Complex	Metal charge	3d population	$K\sigma$	$K\pi$
FeF_6^{3-}	0.87	6.76	1.67	2.10
NiF_6^{4-}	0.78	8.79	1.68	—
$FeCl_4^-$	0.38	6.92	1.66	—
$FeCl_4^{2-}$	0.33	7.01	1.68	—
$CoCl_4^{2-}$	0.32	8.01	1.66	—
$CoBr_4^{2-}$	0.30	8.19	1.56	—
$NiCl_4^{2-}$	0.34	8.99	1.68	—
$PtCl_4^{2-}$	0.24	—	1.98	1.80
$PtBr_4^{2-}$	0.13	—	2.13	1.70
$PdCl_4^{2-}$	0.23	—	2.04	1.75
$PdBr_4^{2-}$	0.09	—	2.22	1.70

In dealing with the complexes of metals containing partly filled d shells, *it has become customary* to label absorption bands as d–d transitions, where the initial and final electronic states can be considered as mainly localized on the metal, and charge-transfer transitions, where one state is mainly localized on the metal and the other on the ligand(s). The classification of spin-allowed and spin-forbidden bands is retained, as are parity restrictions, although these are often termed symmetry restrictions. There are, in addition, several confusing references to the Laporte rule.[76]

A survey of the intensities of d–d transitions of complexes leads to the following general conclusions:

(a) All d–d transitions have oscillator strengths appropriate to parity-forbidden transitions.

(b) The spin quantum number S is near enough to being a good quantum number in that changes in S give rise to much weaker bands, even in the $5d$ series.

(c) Intensities of transitions of $4d$ and $5d$ complexes are several times larger than those of corresponding $3d$ complexes.

(d) Intensities are higher for complexes with higher (formal) positive charge on the central ion.

(e) In a series of complexes of a given central ion with different ligands the intensities of transitions increase with increasing nephelauxetic effect.[107]

(f) The intensities of transitions of variously shaped complexes between the same central ion and ligands increase with the degree of departure from centrosymmetric character.

(g) The selection rule $\Delta J = 0, \pm 1$ of atomic spectroscopy may retain some meaning for the complexed metal ion. [This is one explanation for the fivefold ratio of intensities of $P \leftarrow F$ transitions of $CoCl_4^{2-}$ to $NiCl_4^{2-}$,[76] in that J-mixing from $^2G_{9/2}$ into ^4P of Co(II) is likely to be greater than that of $^1G_4$ into ^3P of Ni(II).]

A temperature increase would be expected to raise the intensity of a centrosymmetric complex by populating more vibrational levels and, possibly, by an average distortion. However, for tetrahedral complexes, say, a temperature increase would change the intensity in either direction or leave it unchanged by a combination of these processes. One might also consider the effect of temperature to act through the nephelauxetic effect but it is not easy to unscramble this from changes of bond length, distortion, or changes of crystal field splitting Δ.

> "Let the goodly empyrean be filled with light."
> Yasna, 31.7

3. GROUP VIII SPECIES IN THE GAS PHASE

3.1. Diatomic Molecules

The spectroscopy of diatomic molecules containing transition metals underwent a thorough review by Cheetham and Barrow[72] and in this section we merely wish to display a list of known molecules and to survey present ideas of their ground states and corresponding electronic configurations. The available data are presented in Table XVII.

As indicated previously, a simple ligand-field treatment[70] of the d^n configuration in $C_{\infty v}$ symmetry is inadequate for a discussion of these molecules because the configurations $d^{n-1}s$ and $d^{n-2}s^2$ are sufficiently low lying, especially in molecules formed formally from metals in the 1+ oxidation state, to participate significantly in the formation of molecular orbitals (cf. Fig. 13).[33,71,72,142] However, we can use the predictions of the separated-atom approach to rationalize an experimentally observed ground state, e.g., for PdH the ground state $^2\Sigma^+$, corresponding to a configuration $\sigma^2\delta^4\pi^4\sigma$, can be derived from $Pd(4d^{10})^1S + H(1s)^2S$ or from $Pd^+(4d^9)^2D + H^-(1s^2)^1S$.

The application of crystal field theory to the (supposed) linear dichlorides of $3d^n$ ions may be of some help in that, for reasonable values of the parameters ($A_2 = \pm 900$ cm^{-1}, $A_4 = \pm 100$ cm^{-1}) the separations of the d orbitals $\sigma_{g^+} - \pi_g$ and $\pi_g - \delta_g$ are 3800 and 4400 cm^{-1}, respectively.[97] The total splitting of 8200 cm^{-1} is like that found for octahedral complexes of divalent $3d^n$ ions and it is tempting to expect similar parameters for the diatomics, making due allowances for the position of the nonmetal ion in the spectrochemical series and for the formal charge on the metal. This would be consistent with Barrow's latest view that the excited states of TiO corresponding to the visible bands are derived from π_{4p} and σ_{4p} and not from π_{3d}.[143]

The scope for further work in this field is obvious and we believe that the extensive use of matrix isolation methods and measurements in the near-infrared region would be very rewarding.[281]

3.2. Other Species

As mentioned previously, we do not propose to discuss warm vapors such as those of anhydrous nitrates, carbonyls, or more esoteric organometallics. Thus this section becomes primarily concerned with halide species.

TABLE XVII. Diatomic Molecules Containing Group VIII Elements

Molecule	Ground state	Ground-state configuration	Ref.
FeH?	—	—	41
FeO	$^7\Sigma$ or $^5\Sigma$	$\sigma\delta^2\pi^2\sigma$ or $\sigma^2\delta^2\pi^2$	108
FeF	—	—	109
FeS	—	—	110, 111
FeCl	—	—	112, 113
FeBr	—	—	112, 114
Co$_2$	—	—	115
CoH	$^3\Phi_4$	$\sigma^2\delta^2\pi^3\sigma^2$	116
CoD	—	—	116
CoO	—	—	112
CoS	—	—	111
CoCl	—	—	117
CoBr	—	—	118
Ni$_2$	—	—	119
NiH	$^2\Delta_{5/2}$	$\sigma^2\delta^3\pi^4\sigma^2$	120
NiD	$^2\Delta_{5/2}$	$\sigma^2\delta^3\pi^4\sigma^2$	120
NiO	—	—	112, 121
NiF	—	—	122
NiS	—	—	111
NiCl	$^2\pi$??	cf. AgO[72]	123, 124, 125
NiGe	—	—	127
NiBr	—	—	123, 124
NiI	—	—	123
RuC	—	—	127, 128
RuO	$^3\Sigma^+$??	cf. Ref. 72	129
RhC	$^2\Sigma$??	cf. Ref. 72	130, 131
RhO	—	—	129, 132
Pd$_2$	—	—	133
PdH	$^2\Sigma^+$	$\sigma^2\delta^4\pi^4\sigma$	134
PdD	$^2\Sigma^+$	$\sigma^2\delta^4\pi^4\sigma$	134
PdO	—	—	135
OsO	—	—	136
IrC	$\Delta_{5/2}$ or $\Phi_{5/2}$	—	127, 137
IrO	—	—	129, 138
PtH	$^2\Delta_{5/2}$	$\sigma^2\delta^3\pi^4\sigma^2$	139, 140
PtD	$^2\Delta_{5/2}$	$\sigma^2\delta^3\pi^4\sigma^2$	139
PtB	—	—	127
PtC	$^1\Sigma$	—	141
PtO	$^1\Sigma$	—	128

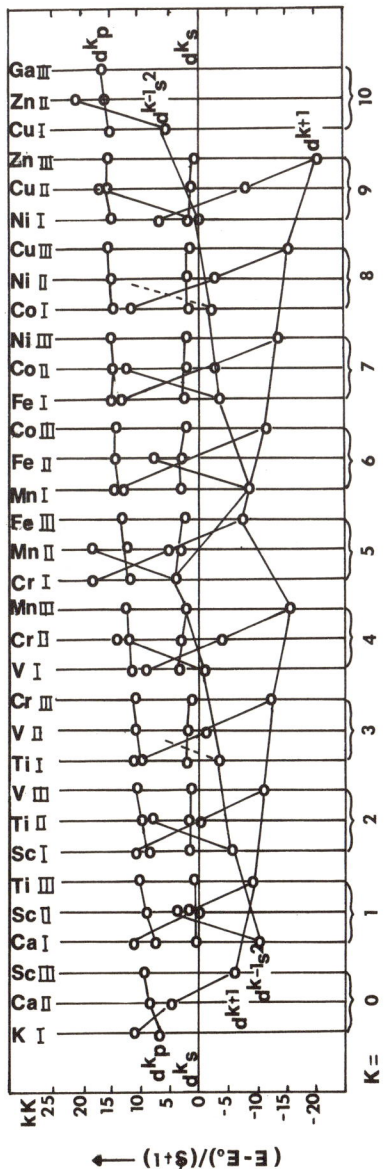

Fig. 13. The normalized, relative energies $(E - E_0)/(\zeta + 1)$ of the lowest terms of the configurations $3d^{k+1}$, $3d^k 4s$, $3d^{k-1} 4s^2$, and $3d^k 4p$, with E_0 the lowest term of $3d^k 4s$.[31,142]

In passing, we note the mass spectrometric evidence for the vapor species RhO_2,[132] and IrO_2 and IrO_3,[138] and the vapor pressure evidence for PtO_2.[144,145]

The halide vapors fall into three groups: monomeric transition metal halides, polymeric transition metal halides, and mixed alkali metal–transition metal halides. The first group has received considerable attention and the last is discussed in some detail elsewhere.[146]

The dipole moments of CoF_2 and NiF_2 were found to be close to zero by mass spectrometric observations of molecular beams,[147] suggesting the linearity of these molecules. However, infrared measurements on the matrix-isolated molecules were interpreted in terms of a distortion of up to 25° from linearity.[148] These deviations were later revised to 10° for CoF_2 and 15° for NiF_2 while FeF_2 was deduced to be linear. The errors in all cases were estimated at 10° and ascribed to lack of detailed knowledge of anharmonic effects.[149,150] Infrared studies were made also of matrix-isolated $NiCl_2$,[151,152] $FeCl_2$,[152] and $NiBr_2$[152] and it was concluded that these molecules were linear.

The electronic spectra of the gaseous dihalides $FeCl_2$,[97] $CoCl_2$,[96,97,153,154] $NiCl_2$,[96–98] $NiBr_2$,[98] and NiI_2[98] have been measured and interpreted on the basis that they are linear molecules. The energy level schemes for d^n systems in a weak axial field were applied (Section 2.3.4) and spin–orbital coupling was allowed for in assigning the spectra of the nickel halides.[98] The results have been critically discussed by Smith.[279]

$FeCl_2$[97] gave two weak infrared bands with vibrational structure and a very weak band close to intense charge-transfer transitions. The infrared bands were assigned as $^5\Pi \leftarrow {}^5\Delta$ at 4600 cm^{-1} and $^5\Sigma \leftarrow {}^5\Delta$ at 7140 cm^{-1}, with $E_\sigma - E_\pi = 2600$ cm^{-1} and $E_\pi - E_\delta = 4550$ cm^{-1}.

The $CoCl_2$ spectrum, reproduced in Fig. 14,[97] was fitted without spin–orbital coupling by taking $A_2 = 9A_4 = 900$ cm^{-1}, $E_\sigma - E_\pi = 3800$ cm^{-1}, $E_\pi - E_\delta = 4400$ cm^{-1}, and $E_P - E_F = 12.52$ kK, i.e., $\beta = 0.86$.[97] Similar spectra and general conclusions were reported in the earlier work[96] but vibrational structure[154] was not observed.

In Fig. 15 are reproduced the electronic spectra of gaseous $NiCl_2$, $NiBr_2$, and NiI_2.[98] They are very thoroughly discussed by the authors and we merely wish to draw attention to a few features. First, the shift to lower energies of the main band system in going to bromide and then iodide is in line with both the spectrochemical and nephelauxetic series for the ligands. Second,[98] the approach and consequent mixing of ^3P and ^1D levels in the same sequence may also be rationalized by the nephelauxetic effect. Why the overall intensity of the spectra should change is not clear. One

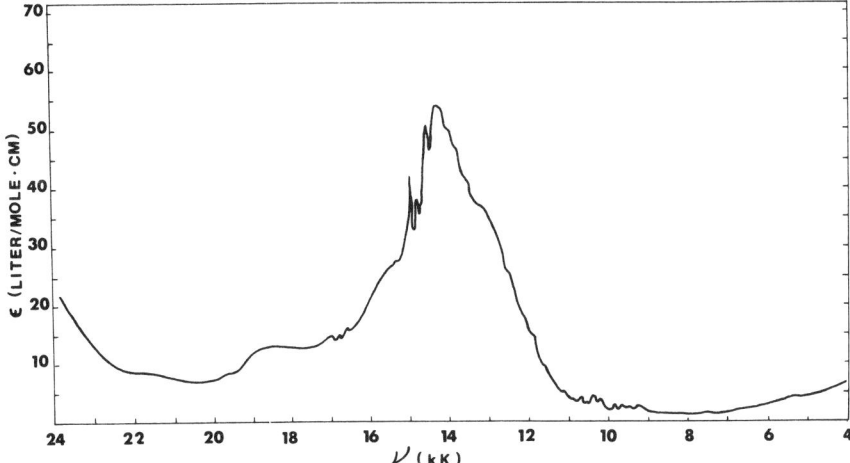

Fig. 14. The absorption spectrum of gaseous $CoCl_2$ at 937°C.[97] (Reproduced by permission from *J. Chem. Phys.*)

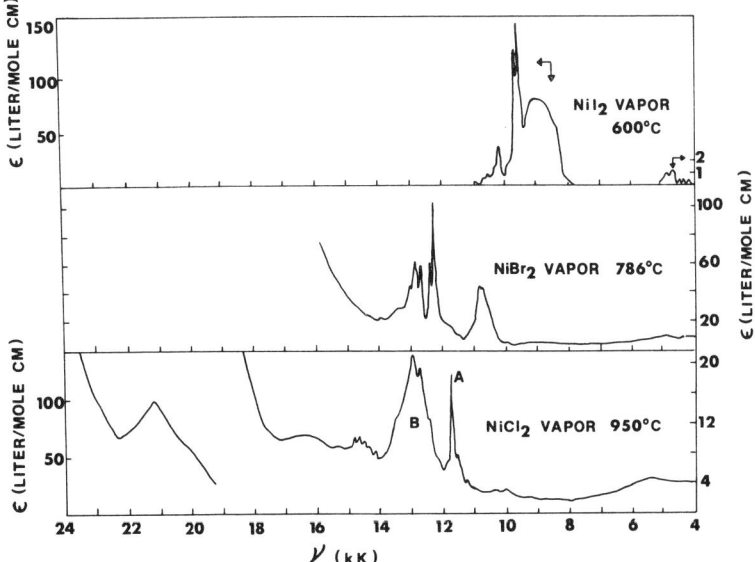

Fig. 15. The absorption spectra of $NiI_2(g)$, $NiBr_2(g)$, and $NiCl_2(g)$.[98] (Reproduced by permission from *J. Chem. Phys.*)

might expect it to decrease in the series on the grounds of the likely decrease of the Ni(II) spin–orbital coupling constant but increase in the series should there be any gross departure from linearity. The authors found that the spectral parameters listed in Table XVIII best fitted the data.

In a subsequent study of $FeCl_2$, $CoCl_2$, and $NiCl_2$ isolated in argon near 4.2°K[155] the ground states of the three molecules were confirmed and, incidentally, the charge-transfer spectra showed good correlation with the ligand-field transitions of gaseous $CoCl_2$, $NiCl_2$, and $CuCl_2$, respectively.

The work on polymeric and heavy metal halides is limited: Fe_2Cl_4, Co_2Cl_4, Ni_2Cl_4, and Ni_2Br_4 isolated in matrices were assigned D_{2h} bridged structures on the basis of infrared measurements.[152] Vapor pressure or mass spectrometric measurements indicated the existence of Fe_2F_6,[156] Fe_2I_4,[157] Fe_2Cl_6,[158] Fe_2Br_6,[158] and Co_2Br_4[157] but the symmetries and electronic structures of these molecules and, incidentally, their monomers remain food for speculation. It was thought that Pd_5Cl_{10} was the stable gaseous molecule below 980°C and $PdCl_2$ the stable one above 980°C,[159] but the polymer has recently been shown to be Pd_6Cl_{12}.[160] This and Pt_6Cl_{12}, which occurs with some Pt_4Cl_{10} in the vapor phase, may well have the chloride-bridged M_6 cluster form which exists in the solid phase in addition to the well-known chains. However, metal–metal bonding as in $Nb_6Cl_{12}^{2+}$[161] is unlikely. $RuCl_4$ is stable to 853°C and $RuCl_3$ above that temperature,[162] while the osmium halides[20] warrant more attention (see Section 4.6).

Molecular beam studies[163] established the existence of $NaFeCl_3$, $KCoCl_3$, $KNiCl_3$, and $CsFeCl_3$ in the vapor phase and large rotational magnetic moments were noted for these molecules. The visible spectrum of $CsCoCl_3$ was deduced from measurements on the vapors in equilibrium with liquids of composition $2CsCl:CoCl_2$ and $CsCl:CoCl_2$.[154] Although the spectrum differs distinctly from that of $CoCl_2$ (Fig. 14),[97] the actual geometry was not assigned.[154] The analysis of the Ni^{2+} spectrum in the $CsCl:CsAlCl_4$ liquid system in terms of planar $NiCl_3^-$ (see p. 167)[247] suggests that a planar $CoCl_3$ unit in the $CsCoCl_3$ vapor is quite likely.

TABLE XVIII. Spectral Parameters of Gaseous Nickel Dihalides

	$E_\sigma - E_\delta$, kK	F_2, cm^{-1}	F_2/F_4	ζ, cm^{-1}	β
$NiCl_2$	−8.2	1260	14	−550	0.79
$NiBr_2$	−7.3	1170	13	−550	0.73
NiI_2	−6.4	900	12	−550	0.56

> *What is this life, if, full of care,*
> *We have no time to stand and stare.*
>
> Leisure, W. H. Davies

4. GROUP VIII SPECIES IN MOLTEN SALTS

4.1. Review of Oxidation States

Studies to date have not revealed many unusual oxidation states of Group VIII metals in molten salt systems.

Much of the spectroscopic work on these systems has depended on the addition of known solid transition metal compounds to the particular solvent and it is conceivable that on some occasions there was an implicit assumption that no redox reaction ensued. In most instances this is a reasonable attitude to take, because the spectra obtained are checked against those of previously studied aqueous and solid systems, but a few curious changes have been noted. Hence it is our view that, whenever possible, the stable oxidation states of a metal in a system should be ascertained by a combination of methods and we thoroughly recommend electrode potential studies as an appropriate supporting technique for electronic spectroscopy. At the same time, the coulometric generation of ions in solution with concomitant determination of coulombic "n" by a standard analytical procedure provides a clean, controlled method of introducing the metal ions into solution. On several occasions the available salt of the cation may contain a different anion from that of the solvent with surprising consequences if this salt is added to the system—the assumption that the foreign anion will be swamped ignores our basic instruction on the subject of equilibrium or stability constants.

In Table XIX we present the oxidation states which have been characterized by electrode potential measurements in four solvents. Theoretical calculations by Hamer et al.[171] support the existence of Ru(III) in chloride systems [$E° = 26$ mV positive with respect to Pt(II)/Pt(0)]* and suggest that Ir(II) could be stabilized, presumably in more acidic chlorides. Further calculations on fluorides[172] suggest that Co(III) may be stable in fluoride melts and the detection of Ta(VI) and Nb(VI) in LiF–NaF–KF[173] indicates the likelihood of finding several more quite high oxidation states in such media. Ni(I) or Ni(0) has been postulated to exist in cyanide melts.[174,175]

The problem of characterizing oxidation states is complicated in oxyanion and similar systems by the ready reduction of the anion by metals

---

* The Ru(III)-Ru(O) potential has just been measured as -107 mV vs. Pt(II)/Pt(O).[282]

TABLE XIX. Relative Potentials (in volts) of Transition Metal Ion Couples in LiCl–KCl, NaCl–KCl, Li$_2$CO$_3$–Na$_2$CO$_3$, and Li$_2$SO$_4$–Na$_2$SO$_4$–K$_2$SO$_4$

Couple	Solvent			
	LiCl–KCl,[165–167, 282] 450°C	NaCl–KCl,[168] 700°C	Li$_2$CO$_3$–Na$_2$CO$_3$,[169] 550°C	Li$_2$SO$_4$–Na$_2$SO$_4$–K$_2$SO$_4$,[170]a 550°C
Fe(II)–Fe(0)	−0.215	−0.196	—	—
Co(II)–Co(0)	0	0	0	0
Ni(II)–Ni(0)	0.162	0.184	0.086	—
Pd(II)–Pd(0)	0.743	—	—	1.208
Rh(III)–Rh(0)	0.761	—	—	1.069
Ir(III)–Ir(0)	0.900	—	—	—
Ru(III)–Ru(0)	0.936	—	—	—
Pt(II)–Pt(0)	1.043	—	—	—
Fe(III)–Fe(II)	1.279	—	—	—

[a] See also Ref. 164.

such as iron or even nickel.[170] This reduces the scope of electrochemical methods and the production of solutions by the addition of salts (premixing of solid solute and solvent is advisable, followed by dilution with more solvent if necessary) must be resorted to.

4.2. Iron(III) $3d^5$

The spectrum of Fe(III) has been recorded in the LiNO$_3$–KNO$_3$ eutectic,[176] the NaCl–KCl–MgCl$_2$ eutectic,[177] the LiCl–KCl eutectic,[178] as molten tetraphenylarsonium tetrachloroferrate (III),[179] and in a tetra-n-butylphosphonium melt.[180]

In the LiNO$_3$–KNO$_3$ eutectic Gruen[176] reported that on addition of FeCl$_3$ no spectral maxima were obtained below 25 kK while there was complete absorption at higher energies.

Silcox and Haendler[177] investigated the UV region of the spectrum of Fe(III) in the NaCl–KCl–MgCl$_2$ eutectic. They reported maxima at ∼28.6, ∼36, and 42 kK and suggested that all these bands were due to charge-transfer transitions, although they had the rather low molar absorptivities of 110, 115, and 174, respectively.

The spectrum of Fe(III) in the LiCl–KCl eutectic was reported to contain a single shoulder located at ∼28.6 kK on a very intense band whose maximum in the UV was not determined.[177]

None of the above investigations were sufficiently detailed to enable the Fe(III) species to be characterized. In this respect the identification and assignment of ligand-field transitions would certainly be helpful. The principal reason why these bands were not observed in the chloride melts is as follows: For a d^5 ion there is one sextet atomic state only, i.e., ^6S and since this term is orbitally nondegenerate, it cannot split, whatever the symmetry of the ligand field. The spectrum of a complex with this ground state will therefore contain only weak bands corresponding to spin-forbidden transitions (cf. Table X).

Balt[179] tackled the problem of locating spin-forbidden transitions by measuring the spectrum of molten tetraphenylarsonium tetrachloroferrate (III) in a 1-mm cell. He reported that the spectrum of this compound in the solid state, in acetone solution, and the molten state were essentially identical and, since the solid has been shown to contain the $FeCl_4^-$ entity as a flattened tetrahedral species having D_{2d} symmetry,[181] it would seem that a similar species is present in the melt. All the observed bands were assigned to sextet–quartet transitions as shown in Table XX.[179,182,183]

The spectrum of $FeBr_3$ dissolved in a tetra-n-butylphosphonium bromide melt as reported by Islam[180] also contains weak bands, in this case in the 20–29-kK region of the spectrum. This spectrum was interpreted in favor of a tetrahedral $FeBr_4^-$ species, all bands again being assigned to sextet–quartet transitions (see Table XX). An interesting feature of these results is that the bands for the bromo complex occur at much higher energies than the corresponding bands for the chloro complex. Although a small shift is expected from the relative positions of the ligands in the spectrochemical series, the very high energies point to charge transfer transitions.

TABLE XX. Spectra of Tetrahedral FeX_4^- Complexes

$FeCl_4^-$,[a] ν, kK	Assignment	$FeBr_4^-$,[b] ν, kK
	$^4T_1(G) \leftarrow {}^6A_1(S)$	
14.60	$^4T_2(G) \leftarrow$	21.05
	$^4E_1, {}^4A_1(G) \leftarrow$	22.47
16.03	$^4T_1(P) \leftarrow$	23.87
16.51	$^4T_2(D) \leftarrow$	25.38
18.22	$^4E(D) \leftarrow$	26.74
22.00	$^4A_2(?) \leftarrow$	

[a] ~180°C.[179,182,183]
[b] 112°C.[180]

Cook and Dunn[15] have interpreted vapor-phase, calorimetric, and phase-diagram studies of the $FeCl_3$–MCl (M = Na, K) systems in favor of an $FeCl_4^-$ species but it should be noted that the presence of small amounts of $Fe_2Cl_7^-$ in molten $KFeCl_4$ has been deduced by chromatographic techniques.[12]

Brown[4] reported that the ESR spectrum of Fe(III) in the LiCl–KCl eutectic did not contain a measurable resonance signal. In the frozen sample, however, a broad line was observed. It was suggested that in the molten state an extremely broad signal which, under the conditions of the experiment, could not be distinguished from the baseline was in fact present. Such a broad signal, it was pointed out, may be interpreted in terms of a very short relaxation time which could be due to the presence of complex ions in the solutions.

4.2.1. Iron(II) $3d^6$

The spectrum of divalent iron has been reported in the LiCl–KCl eutectic,[184,185] an $AlCl_3$ melt,[186] the LiF–NaF–KF eutectic,[187] Li_2BeF_4 melt[188] and in a tetra-n-butyl phosphonium bromide melt.[180]

Gruen and McBeth[184,185] reported that the spectrum in the LiCl–KCl eutectic over the temperature range 400–1000°C consisted (below 28 kK) of a single peak whose maximum moved from 4.8 kK at 400°C to 6.0 kK

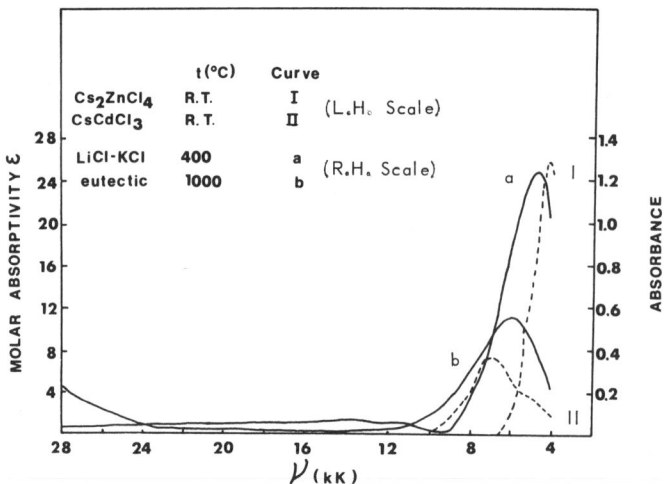

Fig. 16. Absorption spectra of Fe^{2+} in chloride media.[185] I, Cs_2ZnCl_4 at room temperature. II, $CsCdCl_3$ at room temperature (left scale). a, LiCl–KCl at 400°C. b, LiCl–KCl at 1000°C (right scale). (Reproduced by permission from *Pure and Appl. Chem.*)

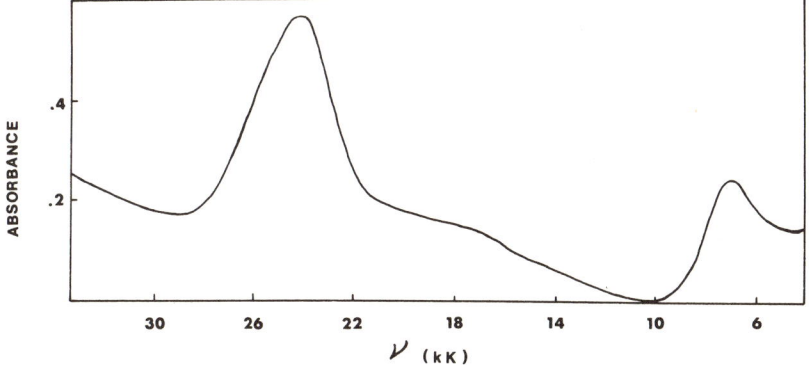

Fig. 17. Absorption spectrum of FeCl$_2$ in molten AlCl$_3$ at 227° and 5.6 atm.[187] (Reproduced by permission from *Inorg. Chem.*)

at 1000°C (Fig. 16). They also observed that as the temperature increased, the intensity of the band decreased. By comparing the melt spectra to the crystal spectra of Fe(II) in both a tetrahedral lattice (Cs$_2$ZnCl$_4$) and an octahedral one (CsCdCl$_3$), it was concluded that at 400°C the melt contained tetrahedral FeCl$_4^{2-}$ species which underwent some form of distortion as the temperature was raised. The single absorption band was assigned to the $^5E \leftarrow {}^5T_2$ transition.

The spectrum of Fe(II) in molten AlCl$_3$ at 227°C is shown in Fig. 17 and the proposed assignment is included in Table XXI.

Figure 17 also shows what appear to be shoulders at ~14 and ~18 kK. This spectrum was assigned by Øye and Gruen[186] to an octahedral Fe(II) species. The tentative assignment of the band at 24.1 kK to a mixture of spin-forbidden transitions is not very satisfactory since the published spectrum of Fe(II) in the octahedral CsCdCl$_3$ lattice[185] does not contain a similar band. It seems unlikely that the band is due to the presence of an Fe(III) impurity because such a species would also be expected to give rise to spin-forbidden transitions in this region of the spectrum (see earlier). Perhaps the excited state derives from $3d^54s$.

TABLE XXI. The Spectrum of Fe(II) in Molten AlCl$_3$

ν, kK	ε	δ	Assignment in O_h
7.1	4.6	2.1	$^5E_g \leftarrow {}^5T_{2g}$
24.1	11.0	4.5	$^3P, {}^3H, {}^3F \leftarrow$

TABLE XXII. Spectra of Fe(II) in Fluoride Melts

LiF–Na–KF (525°C) ν, kK	LiF–BeF$_2$ (66.34 mole %)				LiF–BeF$_2$ (72.28 mole %, 650°C) 650°C	
	540°C		650°C			
	ν, kK	ε	ν, kK	ε	ν, kK	ε
5.5	5.5	3	5.8	2	5.8	2
10.15	9.8	4.5	9.6	5	9.6	5

Although complexes of Fe(I) are not well known, there is a similarity to the spectrum of the isoelectronic Co(II) in the same melt (see Fig. 24A) but this is not very certain evidence because of the expected low energy of terms derived from $3d^6 4s$[142] (cf. Fig. 13). An electrochemical investigation of the Fe–AlCl$_3$ solution might clarify the situation.

Young[187,188] has studied the Fe(II) spectrum in fluoride melts: his results are shown in Table XXII and the spectrum in Li$_2$BeF$_4$ at 540°C in Fig. 18. The results in Table XXII show that the species is relatively insensitive to changes in both temperature and melt composition. The presence of two bands in the spectrum (rather than the single transition $^5E_g \leftarrow {}^5T_{2g}$ expected for an octahedral d^6 ion) was suggested to be due to a Jahn–Teller distortion of the octahedron.

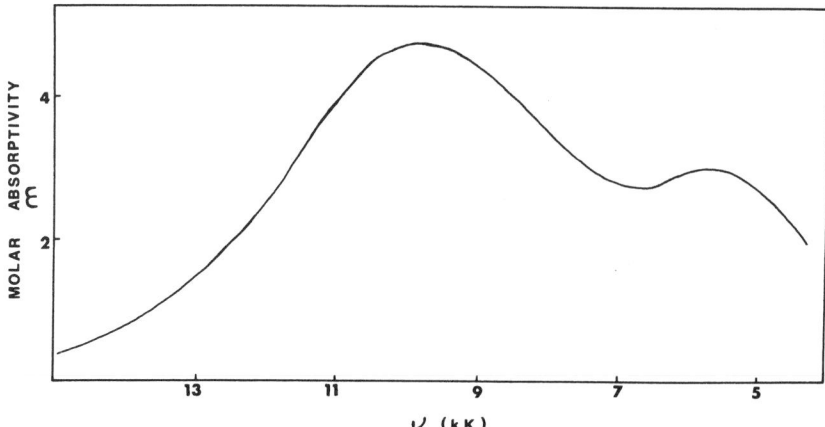

Fig. 18. Spectrum of FeF$_2$ in molten 2LiF·BeF$_2$ at 540°C.[188] (Reproduced by permission from *Inorg. Chem.*)

In the tetra-n-butylphosphonium bromide melt[180] the spectrum of Fe(II) contained a single broad band at 4.5 kK which was assigned to the $^5T_2 \leftarrow {}^5E$ transition of a tetrahedral $FeBr_4^{2-}$ entity. Notice that for Fe(II) the bromo species has its transition at lower energy than the chloro complex.

4.3. Cobalt(II) $3d^7$

Spectral investigations of Co(II) in fused salts have covered a wide range of both inorganic and organic melts. The published literature is summarized in Table XXIII. The discussion that follows will be divided into four sections. The first will consider results obtained in oxyanion melts, the second will be concerned with halide melts; mixed oxyanion–halide melts will be discussed in the third section; and finally we will try to correlate the previous results.

4.3.1. Oxyanion Melts

The spectrum of $CoCl_2$ dissolved in the $LiNO_3$–KNO_3 eutectic was reported by Gruen[175,189] (Fig. 19A). It was suggested that the cobalt species was possibly octahedral $Co(NO_3)_6^{4-}$. Alternative descriptions[209] of the cobalt species have involved coordination by four groups; in one model the 12 equidistant oxygens of the tetrahedrally arranged nitrate groups gave rise to a cubic field, whereas in an alternative model two of the nitrate groups were bidentate and two were monodentate. The possibility of a species involving three bidentate nitrate groups was also considered.[209]

Fung and Johnson[195] have investigated the spectrum of $Co(NO_3)_2$ in both the $LiNO_3$–KNO_3 (Fig. 19B) and the $AgNO_3$–KNO_3 melts, together with that of $Co(ClO_4)_2$ dissolved in dimethylsulfone, with (Fig. 19C) and without (Fig. 19D) the addition of $LiNO_3$–KNO_3. The spectrum in the dimethylsulfone melt (Fig. 19D) can be unequivocally assigned to an octahedral Co(II) species on the basis of band positions, band shapes, and oscillator strengths. In particular, the value of the oscillator strength of the main band (1.1×10^{-5}) is comparable to that of the corresponding band in the $Co(H_2O)_6^{2+}$ spectrum in aqueous solution, namely 8.1×10^{-5}.[210] Addition of $LiNO_3$–KNO_3 to the dimethylsulfone melt produced the spectrum of Fig. 19C, which is almost identical to the spectrum of Fig. 19B and suggests that this latter spectrum is not due to a simple octahedral nitrate complex. The spectra shown in Fig. 19B and 19C and that for the $AgNO_3$–KNO_3 melt were assigned by Fung and Johnson[195] by comparison with the spectrum of $\{CH_3(C_6H_5)_3As\}_2\{Co(NO_3)_4\}$ (Fig. 19E),[211] whose

TABLE XXIII. Systems Used to Study Co(II) Spectra

System	Ref.
$LiNO_3$–KNO_3	175, 190–195
$AgNO_3$–KNO_3	195
Dimethylsulfone	195
Li_2SO_4–Na_2SO_4–K_2SO_4	196, 197
K_2SO_4–$ZnSO_4$	196
Na_2SO_4–K_2SO_4–$ZnSO_4$	196
$NaHSO_4$–$KHSO_4$	198
$Li(CH_3CO_2)$–$Na(CH_3CO_2)$	199
$Tl(CH_3CO_2)$	200
$Na(CH_3CO_2)3H_2O$	200
$Na(CH_3CO_2)$–$K(CH_3CO_2)$	200
LiF–NaF–KF	201
LiCl–KCl	178, 185, 202
LiBr–KBr	202
$CaCl_2$–$MgCl_2$	203
NaCl–KCl–$MgCl_2$	177
KSCN	178, 204
LiCl	193, 203, 205
NaCl	193, 205
KCl	195, 205
RbCl	193
CsCl	193, 205
$MgCl_2$	205
$CdCl_2$	205
$PbCl_2$	205, 206
$SnCl_2$	206
$BiCl_3$	206
$AlCl_3$, $AlCl_3$–KCl, $AlCl_3$–$ZnCl_2$	187, 205, 207
$GaCl_3$	204
$HgCl_2$	204
Pyridine hydrochloride	175
(n–Bu_4P)Cl and (n–Bu_4P)Br	180

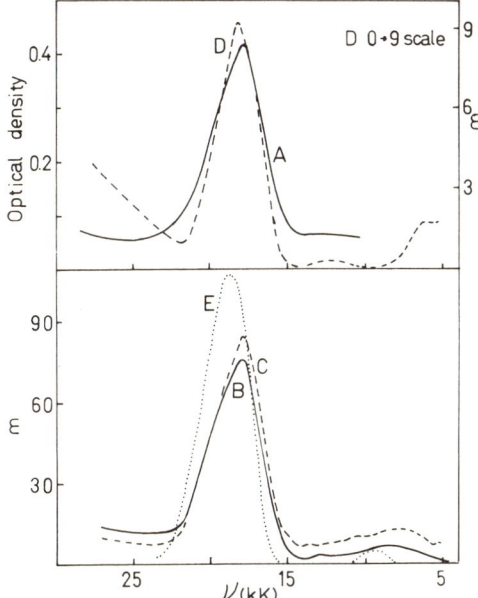

Fig. 19. Absorption spectra of Co(II) in nitrate media: (A) $CoCl_2$ in $LiNO_3$–KNO_3 at 184°C.[176] (B) $Co(NO_3)_2$ in $LiNO_3$–KNO_3 at 156°C.[195] (C) $Co(ClO_4)_2$ in $DMSO_2$ with added $LiNO_3$–KNO_3 at 128°C.[195] (D) $Co(ClO_4)_2$ in $DMSO_2$ at 128°C.[195] (E) $\{CH_3(C_6H_5)_3As\}_2\{Co(NO_3)_4\}$ in nitromethane.[211]

structure is known to contain eight coordinate dodecahedral Co(II),[212] to a complex having the same arrangement of four nitrate groups. A comparison of Fig. 19A and 19B seems to indicate that the spectrum of Co(II) in the $LiNO_3$–KNO_3 melt depends on the particular cobalt salt used in preparing the solutions. The assumption made by Gruen[175,209] (namely that the $CoCl_2$ would produce simple nitrate complexes) is apparently invalid and it would seem that the spectrum shown in Fig. 19A is in fact due to a mixed chloronitrate complex. The following evidence can be cited to support this conclusion: When $Co(NO_3)_2$ was added to a dimethylsulfone melt the spectrum differed from that obtained with $Co(ClO_4)_2$[195]; however, when excess nitrate was added, in the form of $LiNO_3$–KNO_3, the resulting spectrum was almost identical to that of the $LiNO_3$–KNO_3 melt (Fig. 19B). In addition, cryoscopic studies[213,214] of $CoCl_2$ solutions in $LiNO_3$, $NaNO_3$, and KNO_3 melts have allowed calculation of the values for the stability constant K_1 for the reaction

$$CoCl^+ + Cl^- = CoCl_2$$

The results are listed in Table XXIV.

The spectrum of Co(II) in the Li_2SO_4–Na_2SO_4–K_2SO_4 melt was initially interpreted by Johnson and Piper[147] in favor of a distorted tetrahedral arrangement about the Co^{2+} ion. This interpretation was questioned[209] and it was suggested, on the basis of the extinction coefficients, that an octahedral complex was present. Duffy et al.[198] have investigated the spectrum of Co(II) in a K_2SO_4–$ZnSO_4$ glass, a $NaHSO_4$ melt and glass and in sulfuric acid solutions at 25 and 190°C. The sulfuric acid and the $NaHSO_4$–$KHSO_4$ glass spectra, at room temperature, indicate octahedral coordination. A similar geometry was assigned to Co in the K_2SO_4–$ZnSO_4$ glass but this spectrum differed from the $NaHSO_4$–$KHSO_4$ glass and the sulfuric acid solution spectra. It was claimed that, on heating, both the bisulfate glass and the sulfuric acid solution gave spectra resembling that of the K_2SO_4–$ZnSO_4$ glass at 25°C. The pertinent spectra are shown in Fig. 20. These results were interpreted in terms of two octahedral Co(II) complexes which differed in the way that the sulfate ligands were coordinated. Thus it was suggested that in the K_2SO_4–$ZnSO_4$ glass (Fig. 20A) the $NaHSO_4$–$KHSO_4$ melt at 230°C (Fig. 20B) and the sulfuric acid solution at 190°C (Fig. 20E) a high-temperature, or violet, form, involving bidentate sulfate ligands, was present. On cooling the bisulfate melt and the sulfuric acid solution, a low-temperature, or pink, form in which one or more sulfate ligands had converted from bidentate to unidentate, was produced. The inability of the K_2SO_4–$ZnSO_4$ glass to undergo this transition was postulated to be due to either the rigidity of the medium or the absence of protons which could counteract the increase of electrostatic repulsion by protonation.

TABLE XXIV. Stability Constants for $CoCl^+ + Cl^- \rightarrow CoCl_2$ in Nitrate Melts

Melt	K_1
$LiNO_3$	9 ± ?
$NaNO_3$	404 ± 68
KNO_3	910 ± 51

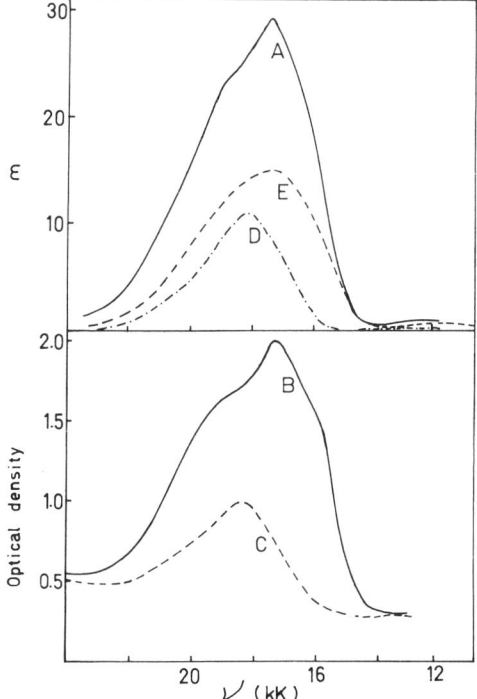

Fig. 20. Absorption spectra of Co(II) in sulfate media: (A) K_2SO_4–$ZnSO_4$ glass at 25°C.[198] (B) $NaHSO_4$–$KHSO_4$ melt at 230°C.[198] (C) $NaHSO_4$–$KHSO_4$ glass at 25°C.[198] (D) 98% H_2SO_4 at 25°C.[210] (E) 98% H_2SO_4 at 190°C.[210]

The present authors have reinvestigated the spectrum of Co(II) in the Li_2SO_4–Na_2SO_4–K_2SO_4 melt and in the K_2SO_4–$ZnSO_4$ glass as a function of temperature in the range 14–191°C.[196] Also investigated in this study were the spectra in the K_2SO_4–$ZnSO_4$ melt and the Na_2SO_4–K_2SO_4–$ZnSO_4$ melt. The results are shown in Fig. 21.

A comparison of Fig. 20B and 21B indicates that in band shapes and positions the bisulfate melt spectrum bears a closer resemblance to that of the K_2SO_4–$ZnSO_4$ melt than to that of the glass. Furthermore, the bisulfate melt (Fig. 20B) is also similar to the other sulfate melts (Fig. 21A–C). The spectrum in the Li_2SO_4–Na_2SO_4–K_2SO_4 melt was assigned to a cobalt species having a dodecahedral arrangement of sulfate ligands.[196] A similar arrangement of ligands in the other sulfate melts and in the bisulfate melt was suggested on the grounds of similar spectra. An interesting feature of

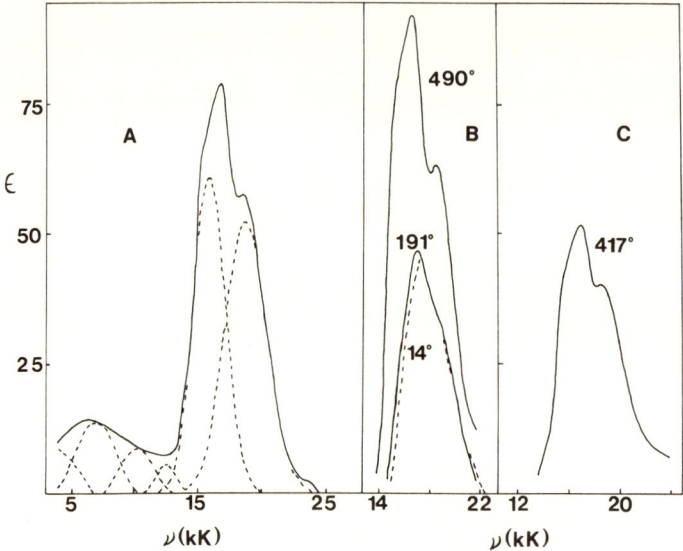

Fig. 21. Absorption spectra of Co(II) in sulphate media: (A) Li_2SO_4–Na_2SO_4–K_2SO_4 melt at 550°C.[196] (B) K_2SO_4–$ZnSO_4$ melt (490°C) and glass (14, 191°C).[196] (C) Na_2SO_4–K_2SO_4–$ZnSO_4$ melt at 417°C.[196] (Reproduced by permission from *J. Mol. Spectry.*)

these spectra was the variation in intensity which occurs for the sulfate and bisulfate melts. The estimated oscillator strengths are shown in Table XXV.[196] It can be seen that the intensity increases with temperature for the sulfate media, which is contrary to the results of Gruen *et al.*[185] for tetrahedral $CoCl_4^{2-}$ (see later) and suggests that the vibronic intensity

TABLE XXV. Oscillator Strengths of Visible Bands of Cobalt(II) in Oxyanion Melts

Melt	T, °C	$f \times 10^5$	Ref.
Li_2SO_4–Na_2SO_4–K_2SO_4	550	176	196
Na_2SO_4–K_2SO_4–$ZnSO_4$	417	125	196
K_2SO_4–$ZnSO_4$	580	197	196
$NaHSO_4$–$KHSO_4$	230	90	198
CH_3COOLi–CH_3COONa	180	~475	199
CH_3COOTl	150	~186	200

mechanism is occurring. Alternatively, the cobalt species may undergo a distortion to a less symmetric species as the temperature is raised (cf. Section 2.3.6).

Further evidence that can be cited in favor of an eight-coordinate Co(II) species are the spectra shown in Fig. 22A and 22B. Both these spectra refer to Co(II) substituted in a fluoride lattice in which the divalent ion is surrounded by eight fluoride ions arranged at the corners of a cube.[215,216]

Recent investigations of the spectrum of Co(II) in acetate melts[199,200] were interpreted in terms of a tetrahedral tetraacetato Co(II) complex. The results of Bailey et al.[200] are given in Table XXVI together with the proposed band assignments. These workers noted the similarity in band shape between their results and the sulfate melt spectrum of Johnson and Piper[197] for the main band in the visible region. Furthermore, they pointed out that the most unequivocal means of identification of a tetrahedral Co(II) spectrum was afforded by the presence of a band in the infrared region ($\sim$4 kK) due to the $^4T_2 \leftarrow {}^4A_1$ transition. The results of Duffey et al.[199] for the Li(CH-CO_2)–Na(CH_3CO_2) melt are in substantial agreement with those of Bailey and co-workers except for the larger oscillator strength of

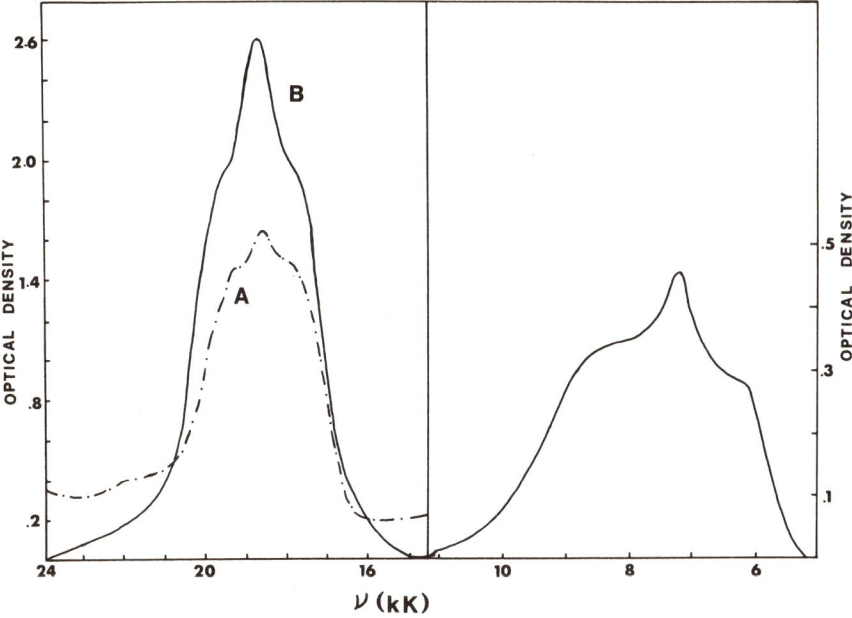

Fig. 22. Absorption spectra of Co(II) in fluoride crystals: (A) CaF_2 at 90°K.[215] (B) CdF_2 at 298°K.[216]

TABLE XXVI. Spectra of CO(II) in Acetate Melts

Melt	ν, kK	ε	Assignment	Δ, kK	B, cm^{-1}
Tl(CH$_3$CO$_2$) (150°C)	3.90	~15	$^4T_2 \leftarrow {}^4A_1$	4.15	755
	7.20	18	$^4T_1(F) \leftarrow$	—	—
	16.10	—	—	—	—
	17.30	95	$^4T_1(P) \leftarrow$	—	—
	18.70	—	—	—	—
Na(CH$_3$CO$_2$)–K(CH$_3$CO$_2$) (250°C)	3.94	~15	$^4T_2 \leftarrow {}^4A_1$	4.05	770
	7.00	20	$^4T_1(F) \leftarrow$	—	—
	16.10	—	—	—	—
	17.20	100	$^4T_1(P) \leftarrow$	—	—
	18.80	—	—	—	—
Na(CH$_3$CO$_2$) · 3H$_2$O (90°C)	3.87	<10	$^4T_2 \leftarrow {}^4A_1$	420	727
	7.50	25	$^4T_1(F) \leftarrow$	—	—
	17.40	120	$^4T_1(P) \leftarrow$	—	—
	18.70	—	—	—	—

the main visible band (see Table XXV). Otherwise, the intensity of the bands and the precise location of the $^4T_2 \leftarrow {}^4A_1$ transition place acetate in the same class as sulfate and nitrate, i.e., a ligand which adopts eightfold coordination about Co^{2+} but which exerts a weaker field than the other two.

4.3.2. Halide Melts

Young and White [201] have reported the spectrum of CoF$_2$ in the LiF–NaF–KF eutectic melt (Fig. 23) for which both octahedral[217] and tetrahedral[218] coordination geometries have been suggested (see Section 4.3.4 for further discussion).

The UV spectrum of Co(II) in the NaCl–KCl–MgCl$_2$ eutectic contained a single peak at 37.6 kK with a molar absorptivity of 203[177] which was assigned to a charge transfer transition. Again, as with the corresponding Fe(III) spectrum, the intensity would appear to be rather low for a charge transfer band.

The spectrum in the LiCl–KCl eutectic has been investigated by Harrington and Sundheim,[178] Gruen and McBeth,[185] and Sundheim and Kukk.[202] The latter workers obtained a spectrum consisting of three well-resolved components on an intense band spanning the region 13–18 kK

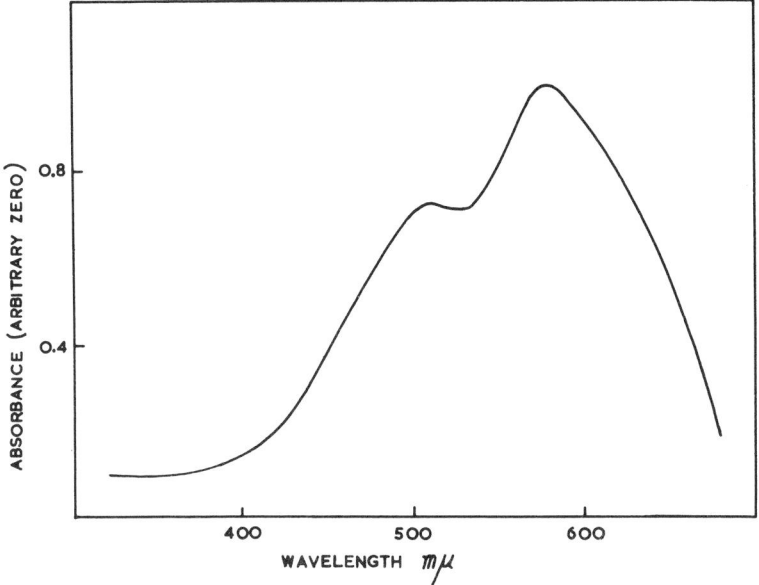

Fig. 23. Spectrum of cobalt fluoride in LiF–NaF–KF at ~500°C.[201] (Reproduced by permission from *Anal. Chem.*)

which they assigned to the components of the $^4T_1(P) \leftarrow {}^4A_2(F)$ transition of tetrahedral $CoCl_4^{2-}$. The more detailed investigation of Gruen and McBeth[185] revealed a further band located in the near-infrared at 5.6 kK. The results and proposed assignments of these workers are shown in Table XXVII.

It was also observed by Gruen and McBeth that as the temperature of the melt increased from 400 to 1000°C, the oscillator strength of the main visible band decreased (Table XXVIII). A similar effect was also reported by Sundheim and Kukk.[202]

TABLE XXVII. Spectrum of Co(II) in LiCl–KCl at 400°C[185]

v, kK	ε	Assignment
5.6	50	$^4T_1(F) \leftarrow {}^4A_2(F)$
14.3	365	$^4T_1(P) \leftarrow$
15.1	360	
16.4	230	$^2E_1, {}^2T_1(^2G) \leftarrow$

TABLE XXVIII. Temperature Dependence of the Oscillator Strength of $CoCl_4^{2-}$ [185]

$f \times 10^3$	T, °C
4.5	400
4.3	600
3.5	800
3.2	1000

As discussed earlier, if the intensity-producing mechanism was vibronic in origin, then an increase in temperature should lead to an increase in oscillator strength. The observed decrease was explained using the covalent mechanism of Liehr and Ballhausen[219]: As the temperature increased, the decrease in intensity was correlated with an increase in the ionicity of the Co—Cl bonds, assuming that both the overlap integral (between Co and Cl) and the ligand coupling integral were temperature invariant.[185]

In the LiBr–KBr eutectic[202] exactly the same type of spectrum and variation with temperature were observed as for the corresponding chloride melt. The only difference was that the spectrum was shifted to slightly lower energies, as would be expected from the relative positions of Br⁻ and Cl⁻ in the spectrochemical series.

The tetrahedral $CoCl_4^{2-}$ species also exists in a pyridine hydrochloride melt.[175] A thorough study of the spectrum of Co(II) in molten tetra-*n*-butylphosphonium bromide and chloride has been carried out by Islam,[180] who found that CoX_4^{2-} also exists in these melts. From this study, which covered the spectrum from 4 to 40 kK, several interesting points emerged. First, the main band in the visible region (13–16 kK) showed *four* very well-resolved components; the infrared band also showed an unusual amount of structure (for a molten salt spectrum). The most striking result, however, was the amount of detail obtained for the spin-forbidden bands occurring in the 17–25-kK region of the spectrum. The spectral band positions, extinction coefficients, and assignments are shown in Table XXIX.

Islam[180] reports that over the temperature range 112–158°C the molar absorptivities of the bands decreased with increasing temperature while the width increased. If the variation in the density of the melt with temperature is taken into account, it is found that the oscillator strength is independent of temperature.

TABLE XXIX. Spectra of Molten $(n\text{-Bu}_4P)_2\text{CoBr}_4$ and $(n\text{-Bu}_4P)_2\text{CoCl}_4$ [180]

$(n\text{-Bu}_4P)_2\text{CoBr}_4$		$(n\text{-Bu}_4P)_2\text{CoCl}_4$	Assignment
ν, kK	ε		
4.325	(sh)	—	—
4.300	(sh)	—	—
4.400	83	—	$^4T_1(F) \leftarrow {}^4A_2(F)$
4.545	87	—	—
4.900	92	—	—
5.435	90	—	—
13.736	910	14.367	—
14.245	819	14.947	—
15.000	520	15.798	$^4T_1(P) + {}^2G \leftarrow$
15.552	280	16.286	—
16.529	—	—	—
16.949	—	—	—
17.271	—	—	$^2G \leftarrow$
17.730	—	—	—
18.587	—	—	—
20.704	—	—	—
21.142	—	—	—
21.505	—	—	$^2P + {}^2H \leftarrow$
21.786	—	—	—
22.910	—	—	$^2T_2(D) \leftarrow$
23.364	—	—	$^2E(D) \leftarrow$
35.336	3090	—	—
37.313	2674	—	Charge transfer

The spectrum of Co(II) in an $AlCl_3$ melt at 227° and 6 atm is shown in Fig. 24A.[187] The main band, occurring at 16 kK, has a molar absorptivity of 76 and was assigned to the $^4T_{1g}(P) \leftarrow {}^4T_{1g}(F)$ transition of an octahedral Co(II) species. The two shoulders occurring at higher energy were suggested to be due to doublet ← quartet transitions.[220] The two other spin-allowed transitions, $^4A_{2g}(F) \leftarrow {}^4T_{1g}(F)$ and $^4T_{2g}(F) \leftarrow {}^4T_{1g}(F)$, which should occur at lower energies, were not reported. Øye and Gruen[187] discuss various pieces of evidence which suggest that addition of alkali halide to an $AlCl_3$ melt results in the formation of $Al_2Cl_7^-$ ions initially,

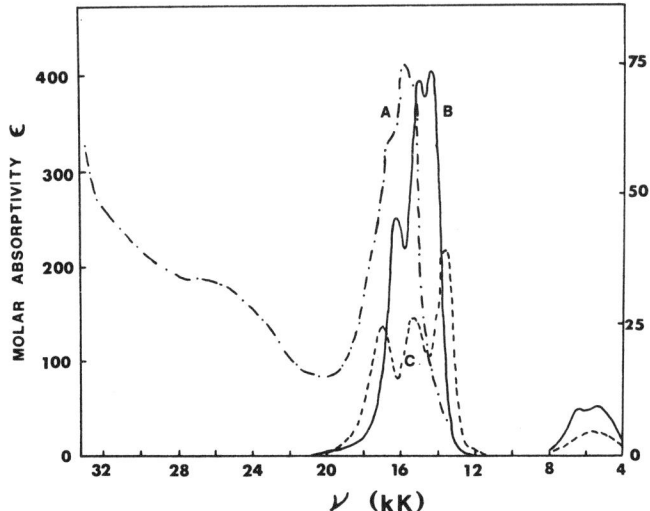

Fig. 24. Absorption spectra of Co(II) in chloride media: (A) Molten AlCl$_3$ at 227° and 5.6 atm (right-hand scale).[187] (B) 50.3 mole % KCl, 49.7 mole % AlCl$_3$ at 300°C.[208] (C) 49.9 mole % KCl, 50.1 mole % AlCl$_3$ at 300°C.[208]

while further alkali halide produces AlCl$_4^-$ ions. It is suggested that on addition of a transition metal dichloride the following reaction occurs:

$$MCl_2 + 2Al_2Cl_6 = M(Al_2Cl_7)_2$$

The structure of this species was believed to have been

```
      Cl           Cl   Cl            Cl
       \          /      \            /
  Cl—Al—Cl—Al—Cl—M—Cl—Al—Cl—Al—Cl
       /          \      /            \
      Cl           Cl   Cl            Cl
```

in which the transition metal ion is surrounded octahedrally by six chloride ions.

On addition of KCl to the AlCl$_3$ melt containing Co(II), it was reported that in the range 0–35.5 mole % KCl the spectra remained virtually unchanged.[205] Between 35.5 and 49.9 mole % KCl a completely new spectrum (Fig. 24C) was produced, while at compositions above 50 mole % KCl the spectrum was characteristic of tetrahedral CoCl$_4^{2-}$ (Fig. 24B), although species such as (CoCl$_3$AlCl$_4$)$^{2-}$ were not completely ruled out. A detailed analysis of the spectra obtained in the 35.5–49.9 mole % KCl region in terms of the two equilibria

$$Al_2Cl_7^- = 2AlCl_4^-$$
$$Co(Al_2Cl_7)_2 + n_1AlCl_4^- = Co(Al_2Cl_7)_{2-n_1}(AlCl_4)_{n_1}^{n_2-n_1} + n_2Al_2Cl_7^-$$

led to a value of n_2 equal to one but was unable to distinguish between $n_1 = 1$ and $n_1 = 2$. The authors assumed that the simplest case, $n_1 = 1$, was true. The stoichiometry of the species responsible for the spectrum of Fig. 24B was, therefore, $Co(Al_2Cl_7AlCl_4)$, in which the cobalt is still six-coordinate but with a severely distorted arrangement.

In $ZnCl_2$–$AlCl_3$ mixtures[207] the spectra were interpreted in terms of a progressive conversion from octahedral (31.5 mole % $ZnCl_2$) to tetrahedral (~50 mole % $ZnCl_2$) coordination about the cobalt. The spectrum in a melt of 61.5 mole % $ZnCl_2$ was studied as a function of temperature over the range 25–400°C and the spectral changes explained on the basis of an octahedral–tetrahedral equilibrium.

Øye and Gruen[205] have also examined the spectra of Co(II) in a wide range of pure chloride melts and assigned them to one of three types: distorted tetrahedral, octahedral, or strongly distorted octahedral, on the basis of the breadth and oscillator strength of the main visible band. Some of their data, together with recent results[206] for $CoCl_2$ dissolved in $PbCl_2$, $BiCl_3$, and $SnCl_2$, are presented in Table XXX. In molten pyridinium chloride, an ionic melt, there is little doubt that the cobalt species is tetrahedral. Comparable oscillator strengths but broader bands are obtained for $PbCl_2$, $BiCl_3$, and $SnCl_2$, perhaps because of distortion or an averaging of essentially tetrahedral sites in the polymeric melts. The spectra in $GaCl_3$ and $AlCl_3$ suggest, rather than octahedral coordination,[205] that we are close to that found in the oxyanion systems. The descriptions "strongly distorted octahedral" (for $HgCl_2$) and "distorted tetrahedral" (for the remainder) are also questionable in the light of the oxyanion results (Table XXV).

Harrington and Sundheim[178] report that in molten KSCN, $CoCl_2$, $CoBr_2$, and $Co(SCN)_2$ produce identical spectra. In fact, a Beer–Lambert plot of data obtained with the three salts produced a straight line. It was concluded that, since no evidence for mixed complexes was obtained, the species present was a tetrahedral $Co(SCN)_4^{2-}$ complex. Recent work by Egghart[204] has confirmed the tetrahedral nature of the cobalt thiocyonate complex which is suggested to involve Co–N bonding.

Inman et al.[11] have used a potentiometric method to demonstrate the formation of cyano complexes in solutions of $CoCl_4^{2-}$ in the LiCl–KCl

TABLE XXX. The Visible Band ($^4P \leftarrow {}^4F$ Mainly) of the Co(II) Spectrum in Chloride Melts

Melt	Temperature, °C	Band components						$\Sigma f \times 10^3$	Ref.
		ν, kK	ε	ν, kK	ε	ν, kK	ε		
$MgCl_2$	800	14.4	154	15.1	156	16.8	101	2.54	205
$CdCl_2$	650	14.2	173	15.0	176	16.6	121	2.83	205
LiCl	800	14.3	173	15.1	179	16.7	117	2.89	205
NaCl	820	14.2	223	16.0	223	16.5	140	3.25	205
KCl	800	14.2	249	14.8	245	16.4	152	3.35	205
CsCl	800	14.2	292	14.8	298	16.4	176	3.82	205
$PbCl_2$	532	13.9	240	14.7	250	16.1	180	4.06	206
$BiCl_3$	270	14.0	246	14.9	255	16.3	178	4.10	206
$SnCl_2$	300	14.0	271	14.8	284	16.2	202	4.70	206
$GaCl_3$	150	14.8	—	15.8	58	16.7	—	0.78	205
$AlCl_3$	227	14.8	—	15.7	76	16.7	—	0.92	205
$HgCl_2$	350	13.6	228	15.4	197	16.8	188	3.8	205
PyCl	150	14.3	630	14.9	570	15.9	360	5.42	205

eutectic to which KCN was added. The complexes were postulated to be four-coordinate having a square-planar geometry. Spectral studies on such solutions may allow a characterization of the geometry to be made.

4.3.3. Oxyanion–Halide Melts

Gruen[175] has examined in some detail the spectral changes obtained by adding excess chloride to a solution of Co(II) in $LiNO_3$–KNO_3. The results are shown in Fig. 25 and have been interpreted as a gradual change from a species uncomplexed with respect to chloride (Fig. 24A) to the fully complexed $CoCl_4^{2-}$ species (Fig. 24E). The following scheme was postulated to represent the spectral changes,[209] assuming the existence of an octahedral two mono–two bidentate nitrate cobalt complex in the pure $LiNO_3$–KNO_3 melt, as discussed earlier:

$Co(NO_3)_4^{2-} \rightarrow Co(NO_3)_3Cl^{2-} \rightarrow Co(NO_3)_2Cl_2^{2-} \rightarrow Co(NO_3)Cl_3^{2-} \rightarrow CoCl_4^{2-}$

It was suggested that the monodentate nitrate groups are replaced first. Similar spectral results have been obtained by Tananaev and Dzhurinskii,[191–194] who have found the same type of behavior in the corresponding

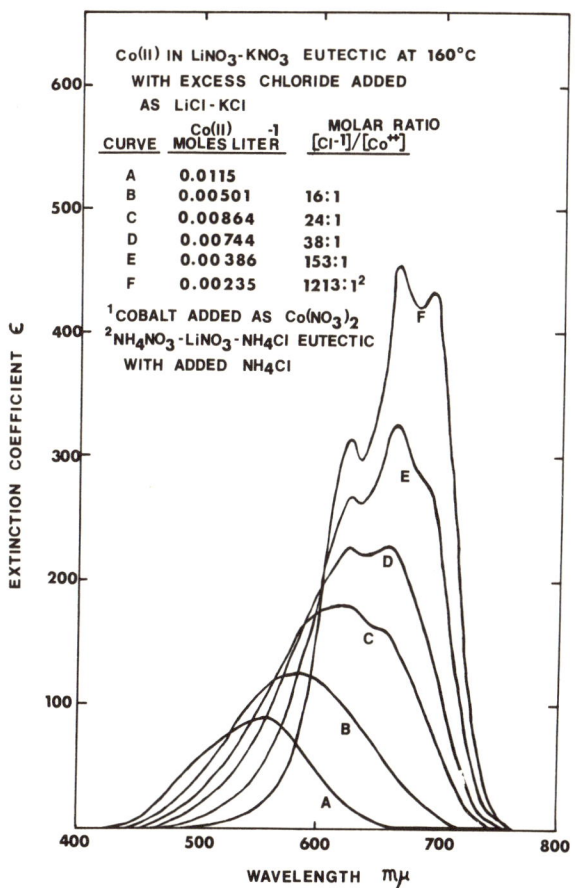

Fig. 25. Changes in the absorption spectrum of Co(II) in LiNO$_3$–KNO$_3$ on chloro complex formation.[209] (Reproduced by permission of McGraw-Hill.)

bromide and iodide systems. Some modification of this reasoning seems necessary in the light of the most recent results (see Section 4.3.1 and Ref. 195).

4.3.4. Discussion of Cobalt(II) Results

Among the spectra discussed in the three previous sections, three distinctive types are apparent. First, those that are undeniably due to an octahedral species, of which Co(ClO$_4$)$_2$ in dimethylsulfone (Fig. 18D) is perhaps the best example. Second, there are those produced by a tetrahedral Co(II) species, for which the LiCl–KCl eutectic spectrum is characteristic.

Third, there is a group of spectra which are remarkably alike but which have been assigned in the published literature, for various reasons, as both octahedral and tetrahedral. This last group includes the sulfate and bisulfate melts (Figs. 20 and 21), the LiF–NaF–KF system (Fig. 23), fluoride crystals (Fig. 22), the $AlCl_3$ melt (Fig. 24A), the $LiNO_3$–KNO_3 melt to which excess Cl^- has been added (Fig. 25E), and the acetate melts (not shown).

The evidence which led the present authors to postulate the existence in the sulfate melt of a species having approximate D_{2d} symmetry has already been discussed. We should like to suggest that a similar type of species is present in each of the systems listed in the third classification above. However, before considering each of these melts in turn, we would like to discuss our model in more detail. As already mentioned, it is based on the structure of the $Co(NO_3)_4^{2-}$ ion, in which four nitrate groups are arranged approximately tetrahedrally around the cobalt ion. However, the crystal structure showed that each nitrate ion was in fact bidentate, although the Co—O distances were not found to be identical.[212] Two sets of distances were found, the first in the 2.00–2.05-Å range and the second in the 2.41–2.69-Å range. Our model then consists of a cobalt ion surrounded by four ligands approximately tetrahedrally with each of the ligands potentially bidentate, $(CoL_4)^{x-}$. That is, the first atom of each ligand is positioned at a vertex of the tetrahedron, while the second lies sufficiently close to the cobalt atom for interaction to occur. We suggest that the oscillator strength of the main visible band is a reasonable indicator of this interaction, as the following argument implies. If the second atom does not interact at all, the cobalt species will be tetrahedral, and the oscillator strength should be of the order of 10^{-3}. Alternatively, if the second atom occupies a site on a perfect cube, then, since this species will have a center of symmetry, the oscillator strength should be reduced to approximately 10^{-5}.

Let us now apply our model to the results of Øye and Gruen[187,205] for the $AlCl_3$–KCl system. In pure $AlCl_3$ we suggest that the cobalt species $Co(Al_2Cl_7)_2$ exists with 2Cl of each $AlCl_3$ unit giving eightfold coordination to Co. Addition of potassium chloride, up to 35.45 mole %, produces no change. In the range 35.45–49.9 mole %, however, a major change in spectrum is observed (Fig. 24B) which can be interpreted by a change to the stoichiometry $Co(Al_2Cl_7)(AlCl_4)_2$, i.e., we have $n_1 = 2$ rather than the value of one chosen by Øye and Gruen (see p. 147). This species has approximately C_{2v} symmetry and, therefore, the band in the visible region $[^4T_{1g}(P) \leftarrow {}^4T_{1g}(F)$ for O_h and $^4T_1(P) \leftarrow {}^4A_2(F)$ for $T_d]$ should split into three components: $^4A_2$, $^4B_1$, and $^4B_2$. Addition of further KCl then causes the second Al_2Cl_7 unit to break down and distort the cube, eventually producing a tetrahedral $CoCl_4^{2-}$ species. At KCl concentrations greater than 50 mole % the system will contain free chloride ions which, it can be supposed, are preferred ligands to $AlCl_4^-$.

In the bisulfate glass at 25°C (Fig. 20C), cobalt is probably octahedrally coordinated. With increase in temperature, the species distorts until, at 230°C (Fig. 20B), the dodecahedral-type spectrum is obtained. A distortion of this type is easily visualized for any combination of bidentate and monodentate sulfate ligands except the one involving three bidentate sulfate groups. In this case a perfect tetrahedral arrangement has not been obtained, even at the highest temperatures examined so far.

The nitrate melts have been shown to contain the dodecahedral $Co(NO_3)_4^{2-}$ entity, as discussed earlier. Addition of chloride results in the eventual production of tetrahedral $CoCl_4^{2-}$, apparently by stepwise replacement. However, this system is not at all straightforward. Goffman[221] reports the isolation and characterization, by means of visible and infrared spectroscopy, powder diffraction studies, and magnetic susceptibility measurements, of $K_4\{CoCl_4(NO_3)_2\}$, which was obtained from a nitrate melt containing Co(II) by the addition of KCl. He concludes that although the nitrates are within the coordination sphere, all the properties are due to a tetrahedral $CoCl_4^{2-}$ entity! It is suggested that the two nitrate groups are located on the C_2 axis but that they have no effect on the absorption spectrum. A single-crystal X-ray and absorption spectrum study of this compound would be very interesting. However, Cotton and Bergman[212,222] found that $\{(C_6H_5)_4As\}_2\{Co(CF_3CO_2)_4\}$ has a structure very similar to that of the corresponding nitrate complex except that in this case the "second" oxygen of each trifluoroacetato group is almost at the limit for bonding to occur. The solution spectrum of this compound is similar to that of the acetate melts and has a value of 2.62×10^{-3} for the oscillator strength of the main visible band.

The model that we have been discussing is (apparently) capable of describing several different systems. It would appear that there is evidence to suggest that in the systems under discussion changes in temperature and/or composition produce a change in coordination from octahedral to tetrahedral, passing through a symmetry which can be described as D_{2d} or dodecahedral.*

---

* *Note added in proof*: The spectra of a whole variety of sulfate systems containing Co(II) have been accounted for[283] in terms of pink, monodentate, octahedral $Co(OSO_3)_6^{10-}$, favored at low temperatures and blue, bidentate, dodecahedral $Co(O_2SO_2)_4^{6-}$ favored at high temperatures.

TABLE XXXI. Investigations of Ni(II) Spectrum in Melts

System	Ref.
$LiNO_3$–KNO_3	175, 190, 223
$LiNO_3$–$NaNO_3$–KNO_3	224
Li_2SO_4–Na_2SO_4–K_2SO_4	197, 225
$Li(CH_3CO_2)$–$Na(CH_3CO_2)$	199
$Tl(CH_3CO_2)$	200
$Na(CH_3CO_2)$–$K(CH_3CO_2)$	200
Dimethylsulfone	224, 226, 227
LiF–NaF–KF	187, 201
$2LiF$–BeF_2	188
LiCl	203, 223, 228
NaCl	228
KCl	228
RbCl	228
CsCl	223, 228–231
$MgCl_2$	228
LiCl–KCl	178, 203, 232–234
KCl–$MgCl_2$	235, 326
NaCl–KCl–$MgCl_2$	177
$CaCl_2$–$MgCl_2$	203
KCl–$ZnCl_2$	237
CsCl–$ZnCl_2$	223, 238, 239
$AlCl_3$	187
$PbCl_2$	206
$SnCl_2$	206
Phosphonium and arsonium halides	180, 225, 240
Pyridine hydrochloride	223
LiBr, KBr, CsBr	228
LiBr–KBr	203, 241
LiI–KI	241
KSCN	178, 204

4.4. Nickel(II) $3d^8$

The list of melts in which the spectrum of Ni(II) has been studied is shown in Table XXXI. Once again we will consider the oxyanion melts first, then the halide systems, and finally the mixed melts.

4.4.1. Oxyanion Melts

The results of spectral investigation of solutions of Ni(II) in nitrate melts are compared in Table XXXII with the data for $Ni(NO_3)_4^{2-}$ in nitromethane[211] and $Ni(H_2O)_6^{2+}$.[242]

TABLE XXXII. The Spectrum of Ni(II) in Nitrate Systems and Water

$LiNO_3-KNO_3^a$		$LiNO_3-NaNO_3-KNO_3^b$		$[CH_3(C_6H_5)_3As]_2[Ni(NO_3)_4]$ in nitromethane^c		$Ni(H_2O)_6^{2+c}$	
v, kK	ε	v, kK	ε	v, kK	ε	v, kK	ε
—	—	7.67	6.25	8.0	1.8	8.5	2
12.9	11	13.0	10.0	13.15	(sh)	13.5	1.8
—	—	14.16	9.2(sh)	14.3	4.05	15.4	1.5(sh)
23.7	30	23.74	30.83	24.1	11.0	23.5	5.2

^{a} 160°C.[175,223]
^{b} 125°C.[224]
^{c} Room temperature.[211]
^{d} Room temperature.[242]

Gruen and McBeth[175,190,223] found two bands in the binary nitrate eutectic, the one at lower energy being asymmetric with an ill-defined shoulder on the high-energy side. They identified the spectrum as being due to an octahedral complex involving four nitrate groups arranged tetrahedrally about the nickel such that the 12 oxygen atoms gave rise to a cubic field. Johnson and Piper[197] pointed out that such an arrangement would give rise to an energy level splitting that is the inverse of that obtained for a six-coordinate octahedral field and suggested an alternative model involving three bidentate nitrate groups. A further possibility suggested by Gruen[209] involves two monodentate and two bidentate nitrate groups. Smith et al.[224] examined the Ni(II) spectrum in the ternary nitrate melt and also a dimethylsulfone melt to which excess nitrate had been added. In dimethylsulfone the changes observed on addition of excess nitrate, both potentiometrically and spectrophotometrically, were shown to be compatible with the formation of dinitrato $Ni(NO_3)_2$ and trinitrato $Ni(NO_3)_3^-$ complexes. Values of the successive formation constants K_2 and K_3 were found to be in fairly good agreement for both types of measurements. The values found were as follows:

	K_2	K_3
potentiometrically	2.7×10^3	6.0
spectrophotometrically	$3.6 \pm 1.2 \times 10^3$	3.5 ± 1

A comparison of the spectrum of $Ni(NO_3)_3^-$ in dimethylsulfone (obtained by extrapolation) with that of the ternary nitrate melt led to the suggestion that the same trinitrato species was present in both systems. It was pointed out, however, that the analysis of the data did not completely rule out the possibility of a tetranitrato complex $Ni(NO_3^-)_4^{2-}$. In Fig. 26 the spectra of the di-, tri-, and tetranitrato Ni(II) complexes are shown, each of which appears to involve octahedral coordination about the nickel ion. The dinitrato complex (Fig. 26B) was postulated to involve either unidentate or bidentate nitrate coordination with, presumably, solvent molecules occupying the remaining octahedral positions. The trinitrato species, identified in both the ternary nitrate melt (Fig. 26A) and the dimethylsulfone melt (Fig. 26B), apparently involves three tridentate nitrates. The tetranitrato complex (Fig. 26C) has been thought to contain both unidentate and bidentate nitrates but no detailed model has been proposed.[211] The close resemblance between all these spectra is somewhat unusual but arises because of the very similar effects of the dimethylsulfone and nitrate ligands on the nickel ion.

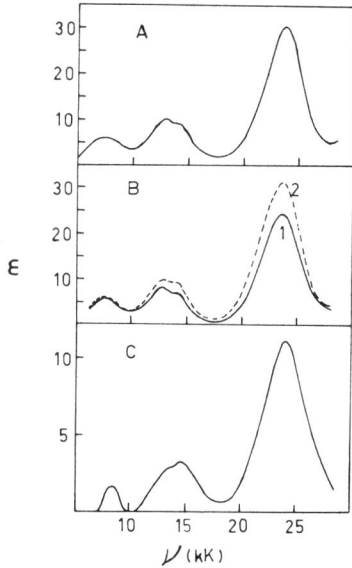

Fig. 26. Absorption spectra of Ni(II) in nitrate media: (A) Molten LiNO$_3$–NaNO$_3$–KNO$_3$ eutectic at 125°C.[224] (B) Molten dimethylsulfone with added nitrate at 125°C: Curve 1, the Ni(NO$_3$)$_2$ species; curve 2, the Ni(NO$_3$)$_3^-$ species.[224] (C) 0.01 M {CH$_3$(C$_6$H$_5$)$_2$As}$_2$ × {Ni(NO$_3$)$_4$} in CH$_3$NO$_2$.[211]

The spectrum obtained in the Li$_2$SO$_4$–Na$_2$SO$_4$–K$_2$SO$_4$ eutectic melt[197,225] was interpreted originally in favor of an octahedral arrangement about the nickel, possibly involving three bidentate sulfate groups. The composite non-Gaussian nature of the principal visible band was pointed out[243] and the possibility of distortion from pure octahedrality raised. A reinvestigation[225] of this spectrum (Fig. 27), including a test of Beer's law for the range 5–25 × 10^{-3} M confirmed most of the original features, as indicated in Table XXXIII. Although a resemblance to the spectra in nitrates is clear (cf. Table XXXII and Ref. 243), the relative intensities of the bands and the skewed nature of the main one are prominent features which arise in the spectra of several halide melts containing nickel(II). We shall return to this subject later (p. 167). The resolution of the spectrum into Gaussian components gave neither a unique nor a decipherable result.[225]

In acetate melts[199,200] nickel(II) adopts a simple octahedral coordination.

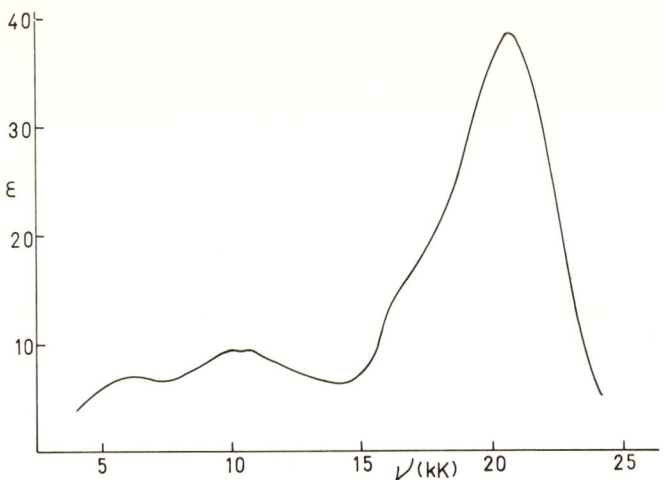

Fig. 27. Absorption spectrum of Ni(II) in Li_2SO_4–Na_2SO_4–K_2SO_4 melt at 550°C.[225]

4.4.2. Halide Melts

In the LiF–NaF–KF melt the Ni(II) spectrum has been assigned to an octahedral NiF_6^{4-} species.[187,201] The ligand field parameters were found to have the values $\Delta = 6.5$ kK and $B = 950$ cm^{-1}. In crystalline $KNiF_3$ the corresponding values are 7.25 kK and 955 cm^{-1}.[244] The decrease in the value of Δ is compatible with the difference in temperature between the crystal and the melt. A similar spectrum is obtained in the 2LiF–BeF$_2$

TABLE XXXIII. Spectrum of Ni(II) in Li_2SO_4–Na_2SO_4–K_2SO_4

At 550°C[197]		At 700°C[197]		At 550°C[225]	
ν, kK	ε	ν, kK	ε	ν,^{a} kK	ε
—	—	—	—	5.2	4.8
9.9	11.3	10.4	11.6	10.4	9.7
—	—	—	—	18.1	14.5
—	—	—	—	19.5	4.0
21.6	42.	22.0	44.6	22.0	35.0

^{a} Best Gaussian fit.

TABLE XXXIV. Spectrum of NI(II) in Fluoride Melts

LiF–NaF–KF[a]		2LiF–BeF$_2$[b]		Assignment
v, kK	ε	v, kK	ε	
6.5	0.8	6.0	1	$^3T_{2g}(F) \leftarrow {}^3A_{2g}(F)$
11.1	2	10.8	2	$^3T_{1g}(F) \leftarrow$
23	12	23.1	11	$^3T_{1g}(P) \leftarrow$

[a] 525°C.[227]
[b] 550°C.[188]

melt,[188] as can be seen from the results shown in Table XXXIV. Variation in the melt composition over the range 72–60 mole % LiF produced no change in the Ni(II) spectrum. It was suggested that in the BeF$_2$-containing melts the coordinating species was the BeF$_4^{2-}$ ion.[188]

The ultraviolet and visible absorption spectrum of a solution of Ni(II) in LiCl–KCl eutectic was first reported by Boston and Smith (Fig. 28).[233]

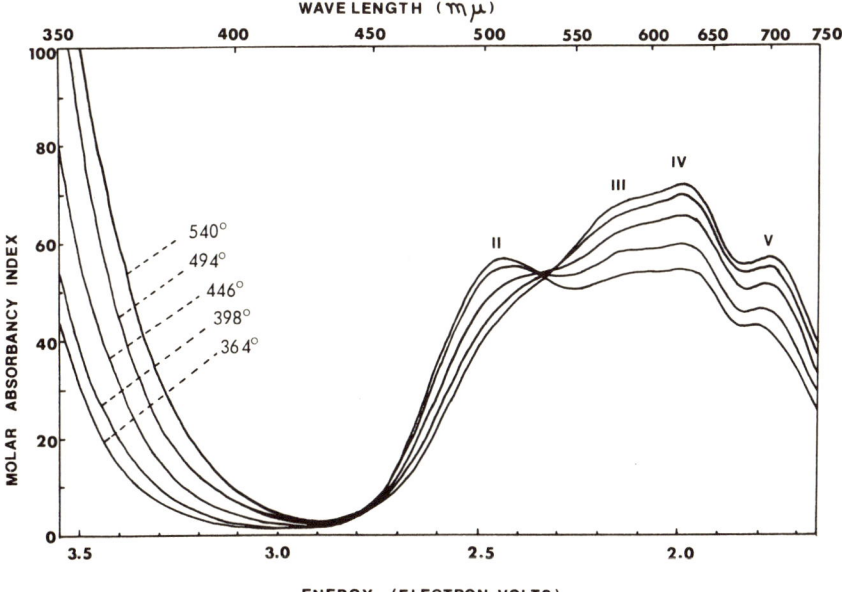

Fig. 28. Visible absorption spectra of a solution of NiCl$_2$ in a fused LiCl–KCl mixture at several temperatures. [233] (Reproduced by permission from *J. Phys. Chem.*)

The spectrum was studied as a function of temperature over the range 360–540°C and it was tentatively suggested that the spectral changes were due to equilibria between two or more Ni(II) species. Jørgensen[245] suggested that octahedral $NiCl_6^{4-}$ and tetrahedral $NiCl_4^{2-}$ were present and that the increase in temperature favored the tetrahedral species. Band II (Fig. 28) was assigned to the octahedral species and bands III, IV, and V to the tetrahedral one. The value of $\varDelta$ for the tetrahedral species was estimated to be $\sim$3.2 kK.[245] This interpretation was challenged by Sundheim and Harrington[234] and Gruen and McBeth,[223] who maintained that the results could be explained by means of a gradual distortion model. At the higher temperatures (Fig. 28) it was suggested that the Ni species is tetrahedral and that the outer coordination shell contains a nearly random arrangement of Li^+ and K^+ ions. As the temperature is lowered, the distribution in the outer coordination shell shifts in favor of Li^+ and these more strongly polarizing ions induce a tetragonal distortion of the $NiCl_4^{2-}$ tetrahedron. Gruen and McBeth[223] proved that in the C_5H_5NHCl, $CsCl$, and Cs_2ZnCl_4 melts Ni(II) exists as a tetrahedral $NiCl_4^{2-}$ species. This was done by comparing these melt spectra with the spectrum of a crystal of Cs_2ZnCl_4 doped with Ni(II) (Fig. 29A). To support their argument against the equilibrium theory of Boston and Smith[233] and Jørgensen,[245] they claimed that the spectrum in LiCl–KCl at 360°C showed no similarities with that of Ni(II)

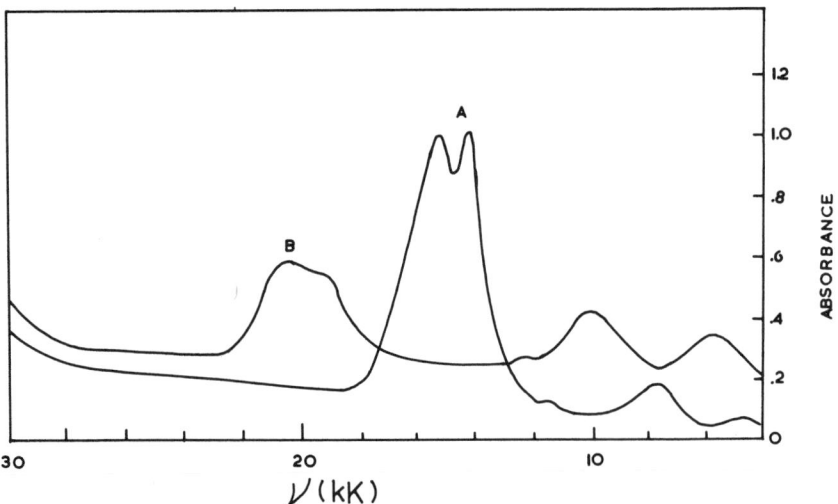

Fig. 29. Absorption spectra of Ni(II) in chloride crystals: (A) Cs_2ZnCl_4 at room temperature, (B) $CsCdCl_3$ at room temperature.[185] (Reproduced by permission from *Pure Appl. Chem.*)

TABLE XXXV. The Spectrum of NiCl$_4^{2-}$ in the CsCl Melt

ν, kK	ε	Assignment
4	13	$^3T_2(F) \leftarrow {}^3T_1(F)$
7.54	15	$^3A_2(F) \leftarrow$
14.0	90	—
15.7	105	$^3T_1(P) \leftarrow$
16.8	85	—

^{a} Δ = 3.70 kK. All values estimated from Ref. 229.

doped in a CsCdCl$_3$ crystal (Fig. 29B).[185] That the spectrum in the CsCl melt was in fact due to tetrahedral NiCl$_4^{2-}$ was confirmed by Boston and Smith,[229] who obtained, for the first time, the complete spin-allowed spectrum of this ion. Their results and assignments are shown in Table XXXV.

Smith and Boston have extended their studies of Ni(II) spectra to include single salt melts of all the alkali chlorides and of magnesium chloride (Fig. 30).[228,246] The CsCl, RbCl, and KCl melt spectra were all assigned to tetrahedral species, that of CsCl being the least distorted. In the NaCl, LiCl, and MgCl$_2$ melts the results are similar to one another and to the 360°C spectrum in the LiCl–KCl eutectic (Fig. 28). The results obtained in this study were such that it was not possible to distinguish between the continuous distortion and the equilibrium models discussed earlier. In a study of the CsCl–NiCl$_2$ system over the composition range 2–60 mole % NiCl$_2$, Smith et al.[230] found that at compositions up to 20 mole % NiCl$_2$ the spectra were due to tetrahedral NiCl$_4^{2-}$ chromophores. At higher NiCl$_2$ compositions gross deviations from a simple tetrahedral spectrum were noted (Fig. 31); these were interpreted in terms of an equilibrium between two species, one of which was tetrahedral while the other was not identified. The similarity between the 40 mole % NiCl$_2$ spectrum (curve C in Fig. 31) and the NaCl melt spectrum (Fig. 30) led the authors to suggest that a similar type of equilibrium was present in the NaCl melt. A similarity was also noted between the LiCl and MgCl$_2$ spectra (Fig. 30) and the spectrum of a 60 mole % NiCl$_2$ melt (curve E in Fig. 31).

In an attempt to identify the low-temperature species of their two-center equilibrium model, Smith et al.[232] undertook an extremely detailed investigation of the LiCl–KCl system, including studies of the effect of melt composition over the entire range 0–100% LiCl and of temperature in the range 363–1070°C. The effects of melt composition at 526°C are shown in Fig. 32. These eight spectra were shown to have the property of internal linearity with respect to the mole fraction of KCl. This means that, at any fixed temperature for the spectrum corresponding to a KCl mole fraction

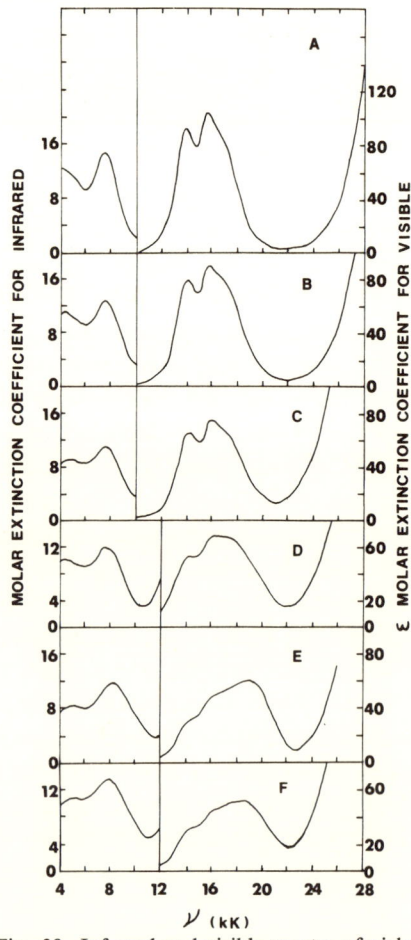

Fig. 30. Infrared and visible spectra of nickel centers in molten chloride salts. Molten salt solvents and temperatures for each spectrum are as follows: (A) CsCl, 680°C. (B) RbCl, 730°C. (C) KCl, 815°C. (D) NaCl, 822°C. (E) LiCl, 640°C. (F) $MgCl_2$, 740°C.[228,246] (Uncorrected for errors in ε.)

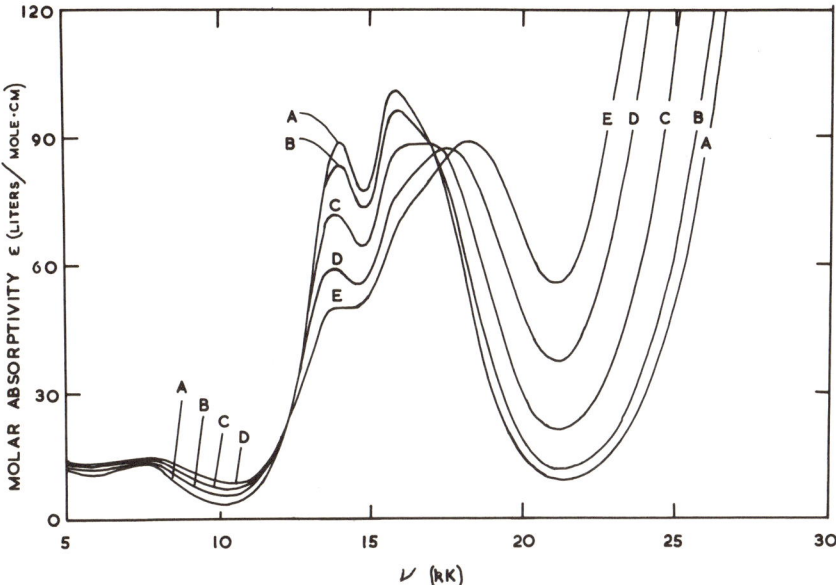

Fig. 31. Infrared and visible spectra of CsCl–NiCl$_2$ mixtures at 864°C. Compositions in mole per cent NiCl$_2$ were (A) 20.0, (B) 30.0, (C) 39.9, (D) 49.9, (E) 60.0.[230] (Reproduced by permission from *J. Chem. Phys.*)

N_{K3}, the molar absorptivity over the entire wavelength region is given by a linear combination of two other spectra corresponding to the mole fractions N_{K2} and N_{K1}, i.e., for each wavelength the following expression holds true:

$$(N_{K3}) = (1 - b)(N_{K2}) + b(N_{K1})$$

where b is a number which varies monotonously with N_{K3}. Similar behavior was also observed when the melt composition was held constant and the temperature varied above 700°C. This type of behavior was shown to be compatible with a two-species equilibrium. By a linear extrapolation, using the above equation, the two spectra shown in Fig. 32B (curve 1) and Fig. 32C (curve 1), which are due to the two nickel species present in the melt, were obtained. In Fig. 32B a comparison of curve 1 with the extrapolated spectrum of tetrahedral $NiCl_4^{2-}$ in KCl (curve 2) leaves little doubt that this tetrahedral species is in fact one of the components in the LiCl–KCl melt. The resemblance between Fig. 32C (curve 1) and the spectrum of nickel-doped KMgCl$_3$ (curve 2, Fig. 32C), which contains octahedral $NiCl_6^{4-}$ units, is not as good. It was suggested by Smith *et al.*[232] that this

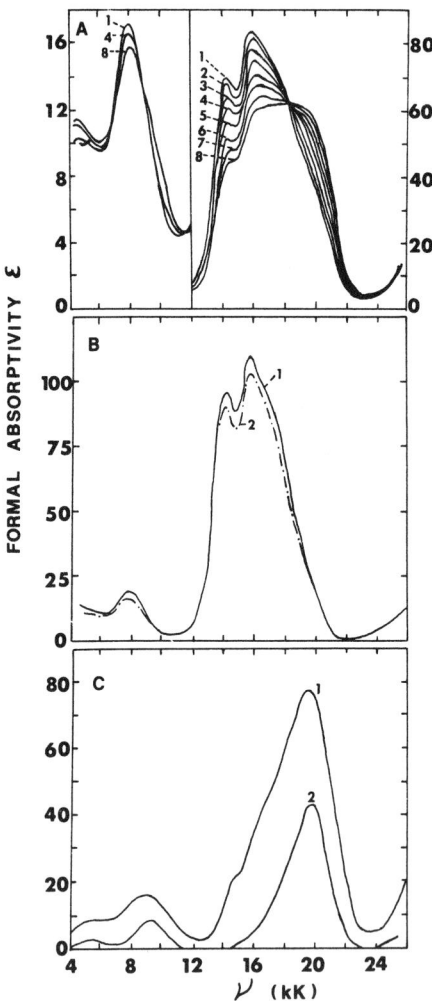

Fig. 32. Effect of solvent composition on the spectrum of Ni(II) in LiCl–KCl at 526°C. (A) Mole per cent KCl for these spectra are (1) 55.0, (2) 50.1, (3) 45.0, (4) 40.9, (5) 35.0, (6) 30.0, (7) 25.0, (8) 20.0. (B) Extrapolated spectrum for T centers (curve 1) compared with 100% KCl spectrum (curve 2) extrapolated to 526°C. (C) Extrapolated spectrum for O centers (curve 1) compared with the spectrum of nickel-doped $KMgCl_3$ at 490°C.[232] (Reproduced by permission from *J. Chem. Phys.*)

second spectrum (Fig. 32C, curves 1 and 2) was due to **O** centers (as opposed to the **T** centers giving rise to Fig. 32B, curves 1 and 2) which have a geometry that is based on the octahedron. However, in order to explain the rather high intensity of the bands in Fig. 32C, curves 1 and 2, it was suggested that the **O** centers consisted of a fairly wide distribution of geometries about an octahedron, a large proportion of them having a noncentrosymmetric arrangement. The difference between the well-defined **T** centers and the much less well-defined **O** centers was rationalized on the basis of differences in the outer-shell compositions. The outer shells of the **T** centers were thought to be composed of K^+ ions. The **O** centers, on the other hand, had a preponderance of Li^+ ions which strongly polarize the coordination-shell chloride ions and lead to a wider range of geometries. As the temperature was raised, the outer shell became less exclusive and also the equilibrium shifted in favor of the tetrahedral species. At high temperatures Smith et al.[232] concluded that the spectral changes caused by changing melt composition were best explained as due to a continuous distribution of **T** centers. Their extremely detailed paper thus proposed that at low temperatures there was an equilibrium between ill-defined **O** centers and well-defined **T** centers, while at high temperatures less-defined **T** centers remain.

Smith et al.[231] continued their studies of the $CsCl-NiCl_2$ system by examining the effect of melting on the two compounds Cs_3NiCl_5 and $CsNiCl_3$. In the solid state Cs_3NiCl_5 contains tetrahedral $NiCl_4^{2-}$ units[17] while $CsNiCl_3$ contains octahedral $NiCl_6^{4-}$. On melting, the spectrum of Cs_3NiCl_5 showed very little change from the high-temperature crystalline spectrum. In the case of $CsNiCl_3$ a major change in the spectrum was interpreted in favor of a single, broad distribution of geometries in the melt.

The results of temperature and composition studies of the $KCl-MgCl_2$ system containing Ni(II)[236] were found to be qualitatively similar to those obtained in the LiCl–KCl melt and the same model was used to interpret them. Once again the outer shells of the **T** centers were supposed to consist mainly of K^+ ions while those of the **O** centers were thought to be mainly Mg^{2+} ions. At high temperatures (790°C) only one species appeared to be present. The results in the 50 mole % KCl melt (corresponding to $KMgCl_3$) were very similar to those obtained for $CsNiCl_3$.[231]

Angell and Gruen[237] have studied the effect on the Ni(II) spectrum of varying both the temperature and melt composition in the $ZnCl_2$–KCl system. At compositions above 52.5 mole % KCl the spectral changes were interpreted on the basis of an octahedral–tetrahedral equilibrium, while below 45.3 mole % KCl the single-species continuous distortion model was favored.

An extensive examination of the $ZnCl_2$–CsCl system has been reported by Smith *et al.*[238,239] A preliminary note suggested that in these melts two different octahedral–tetrahedral (actually **O** center–**T** center) equilibria were present, the first occurring in the 50–70 mole % $ZnCl_2$ composition range and the second in the 92–100 mole % range. At intermediate composition the Ni(II) spectrum changed in minor ways only. The second detailed paper on this system[239] covered the entire composition range 0–100 mole % $ZnCl_2$ and led to the suggestion that no less than six different Ni(II) centers were present at low temperatures (further unidentified centers were thought to exist at higher temperatures). The results at low temperatures (100–300° above the liquidus temperature) are summarized in Table XXXVI.

In contrast to the nickel centers in the LiCl–KCl melts discussed earlier, where the **O** centers were found to be rather ill-defined whereas the **T** centers were very close to regular tetrahedra, in the $ZnCl_2$–CsCl melts all of the nickel centers have reasonably well-defined coordination geometries. At compositions between 92 and 100 mole % $ZnCl_2$ the spectra were found to be internally linear with respect to composition and were interpreted on the basis of an $\mathbf{O}_N$–$\mathbf{T}_N$ equilibrium. It was suggested that the $\mathbf{T}_N$ centers were species where the Ni(II) had substituted at a Zn(II) site in the three-dimensional polymeric chlorozincate agglomerates which are present in these $ZnCl_2$-rich melts. The $\mathbf{O}_N$ centers, on the other hand, were thought to be due to nickel ions trapped in octahedral holes present in the network structure of the melts. This three-dimensional chlorozincate network structure is broken down by the addition of further CsCl into polymeric chain type species. The $\mathbf{T}_P$ centers were postulated to involve species in which the number of bridging chlorides (which was believed to be four in the case of the $\mathbf{T}_N$ centers) was reduced to three or two. Apparently it is

TABLE XXXVI. Low-Temperature Ni(II) Species Present in the $ZnCl_2$–CsCl System

Mole % $ZnCl_2$	Species present
100–92	$\mathbf{T}_N \rightleftarrows \mathbf{O}_N$ equilibrium
92–66	Mainly $\mathbf{O}_P$ centers but some $\mathbf{T}_P$ centers present
66–50	$\mathbf{T}_P \rightleftarrows \mathbf{O}_P$ equilibrium
40–0	$(Cl_3NiClZnCl_3)^{3-} \rightleftarrows NiCl_4^{2-}$ equilibrium

possible to fold a chlorozincate polymer chain around a Ni(II) ion so that the terminal chlorides form a regular octahedral array. In melts rich in CsCl the main zinc species is $ZnCl_4^{2-}$. However, as the amount of $ZnCl_2$ increases, dimeric chlorozincates of the type $Zn_2Cl_7^{3-}$ are formed. It was suggested that in melts containing substantial amounts of $Zn_2Cl_7^{3-}$ the following equilibrium might be established:

$$NiCl_4^{2-} + Zn_2Cl_7^{3-} \rightleftharpoons Cl_3NiClZnCl_3^{3-} + ZnCl_4^{2-}$$

This particular nickel species ($Cl_3NiClZnCl_3^{3-}$) involves only a single chloride bridge (cf. the T_N and T_P centers). The extrapolated spectra of all these species are shown in Fig. 33 and the band maxima are listed in Table XXXVII.

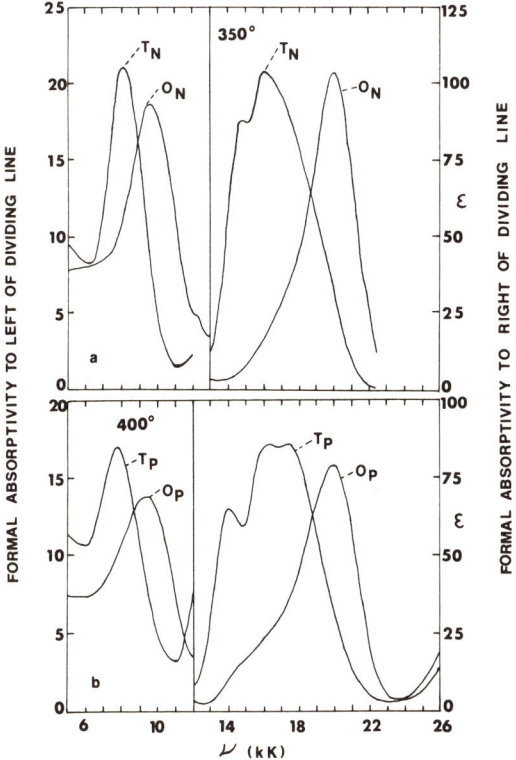

Fig. 33. Extrapolated spectra for (a) T_n and O_n centers at 350°C and (b) T_p and O_p centers at 400°C in liquid mixtures of zinc and cesium chlorides.[239] (Reproduced by permission from *J. Chem. Phys.*)

TABLE XXXVII. Spectra of Low-Temperature Ni(II) Species in the $ZnCl_2$–CsCl System

A. Tetrahedral Species

Assignment	T_N, 350°C		T_P, 400°C		$Cl_3NiClZnCl_3^-$, 600°C	$NiCl_4^{2-}$, 700°C	
	ν, kK	ε	ν, kK	ε	ν, kK	ν, kK	ε
$^3A_2(F) \leftarrow {}^3T_1(F)$	8.1	21	7.7	17	7.5	7.45	13
$^3T_1(P) \leftarrow$	14.7	88	14.0	66	14.0	14.0	93
	16.0	104	16.0	85	15.9	15.6	106
	~17.5	sh	17.4	86	~12.3(sh)	~17.0	sh

B. Octahedral Species

Assignment	O_N, 350°C		O_P, 260°C	
	ν, kK	ε	ν, kK	ε
$^3T_{2g}(F) \leftarrow {}^3A_{2g}(F)$	~6.0	sh	~6.3	8
$^3T_{1g}(F) \leftarrow$	9.7	19	10.1	14
$^1E_g(D) \leftarrow$	12.0	<1	12.2	<1
$^3T_{1g}(P) \leftarrow$	20.0	104	20.3	69

The above discussion was limited to effects observed at "low" temperatures, i.e., up to 300° above the liquidus temperatures. At higher temperatures the spectral changes became much more complicated and are less easily explained, partly because little is known about the melt structure at these temperatures. For the composition range covering the T_N–O_N equilibrium the coordination geometries cannot be identified above about 450°C. In the T_P–O_P range, however, an increase in temperature up to 800°C favors the formation of the T_P centers but above this temperature the spectra resemble those observed in the $MgCl_2$–KCl system, where no definite coordination geometry could be assigned to the nickel species.

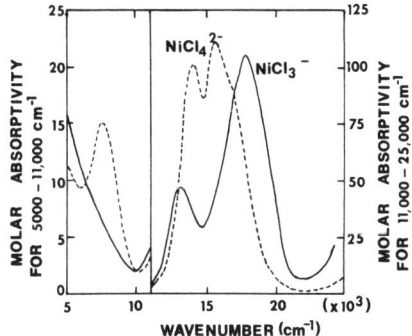

Fig. 34. Spectra of $NiCl_4^{2-}$ and $NiCl_3^-$ in molten $CsAlCl_4$ containing slight excess CsCl at 600°C.[247] (Reproduced by permission from *J. Am. Chem. Soc.*)

At 227°C the spectrum of Ni(II) in the $AlCl_3$ melt appeared to be due to an octahedral $Ni(Al_2Cl_7)_2$ entity.[187] A temperature and composition dependence study of the Ni(II) spectrum in the $CsCl-AlC_3$ systems showed[247] that $CsNiCl_3$ dissolved in $CsAlCl_4$ containing a little excess CsCl to give $NiCl_4^{2-}$ and $NiCl_3^-$ in solution over the range 500–750°C. The spectrum for the latter entity (Fig. 34) was rationalized in terms of D_{3h} symmetry.

The spectra of Ni(II) in molten $PbCl_2$ and $SnCl_2$[206] show great resemblance to those of the **O** in LiCl–KCl and $\mathbf{O}_N$ in $ZnCl_2$–CsCl centers, respectively. Details of these spectra appear in Table XXXVIII and Figs. 35 and 36, which also include the spectrum of Ni(II) in the sulfate melt; apart from the order of the absorptivity, the sulfate spectrum obviously belongs to the same family.

Although the alkali bromide and iodide melts have not been studied in as much detail as the corresponding chloride melts, the data at present available[203,228,241] suggest that similar behavior occurs.

Extensive studies of nickel halides in molten organic salts have shown that in most cases a virtually undistorted tetrahedral halonickel complex is formed, although Griffiths[226] has suggested that both $NiCl_4^{2-}$ and $NiCl_3^-$ species are present in dimethylsulfone melts containing added Cl^-. A detailed ligand field treatment of NiX_4^{2-} complexes in organic melts, using the four-parameter model of Liehr and Ballhausen,[248] has been given by Smith *et al.*[227] Both Δ and B were found to decrease in the order $Cl^- > Br^- > I^-$.

TABLE XXXVIII. Spectra of Ni(II) in $PbCl_2$ at 532°C and in $SnCl_2$ at 300°C[206]

	ν, kK	ε	$f \times 10^4$
$PbCl_2$	5.05	6.9	0.77
	8.92	12.3	2.29
	13.44	9.4	0.66
	15.96	35.3	5.42
	19.42	86.3	14.0
$SnCl_2$	5.51	3.7	1.00
	10.14	7.0	1.06
	16.11	8.3	1.73
	18.54	13.8	2.06
	20.70	58.0	8.84

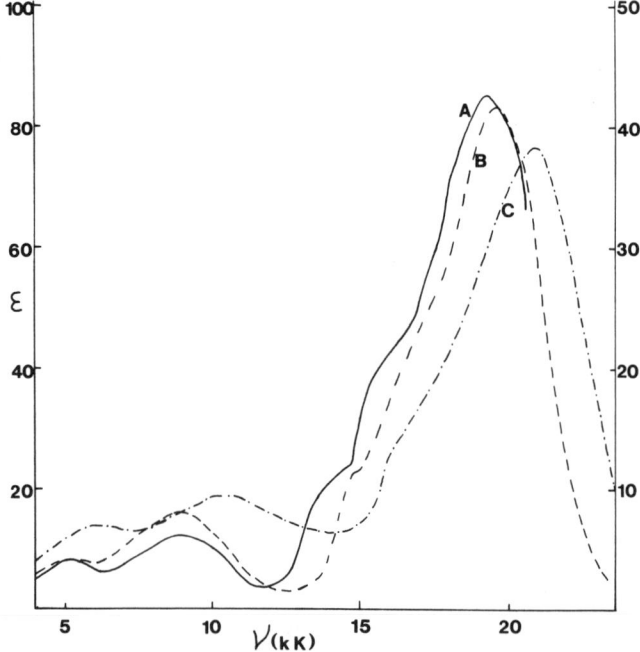

Fig. 35. Absorption spectra of Ni(II). (A) In $PbCl_2$ at 532°C.[206] (B) The O center in LiCl–KCl at 526°C.[232] (C) In Li_2SO_4–Na_2SO_4–K_2SO_4 at 550°C[225]: right hand ε scale.

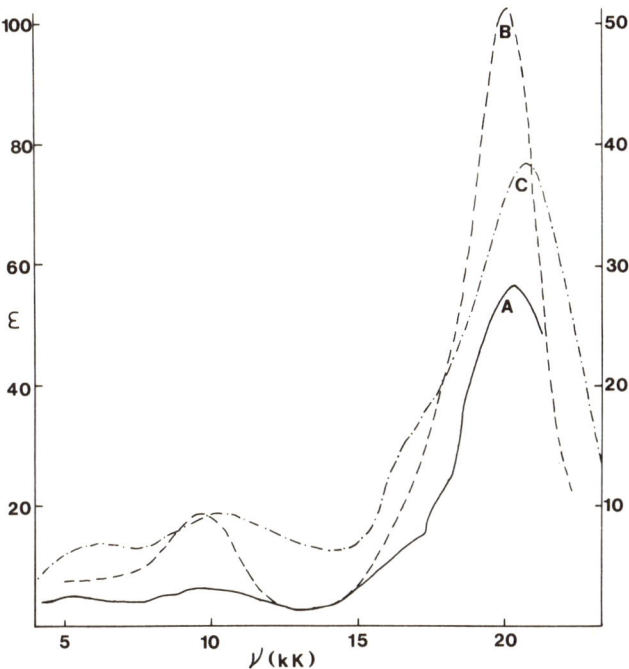

Fig. 36. Absorption spectra of Ni(II). (A) In $SnCl_2$ at 300°C.[206] (B) The O_N center in $ZnCl_2$–CsCl at 350°C.[239] (C) In Li_2SO_4–Na_2SO_4–K_2SO_4 at 550°C: right-hand ε scale.

The systems studied are shown in Table XXXIX.

A potentiometric titration of a solution of $NiCl_2$ in the LiCl–KCl eutectic melt with NaCN led to the suggestion that $Ni(CN)_4^{2-}$ was being produced under these conditions.[11]

Before leaving the halide systems we must mention some rather interesting magnetic susceptibility measurements. These measurements were made on solutions of Ni(II) in the LiCl–KCl eutectic over the temperature range 85–1150°K.[8,249] The results are summarized in Table XL. In the liquid state the value of the magnetic moment lies at the upper end of the range of values expected for tetrahedral Ni(II) complexes. The expected behavior of octahedral and tetrahedral complexes is shown in Table XLI.[250] The results in Table XL were obtained over a temperature range which includes the low-temperature region of Smith et al.,[232] where, from spectral measurements, a mixture of octahedral and tetrahedral complexes was postulated. A system containing such a mixture would surely be expected to have an average magnetic moment between the octahedral and

TABLE XXXIX. Ni(II) Spectra in Organic Melts

Organic salt	Solute	Species present	Ref.
$(CH_3)_2SO_2$	$NiCl_2 + LiCl$	$NiCl_4^{2-} + NiCl_3^-$	226, 227
	$NiCl_2 +$ Hyamine $10x^a$	$NiCl_4^{2-} + NiCl_3^-$	226
$\{C_6H_5CH_2(C_4H_9)_3P\}Cl$	$NiCl_2$	$NiCl_4^{2-}$	227, 240
$\{(2,4-Cl_2C_6H_3CH_2)(C_4H_9)_3P\}Cl$	$NiCl_2$	$NiCl_4^{2-}$	227
$\{C_5H_5N\}HCl$	$NiCl_2$	$NiCl_4^{2-}$	227
$(C_2H_5)_2NH_2Cl$	$NiCl_2$	$NiCl_4^{2-}$	227
$\{(C_6H_5)_4As\}Cl$	$NiCl_2$	$NiCl_4^{2-}$	227
$(CH_3)_2SO_2$	$NiBr_2 + LiBr$	$NiBr_4^{2-}$	227
$\{(C_4H_9)_4N\}Br$	$NiCl_2; NiBr_2$	$NiBr_4^{2-}$	227
$\{(C_5H_{11})_4N\}Br$	$NiCl_2; NiBr_2$	$NiBr_4^{2-}$	227
$(CH_3)_2SO_2$	$NiI_2 + LiI$	NiI_4^{2-}	227
$\{(C_6H_{13})_4N\}I$	NiI_2	NiI_4^{2-}	227
$\{(C_4H_9)_4N\}I$	$NiCl_2$	NiI_4^{2-}	227

^{a} Hyamine $10x$ = benzyldimethyl-2[2-{4-(1,3,3-tetramethylbutyl)-m-tolyloxy}ethoxy]ethylammonium chloride monohydrate.

tetrahedral values. It must be pointed out, however, that the results shown in Table XL were obtained with fairly concentrated solutions. Magnetic susceptibility measurements at concentrations similar to those used in the spectral investigations might be very interesting.

Both Harrington and Sundheim[178] and Egghart[204] have reported the spectrum of Ni(II) in the potassium thiocyanate melt and interpreted it as arising from an octahedral nickel complex.

TABLE XL. Magnetic Studies of Ni(II) in LiCl–KCl

% $NiCl_2$	85–350°K solid		350–600°K solid		800–1150°K liquid		Ref.
	μ	θ	μ	θ	μ	θ	
18.7	3.24	−25	3.54	−90	4.39	−255	249
17.4	—		3.70	−150	4.28	−210	8

TABLE XLI. Susceptibilities Expected for Ni(II) Species

	Ground state	Range of μ observed for room-temperature species
O_h	$^3A_{2g}$	~3.0–3.4
T_d	$^3T_1$	3.2–4.0

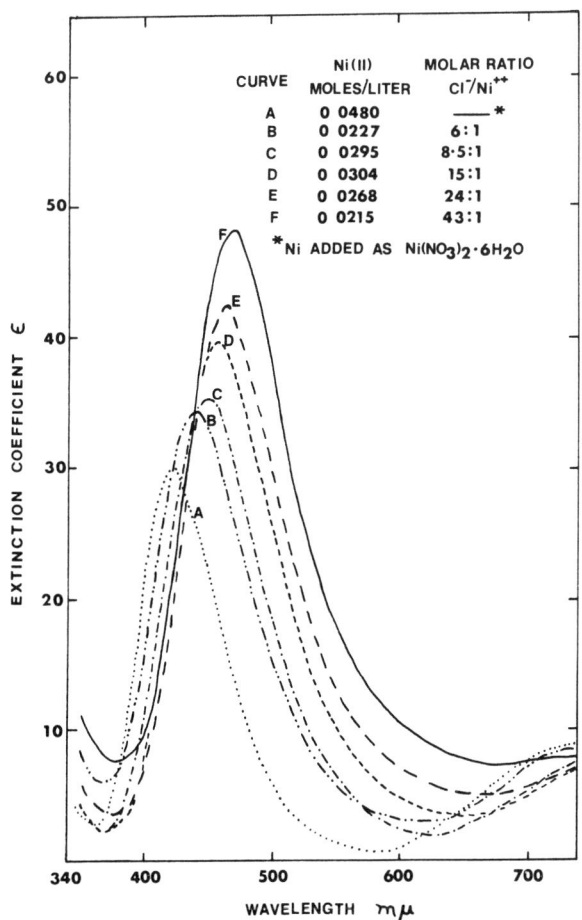

Fig. 37. Absorption spectra of Ni(II) in $LiNO_3$–KNO_3 eutectic at 160°C with increasing amount of LiCl–KCl added.[223] (Reproduced by permission from *J. Phys. Chem.*)

4.4.3. Mixed Halo–Oxyanion Systems

Absorption spectra of solutions of Ni(II) in the $LiNO_3$–KNO_3 eutectic to which excess Cl^- has been added are shown in Fig. 37.[223] The results in this case are not as clear-cut as for the corresponding Co(II) system, partly because of the lower solubility of nickel chloro complexes in the nitrate melt. By increasing the temperature, a higher concentration of nickel complexes could be stabilized and the resulting spectrum closely resembled that of Ni(II) in pure LiCl.[223] The spectral changes shown in Fig. 37 may be due to a conversion to an octahedral chloro complex.

4.5. Ruthenium(III) $4d^5$

The spectrum of $RuCl_3$ in the LiCl–KCl eutectic melt (Fig. 38) has been interpreted in favor of an octahedral $RuCl_6^{3-}$ species.[225,280,284] Table XLII contains the band positions, intensities, and assignments.

The similarities between the melt and HCl solution spectra[251] confirm that the same species is present in both solutions. Using an iterative procedure, it was found that the melt spectrum could be fitted with $\Delta = 18.7$ kK, $B = 0.49$ kK, and $C = 2.3$ kK, which are very reasonable parameters.

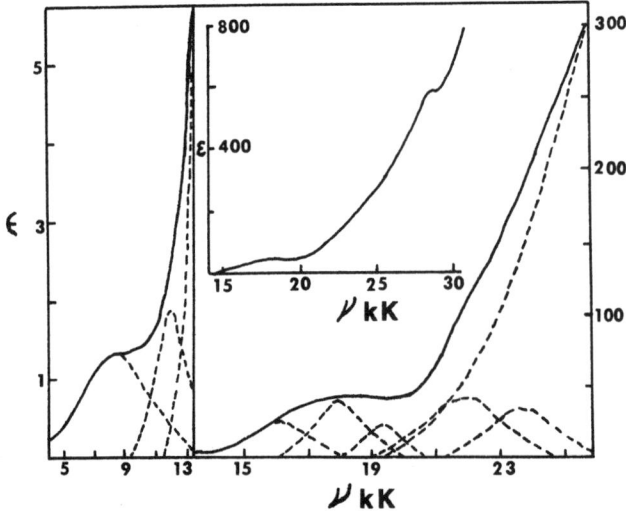

Fig. 38. Absorption spectrum of Ru(III) in LiCl–KCl eutectic at 450°C.[284] (Reproduced by permission from *Inorg. Chem.*)

TABLE XLII. Spectrum of $RuCl_3$ in LiCl–KCl at 450°C[284]

v(found), kK	ε^a	$10^4 f$	δ	Assignment	v(calc), kK
8.42	1.39	0.27	4.75	$^4T_{1g} \leftarrow {}^2T_{2g}$	8.23
				$^5A_{1g} \leftarrow$	8.64
12.02	1.89	0.22	2.54	$^4T_{2g} \leftarrow$	11.66
16.00	26.1	2.47	2.06	$^2A_{2g} \leftarrow$	16.11
				$^2T_{1g} \leftarrow$	16.27
17.87	37.9	3.56	2.04	$^2T_{2g} \leftarrow$	17.48
19.23	22.7	1.54	1.48	$^2E_g \leftarrow$	18.68
21.73	43.9	5.77	2.86	$^2T_{1g} \leftarrow$	21.90
23.56	34.0	4.06	2.60	$^2A_{1g} \leftarrow$	24.02
28.50	580	230	7.60	$\gamma_5 \leftarrow \pi$	—
33.00	~1150	—	—	$\gamma_5 \leftarrow$	—

^{a} Standard error of 0.15 (5%) from 4 to 14.5 kK and 2.8 (6%) from 14.5 to 26 kK.

4.6. Osmium(IV) $5d^4$ and Osmium(III) $5d^5$

The spectrum of Os(IV) was obtained in LiCl–KCl eutectic at 450°C over the 3–35-kK region.[225,253] It consists of two distinct parts: Below about 20 kK there are four relatively weak bands, which may be assigned to transitions within the t_{2g}^4 ground configuration (Fig. 39) and above 20 kK there are several strong bands (Fig. 40) variously assigned as charge-transfer transitions[254–256] and vibronically allowed electric dipole $t_{2g}^3 e_g \leftarrow t_{2g}^4$ transitions.[257,258] The overall spectrum compares very well with that of

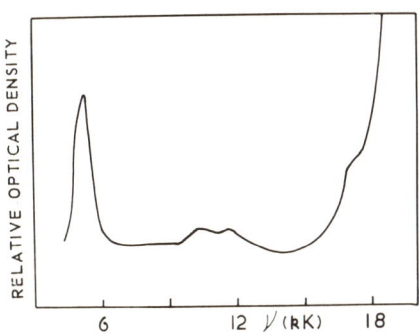

Fig. 39. Absorption spectrum of Os(IV) in the LiCl–KCl eutectic melt at 450°C in the 3–18 kK region.[253] (Reproduced by permission from *Mol. Phys.*)

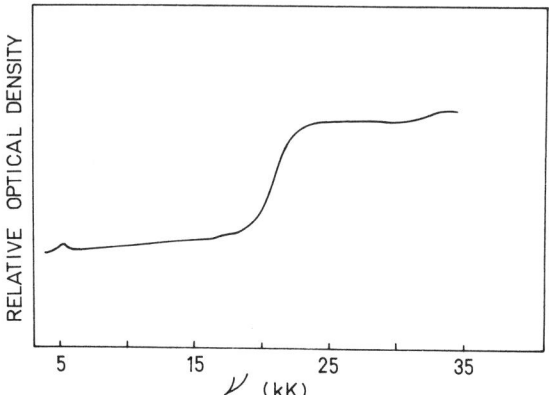

Fig. 40. Absorption spectrum of Na_2OsCl_6 in LiCl–KCl at 450°C.[225]

$OsCl_6^{2-}$ in aqueous HCl[254,259] except that for the melt there is no vibrational structure on the ~17-kK band and a band is observed in the latter at 5.3 kK, whereas it would be obscured by water.

Since we are dealing with a $5d^4$ system, the spectrum may be analyzed approximately in terms of transitions between one-electron spin–orbital configurations, as outlined in Section 2.3.2,[85–87] or the intra-t_{2g}^4 transitions may be assigned in terms of intermediate coupling[85,88,89] using the energy expressions for p^2.[24,31] From the first approach the values $\zeta_{5d} = 3.3$ kK and $\Delta = 26.0$ kK resulted, whereas from the second $\zeta_{5d} = 3.4$ kK and $3B + C = 2.2$ kK were obtained. All these parameters seem reasonable for octahedral $OsCl_6^{2-}$.

It was noted[225,253] that the lime-green melt obtained from Na_2OsCl_6 in LiCl–KCl slowly faded in color (cf. Fig. 41), giving rise to as yet unidentified species although volatile $OsCl_4$ is a likely product. A solution of $OsCl_3$ in LiCl–KCl gave an initial spectrum[284] resembling that of the tris(acetylacetonate),[260] which, therefore, may have been due to trigonally distorted $OsCl_6^{3-}$. This solution decomposed with increasing temperature, the spectrum changing to that of $OsCl_6^{2-}$ and Os metal depositing. However, while Os(III) was present the rate of decomposition of Os(IV) was significantly slowed down.

It was possible to maintain a solution of Os(III) and Os(IV) at a constant ratio for several hours in both LiCl–KCl and $BiCl_3$ provided the temperature was below 500°C (Fig. 42) and the Os(III) spectrum was obtained by subtraction. A distorted octahedral geometry was postulated and a spin–orbital coupling constant of 2.7 kK estimated.

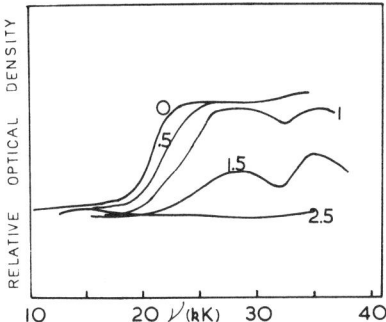

Fig. 41. Absorption spectrum of Na_2OsCl_6 in LiCl–KCl as a function of time in hours.[253] (Reproduced by permission from *Mol. Phys.*)

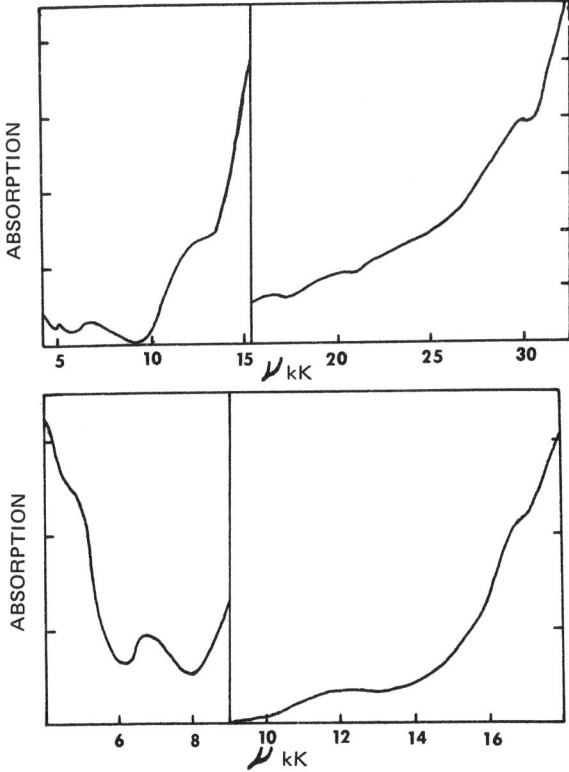

Fig. 42. Spectra of $OsCl_3$ solutions: (top) in LiCl–KCl eutectic at 450°C; (bottom) in $BiCl_3$ at 275°C.[284] (Reproduced by permission from *Inorg. Chem.*)

4.7. Rhodium and Iridium

4.7.1. Rhodium(III) $4d^6$

The spectrum of trivalent rhodium has been reported in the LiCl–KCl eutectic,[261–263] the $NaNO_3$–KNO_3 eutectic,[261] and the Li_2SO_4–Na_2SO_4–K_2SO_4 eutectic.[225]

Ogilvie and Holmes[261] reported that the spectrum of a solution of $RhCl_3$ in the LiCl–KCl eutectic melt at 440°C contained a single peak at 18.2 kK together with an intense absorption band beginning at ~26 kK. The present authors repeated this work and confirmed the results.[262] However, it was observed that a shoulder was present on the edge of the intense absorption band and, using the Gaussian wave-analysis technique of Chatt et al.,[264] it was possible to define this band more precisely (Fig. 43). The results for the chloride melt spectra are listed in Table XLIII, which also includes the results for a solution of $RhCl_6^{3-}$ in aqueous HCl.[265] The similarity of the melt and aqueous HCl spectra was used to identify the rhodium species in the chloride melt as the octahedral $RhCl_6^{3-}$ ion and rules out the existence of chloro-bridged polymeric complexes which Ogilvie and Holmes appeared to favor. The band assignments in the octahedral point group, together with the values of Δ, B, and β, are included in the footnote to Table XLIII. The same general conclusions were reported by Suzuki and Tanaka,[280] who also observed a transition at 38.8 kK.

In the nitrate melt,[261] $RhCl_3$ gave rise to a single, strong, asymmetric band, with a maximum at 28.2 kK ($\varepsilon = 2100$), which had a long tail on the low-energy side. On the addition of $NaNO_2$ to this melt, a spectrum was obtained which approached that of a solution of $K_3Rh(NO_2)_6$ in the melt.

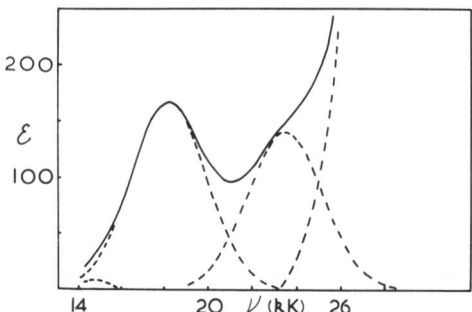

Fig. 43. Absorption spectrum of Rh(III) in the fused LiCl–KCl eutectic at 450°C.[263] (Reproduced by permission from *Can. J. Chem.*)

TABLE XLIII. Spectral Results for Rh(III) in LiCl–KCl Melt and Aqueous HCl

LiCl–KCl				Aqueous HCl[255]c		Assignments
Ref. 261[a]		Ref. 263[b]		ν, kK	ε	
ν, kK	ε	ν, kK	ε			
—	—	14.8	8	14.7	3	$^3T_{1g} \leftarrow {}^1A_{1g}$
18.2	160	18.2	167	19.3	102	$^1T_{1g} \leftarrow$
—	—	23.5	140	24.3	82	$^1T_{2g} \leftarrow$

[a] 400°C.
[b] 450°C, $\Delta = 19.3$ kK, $B_{corr} = 370$ cm^{-1}, $\beta = 0.51$.
[c] 25°C, $\Delta = 20.3$ kK, $B_{corr} = 340$ cm^{-1}, $\beta = 0.47$.

It was suggested that in the nitrate melt the rhodium species was Rh(NO$_2$)$_6^{3-}$ in which nitro coordination was present. With the addition of KCl, pronounced spectral changes were observed until, at a mole ratio of RhCl$_3$:KCl of 1:600, the spectrum contained two bands, at 18.8 and 24 kK, which were almost certainly due to the RhCl$_6^{3-}$ species.

In the ternary sulfate melt[225] the spectrum of Rh(III) contains a single band at 21 kK ($\varepsilon = 150$) together with the low-energy edge of a charge transfer band (Fig. 44).[225] A comparison of Fig. 43 and 44 led us to expect

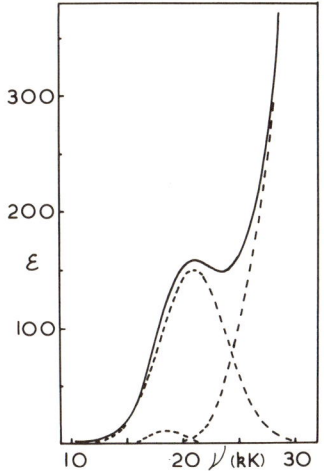

Fig. 44. Absorption spectrum of Rh(III) in fused Li$_2$SO$_4$–Na$_2$SO$_4$–K$_2$SO$_4$ eutectic at 550°C.[225]

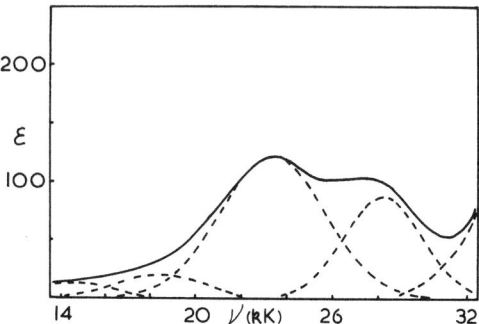

Fig. 45. Absorption spectrum of Ir(III) in fused LiCl–KCl eutectic at 450°C.[263] (Reproduced by permission from *Can. J. Chem.*)

a second, moderately intense band at ~26 kK. Gaussian analysis of this spectrum revealed the presence of a weak band at 18.5 kK ($\varepsilon = 12$) but no indication of the second band. Even so, because of the similarity of the chloride and sulfate melt spectra, it was suggested that an octahedral Rh(III) species was present in the melt, although further characterization of the species was not possible.

4.7.2. Iridium(III) $5d^6$

The spectrum of Ir(III) in the LiCl–KCl eutectic melt has been shown to contain four bands (Fig. 45).[262,263] Table XLIV lists the band maxima and extinction coefficients for both the LiCl–KCl melt and an aqueous HCl solution of $IrCl_6^{3-}$. The similarity between the two results suggests that

TABLE XLIV. Ir(III) Spectra

LiCl–KCl[a]		Aqueous HCl[b]		Assignments
ν, kK	ε	ν, kK	ε	
15.0	12	16.3	7	$^3T_{1g} \leftarrow {}^1A_{1g}$
18.5	20	17.9	10	$^3T_{1g}$ or $^3T_{2g} \leftarrow$
23.4	121	24.1	76	$^1T_{1g} \leftarrow$
28.1	86.5	28.1	64	$^1T_{2g} \leftarrow$

[a] $\Delta = 24.7$ kK, $B_{\text{corr}} = 300$ cm^{-1}, $\beta = 0.48$.[263]
[b] $\Delta = 25.0$ kK, $B_{\text{corr}} = 270$ cm^{-1}, $\beta = 0.41$.[265]

octahedral $IrCl_6^{3-}$ is present in the melt also. Band assignments and values of the ligand field parameters are included in Table XLIV. The one-electron spin–orbital configuration approach gave the reasonable values of ζ_{5d} = 2.85 kK and Δ = 26.0 kK for Ir(III) in the melt.[87]

4.8. Palladium and Platinum

4.8.1. Palladium(II) $4d^8$

The spectrum of Pd(II) has been recorded in the LiCl–KCl[262,263,266] and Li_2SO_4–Na_2SO_4–K_2SO_4[225] eutectic melts (cf. Fig. 46 and Table XLV). An increase in temperature from 400 to 600°C caused a red shift of 0.2 kK[266] in the chloride melt spectrum. The spectrum in the chloride melt was interpreted by both groups of workers[262,266] as being due to an essentially square-planar $PdCl_4^{2-}$ entity.

The original assignment of the transitions[262,263] was in keeping with the scheme of Gray and Ballhausen[92] but this scheme was changed for Pt(II) by Mason and Gray[93] in the light of recent studies of K_2PtCl_4 crystals with polarized light[267] and by magnetic circular dichroism.[268] The assignment of the spectrum of Pd(II) in $2M$ HCl implies a different order of d orbitals for Pd(II) and Pt(II) but it may not be appropriate to the melt spectrum. If the order of d-orbital energies is the same in Pd(II) and Pt(II), then we find for the chloride melt Δ_1 = 23.3 kK and Δ_2 = 6.2 kK, while for the sulfate melt Δ_1 = 25.3 kK.

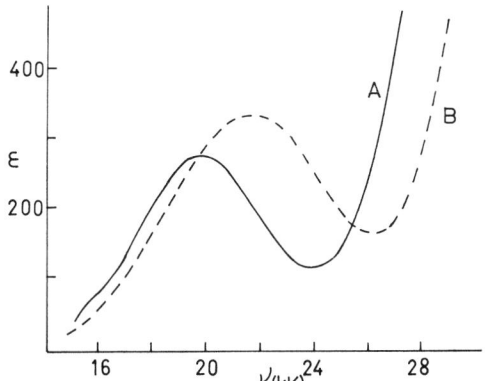

Fig. 46. Absorption spectra of (A) Pd(II) in fused LiCl–KCl eutectic at 450°C.[263] (B) Pd(II) in fused Li_2SO_4–Na_2SO_4–K_2SO_4 at 550°C.[225]

TABLE XLV. Spectrum of Pd(II) in LiCl–KCl and Li$_2$SO$_4$–Na$_2$SO$_4$–K$_2$SO$_4$

LiCl–KCl[a]		LiCl–KCl[b]		Li$_2$SO$_4$–Na$_2$SO$_4$–K$_2$SO$_4$[c]	
ν, kK	ε	ν, kK	ε	ν, kK	ε
		15.8	14	17.8	33
20.3	240	19.8	278	21.8	368
		24.5	65		

[a] 400°C.[266]
[b] 450°C.[263]
[c] 550°C.[225]

4.8.2. Platinum(II) $5d^8$

The spectrum of a solution of K$_2$PtCl$_4$ in the LiCl–KCl eutectic has been reported by two groups.[225,262,263,266] The two sets of results are collected in Table XLVI, while a spectrum is shown in Fig. 47.

Although there has been much discussion in the literature concerning the assignment of both crystal and aqueous solution spectra of Pt(II) compounds,[90,92,93,269] it is generally agreed that the two main bands in the visible region should be assigned as shown in Table XLVI; the energy expressions were given in Table XV. With regard to the weak spin-forbidden bands, agreement has not been reached[93,269] but the assignment of Day et al.[269] has been used above.

Substantial agreement exists between the two groups as far as band positions are concerned. The extinction coefficients of the spin-allowed bands, on the other hand, differ by a factor of ten almost exactly.

TABLE XLVI. Spectrum of Pt(II) in the LiCl–KCl Eutectic

ν,[a] kK	ε^a	ν,[b] kK	ε^b	ν,[225]c kK	$\varepsilon^{(2-3)c}$	Assignment
—	—	—	—	15.0	15	$^3A_{2g} \leftarrow {}^1A_{1g}$
19.5	8	19.0	8	18.8	47	$^3E_g \leftarrow$
25.6	15	25.6	15	24.7	167	$^1A_{2g} \leftarrow$
28.8	15	28.4	15	29.2	146	$^1E_g \leftarrow$

[a] 400°C.[266]
[b] 450°C.[266]
[c] 450°C, Δ_1 = 28.2 kK, Δ_2 = 6.0 kK.

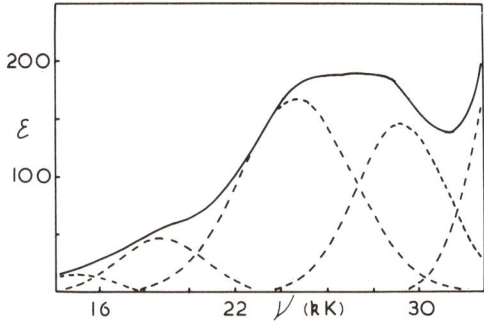

Fig. 47. Absorption spectrum of Pt(II) in fused LiCl–KCl eutectic at 450°C.[263] (Reproduced by permission from *Can. J. Chem.*)

Both groups argued that in the LiCl–KCl eutectic there was evidence of some interaction between the square-planar $PtCl_4^{2-}$ ion and solvent anions oriented above and below the molecular plane. However, further work remains to be done to substantiate this interpretation.

"*The accidents will disappear, what seems episodical and isolated will be absorbed and ranged in the harmonious course of history, in proportion as we understand the ideas which have influenced each separate country and each successive age.*"

Acton, The Rambler ii:*105* (*1859*)

5. WHITHER HIGH-TEMPERATURE COORDINATION CHEMISTRY?

5.1. General Conclusions

A summary of the symmetries observed for the Group VIII metal ions in melts appears in Table XLVII.

Gruen and McBeth[185] made a qualitative correlation of the observed coordination of $3d$ metal ions in chloride melts (mostly LiCl–KCl) with electrostatic and crystal field stabilization energies. Now that many more systems have been examined, it is apparent that their grouping of the ions holds generally but with a broader range of behavior within each group. Thus Ru^{3+}, Rh^{3+}, Os^{4+}, and Ir^{3+} belong to the pure O_h group of Cr^{3+}, while Fe^{3+} joins Mn^{2+} in the pure T_d group.

TABLE XLVII. Symmetries Observed for Group VIII Metal Ions in Melts

Ion	Symmetry	Remarks
Fe^{3+}	T_d	—
Fe^{2+}	T_d, O_h	Some distortion
Co^{2+}	T_d, D_{2d}, O_h	Wide range of symmetries
Ni^{2+}	T_d, O_h, T, O	Wide range of symmetries and equilibria between species
Ru^{3+}, Rh^{3+}	O_h	—
Ir^{3+}, Os^{4+}	—	—
Os^{3+}	O_h	Some distortion
Pd^{2+}, Pt^{2+}	D_{4h}	Considerable axial interaction

Tetrahedral symmetry certainly occurs for Fe^{2+} and Co^{2+} but a weak ligand such as F^- can lead to octahedral Fe^{2+}, and dodecahedral Co^{2+} appears to be quite common. The coordination number four is a very close description for Pd^{2+} and Pt^{2+} with central ion size eliminating tetrahedral in favor of square-planar coordination.[270] Finally, Ni^{2+} spans the octahedral–tetrahedral equilibrium category and that where the system distorts rather than adopt six-coordination.

To date all the symmetries adopted by transition metal ions in melts are very simply related to the octahedron and tetrahedron. The **T** and **O** centers discussed in detail in Section 4.4 have not had specific geometries assigned to them but they are probably distorted versions of T_d and O_h, respectively. In the same way that the dodecahedral symmetry can arise from double of flattened tetrahedra (Section 4.3 and Ref. 196), so a T_d–O_h transformation through a trigonal pyramid (C_{3v}) and a trigonal bipyramid (D_{3h}) is easily visualized (Fig. 48). One might then expect **O**-center spectra to resemble those of Ni(II) in D_{3h} symmetry and indeed a considerable likeness is apparent between the spectrum of a known D_{3h} complex (Fig. 49D),[271] that of the proposed $NiCl_3^-$ (Fig. 34), and the **O**-center spectra depicted in Fig. 33. Unfortunately, no trigonal pyramidal complexes of Ni(II) appear to be known.

For the examples of pure O_h, T_d, $D_{\infty h}$, or D_{4h} symmetry and the systems where the distortion is small, the parameters of crystal field splitting, interelectronic repulsion, and spin–orbit coupling are little removed from those obtained in aqueous and crystalline media with corresponding ligands.[272] A basic tenet of ligand field theory—that only the first coordination sphere

High-Temperature Coordination Chemistry of Group VIII

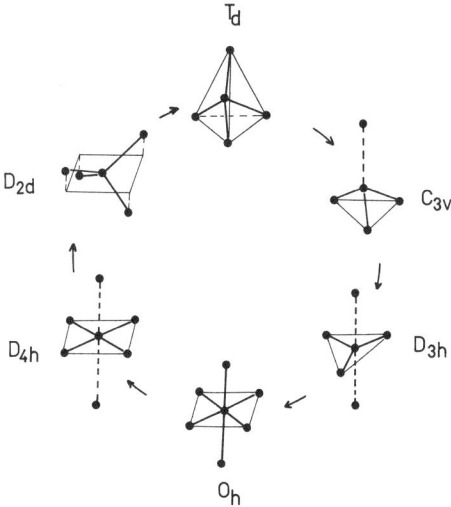

Fig. 48. Interconversion of species of various symmetries.

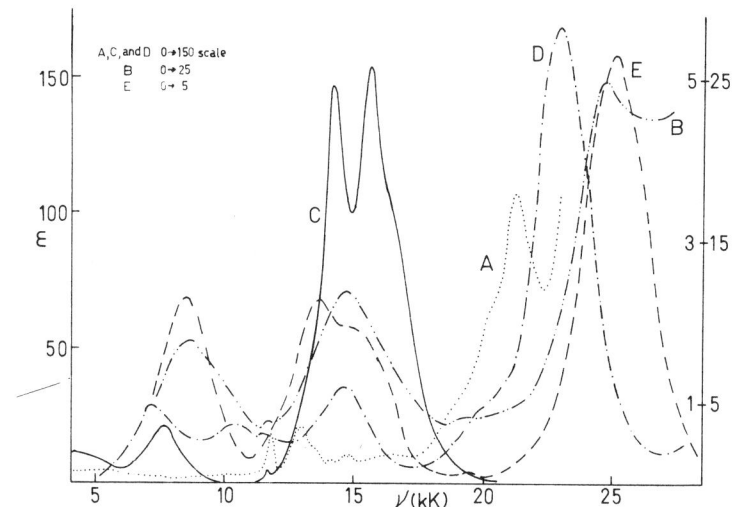

Fig. 49. Absorption spectrum of Ni(II) in various symmetries. (A) $D_{\infty h}$ example $NiCl_2$.[97] (B) D_{4h} example $Ni(\Phi NHCH_2CH_2NH\Phi)_2Cl_2$.[278] (C) T_d example $NiCl_4^{2-}$.[227] (D) D_{3h} example $NiBr(Me_6tren)Br$.[271] (E) O_h example $Ni(H_2O)_6^{2+}$.[242]

really matters as far as the bonding of the transition metal ion is concerned[273]—is breached somewhat if we think of Cl^- as the ligand in all chloride melts, for example, but the derivation of all the spectra of one ion, regardless of coordination, from the free gaseous ion is still perceptible (cf. Fig. 49).

The coordination and the spectrum of an ion in a melt are clearly a compromise between the variety of coordinations favored by the said ion and made available by the solvents; the importance of the precise nature of the cations of the latter cannot be overlooked.

5.2. Suggestions for Further Study

Much more information is needed about the spectra of the diatomic molecules, especially the halides, and it is time to examine the spectra and magnetic properties[250,274] of the double and heavier halides for comparison with the solid and liquid phases.

For molten salt studies it is now clear that the use of just a single salt or a conveniently low-melting eutectic is adequate for exploratory studies but using wide ranges of temperature and solvent composition aids tremendously in interpreting (though sometimes confusing!) the results. This approach is currently lacking in the case of oxyanion systems.

A few specific studies we would like to see made are as follows:

(a) The electronic and Raman spectrum of $KNiCl_3$ in the vapor phase.

(b) The electronic spectra, magnetic susceptibilities, and magnetic circular dichroism of all the palladium halides (including F^-, which favors high spin in solid PdF_2[18]) in the vapor phase, and in alkali halide melts, with a wide temperature range and each alkali metal ion in the solvent in at least one instance.

(c) An electrochemical and electronic spectral investigation of the oxidation states and coordination in cyanide melts.[174,175,272]

(d) The electronic spectra of the Group VIII chlorides in fused pyridinium chloride[275,276] in which Ni(II) is forced into tetrahedral coordination.[223]

(e) Spectra of transition metal halide–organic solvent mixtures where equilibria and distortions can be followed (cf. $NiCl_2$–acetamide[277]).

(f) The Raman spectra of $NiCl_4^{2-}$, $PdCl_4^{2-}$, and $PtCl_4^{2-}$ in several solvents.

(g) Spectra and X-ray crystal structures of single and mixed salts of transition metals as a function of temperature.

APPENDIX A. The Coulomb (Direct) and Exchange Integrals[24]

In the Hamiltonian of the many-electron atom the interelectronic repulsion term has the form $\sum_{i>j=1}^{N} (e^2/r_{ij})$. Two particular matrix elements arise in the evaluation of the diagonal element of interelectronic repulsion:

$$\iint a^*(1)b^*(2)(e^2/r_{12})a(1)b(2)\, d\tau_1\, d\tau_2$$

and

$$\iint a^*(1)b^*(2)(e^2/r_{12})b(1)a(2)\, d\tau_1\, d\tau_2$$

These may be written as $(ab \mid q \mid ab) = J(a, b)$ and $(ab \mid q \mid ba) = K(a, b)$, respectively. $J(a, b)$ is termed the Coulomb or direct integral and corresponds to the classical interaction of the electronic charge distributions ψ_a^2 and ψ_b^2; $K(a, b)$ is termed the exchange integral. They may be written

$$J(a, b) = \sum_{k=0}^{\infty} a^k(l^a m_l^a, l^b m_l^b) F^k(n^a l^a, n^b l^b)$$

and

$$K(a, b) = \delta(m_s^a, m_s^b) \sum_{k=0}^{\infty} b^k(l^a m_l^a, l^b m_l^b) G^k(n^a l^a, n^b l^b)$$

n, l, m_l, and s being the quantum numbers. a^k and b^k are defined thus

$$a^k(l^a m_l^a, l^b m_l^b) = c^k(l^a m_l^a, l^a m_l^a) c^k(l^b m_l^b, l^b m_l^b)$$

and

$$b^k(l^a m_l^a, l^b m_l^b) = \{c_k(l^a m_l^a, l^b m_l^b)\}^2$$

where

$$c^k(lm_l, l'm_l') = [2/(2k+1)]^{1/2} \int_0^{\pi} \Theta(km_l - m_l')\Theta(lm_l)\Theta(l'm_l') \sin\theta\, d\theta$$

The F^k and G^k are integrals dependent on the radial functions and the Slater parameters take the form $F_k = F^k/D_k$ and $G_k = G^k/D_k$ so that

$$F_2 = (1/49)F^2(n^a l^a, n^b l^b) = (1/49)R^2(n^a l^a n^b l^b, n^a l^a n^b l^b)$$

$$= (e^2/49) \int_0^{\infty}\int_0^{\infty} (r_<^2/r_>^3) R_1^2(n^a l^a) R_2^2(n^b l^b)\, dr_1\, dr_2$$

$$F_4 = (e^2/441) \int_0^{\infty}\int_0^{\infty} (r_<^4/r_>^5) R_1^2(n^a l^a) R_2^2(n^b l^b)\, dr_1\, dr_2$$

and

$$G_k = (1/D_k)G^k(n^a l^a, n^b l^b) = R^k(n^a l^a n^b l^b, n^b l^b n^a l^a)$$

$$= (e^2/D_k) \int_0^\infty \int_0^\infty (r_<^k/r_>^{k+1}) R_1(n^a l^a) R_1(n^b l^b) R_2(n^a l^a) R_2(n^b l^b) \, dr_1 \, dr_2$$

$$= (2e^2/D_k) \int_0^\infty dr_2 \int_0^{r_2} (r_1^k/r_2^{k+1}) R_1(n^a l^a) R_1(n^b l^b) R_2(n^a l^a) R_2(n^b l^b) \, dr_1$$

APPENDIX B. Effects of Electric and Magnetic Fields

The splitting of spectral lines in an external electric field is known as the Stark effect. In a weak field each term gives rise to $J + \tfrac{1}{2}$ or $J + 1$ components according to whether J is half-integral or integral. In a strong field, **L** and **S** are uncoupled (an LS coupling scheme has been assumed for this argument) and each term is split into $L + 1$ groups characterized by the quantum number $M_L = L, L - 1, \ldots, 0$, each group being split into $2S + 1$ components (except when $M_L = 0$) characterized by the quantum number $M_S = S, S - 1, \ldots, -S$.[25,39]

A weak magnetic field B splits each spectral term into $2J + 1$ components of energy separation: $g_J B M_J \mu_B$.[25,39]

Here μ_B is the Bohr magneton, M_J the component of total angular momentum in the field direction, and g_J the Lande g factor given in the case of LS coupling by*

$$g_J = 1 + \{[J(J+1) + S(S+1) - L(L+1)]/2J(J+1)\}$$

The consequent appearance of the spectrum is a splitting into several lines unless $S = 0$ for both terms involved, in which case a symmetric triplet results. The special case of singlet–singlet transitions split in a magnetic field is known, for historical reasons, as the normal Zeeman effect. Otherwise we are dealing with the anomalous Zeeman effect.[25,39]

In a magnetic field strong enough to make the split components of adjacent multiplets overlap, we have what is known as the Paschen–Back effect.[25,39] Once again **L** and **S** are uncoupled and the spectral splitting pattern tends toward series of triplets as for the normal Zeeman case with each triplet component itself showing the field-free multiplicity of the transition.

---

* For jj coupling situations g_J is a function of the individual j's and J.

Additional selection rules to those given in Section 2.1.2 arise for the Zeeman effect, namely[25,39]

$\Delta M_J = 0$ components polarized $\parallel$ field (π)

$\Delta M_J = \pm 1$ components polarized $\perp$ field (σ)

and $M_J = 0 \not\leftrightarrow M_J = 0$ for $\Delta J = 0$ unless $S_1 = S_2 = 0$. For the Paschen–Back effect[39] the breakdown is

$$\Delta M_L = 0 \; (\pi), \quad \Delta M_L = \pm 1 \; (\sigma), \quad \Delta M_S = 0$$

APPENDIX C. The Correlation of Molecular and Atomic Electronic States

The Λ values resulting from the combination of L_1 and L_2, the orbital quantum numbers of two unlike atoms, are

$$\Lambda = L_1 + L_2, L_1 + L_2 - 1, L_1 + L_2 - 2, \ldots, 0$$
$$L_1 + L_2 - 1, L_1 + L_2 - 2, \ldots, 0$$

to

$$L_1 - L_2, L_1 - L_2 - 1, \ldots, 0$$

If $\Lambda \neq 0$, each state is doubly degenerate, a degeneracy removed by rotational interaction to give Λ doubling. When $\Lambda = 0$, if $L_1 + L_2 + \sum l_1 + \sum l_2$ is even there will be one more Σ^+ state than Σ^- state and if it is odd, one more Σ^- state than Σ^+ state.

The molecular spin S will take all values from $S_1 + S_2$ to $S_1 - S_2$, where S_1 and S_2 are the total atomic spin quantum numbers.

For homonuclear diatomic molecules the combination of different states proceeds as for heteronuclear molecules except that each molecular state occurs with both parities. However, like atomic states give rise to the same molecular state multiplicities as for heteronuclear molecules but the Λ and parity values are a function of the spin S as follows.

If S is integral, the molecular states resulting from $L_1 = L_2 = L$ are

$$(2L)_g, (2L-1)_g, (2L-2)_g \quad \Pi_g, \Sigma_g^+$$
$$(2L-1)_u, (2L-2)_u \quad \Pi_u, \Sigma_u^-$$
$$\Pi_g, \Sigma_g^+$$

to

$$\Sigma_g^+$$

TABLE XLVIII. Molecular Electronic States Correlated with States of Separated Unlike Atoms[48]

Separated atomic states	Molecular states						
	Σ^+	Σ^-	Π	Δ	Φ	Γ	H
$S_g + S_g$	1	—	—	—	—	—	—
$S_g + S_u$	—	1	—	—	—	—	—
$S_u + P_u$	—	1	1	—	—	—	—
$S_g + P_u$	1	—	1	—	—	—	—
$S_g + D_g$	1	—	1	1	—	—	—
$S_g + F_g$	—	1	1	1	1	—	—
$P_u + P_u$	2	1	2	1	—	—	—
$P_g + D_g$	1	2	3	2	1	—	—
$P_u + F_u$	2	1	3	3	2	1	—
$D_g + D_u$	2	3	4	3	2	1	—
$D_g + F_u$	3	2	5	4	3	2	1

whereas if it is odd, the parities of these states are reversed. Tables XLVIII and XLIX include some examples of correlations.

When A, the spin–orbital coupling constant, is large, Λ and S are not well defined and it becomes appropriate to correlate atomic J values with molecular Ω values. The possible Ω values are

$$J_1 + J_2, J_1 + J_2 - 1, J_1 + J_2 - 2, \ldots, \tfrac{1}{2} \text{ or } 0$$

$$J_1 + J_2 - 1, J_1 + J_2 - 2, \ldots, \tfrac{1}{2} \text{ or } 0$$

to

$$J_1 - J_2, J_1 - J_2 - 1, \ldots, \tfrac{1}{2} \text{ or } 0$$

When $\Omega \neq 0$ we have a double degeneracy of states (Ω doubling corresponding to Λ doubling). When $\Omega = 0$, if J_1 and J_2 are both half-integral, there are the same number of 0^+ as 0^- states, whereas if J_1 and J_2 are both integral, there will be one more 0^+ than 0^- state or the reverse according to whether $J_1 + J_2 + \sum l_1 + \sum l_2$ is even or odd. In the case of like atoms, parity must be considered as above. Tables L and LI summarize some correlations of J and Ω.

TABLE XLIX. Molecular Electronic States Correlated with Like States of Separated Like Atoms[48]

Separated atomic states	Molecular states																			
	$^1\Sigma_g^+$	$^1\Sigma_u^-$	$^1\Pi_g$	$^1\Pi_u$	$^1\Delta_g$	$^1\Delta_u$	$^1\Phi_g$	$^1\Phi_u$	$^1\Gamma_g$	$^3\Sigma_u^+$	$^3\Sigma_g^-$	$^3\Pi_g$	$^3\Pi_u$	$^3\Delta_u$	$^5\Sigma_g^+$	$^5\Sigma_u^-$	$^5\Pi_g$	$^5\Pi_u$	$^5\Delta_g$	$^7\Sigma_u^+$
$^1S + {}^1S$	1	—	—	—	—	—	—	—	—	—	—	—	—	—	—	—	—	—	—	—
$^2S + {}^2S$	1	—	—	—	—	—	—	—	—	1	—	—	—	—	—	—	—	—	—	—
$^4S + {}^4S$	1	—	—	—	—	—	—	—	—	1	—	—	—	—	1	—	—	—	—	1
$^1P + {}^1P$	2	1	1	1	1	—	—	—	—	—	—	—	—	—	—	—	—	—	—	—
$^2P + {}^2P$	2	1	1	1	1	—	—	—	—	2	1	1	1	1	—	—	—	—	—	—
$^3P + {}^3P$	2	1	1	1	1	—	—	—	—	2	1	1	1	1	2	1	1	1	1	—
$^1D + {}^1D$	3	2	2	2	2	1	1	1	1	—	—	—	—	—	—	—	—	—	—	—

TABLE L. Correlation of Molecular Ω with Atomic J's for Unlike Atoms[48]

Atomic J's	Molecular Ω							
	0^+	0^-	$\frac{1}{2}$	1	$1\frac{1}{2}$	2	$2\frac{1}{2}$	3
$0_g + 0_g$	1	—	—	—	—	—	—	—
$0_g + 0_u$	—	1	—	—	—	—	—	—
$0 + \frac{1}{2}$	—	—	1	—	—	—	—	—
$\frac{1}{2} + \frac{1}{2}$	1	1	—	1	—	—	—	—
$0_g + 1_g$	—	1	—	1	—	—	—	—
$\frac{1}{2} + 1$	—	—	2	—	1	—	—	—
$1_g + 1_g$	2	1	—	2	—	1	—	—
$1_g + 1_u$	1	2	—	2	—	1	—	—
$1 + 1\frac{1}{2}$	—	—	3	—	2	—	1	—
$1\frac{1}{2} + 1\frac{1}{2}$	2	2	—	3	—	2	—	1
$0_u + 2_u$	1	—	—	1	—	1	—	—
$1_g + 2_u$	2	1	—	3	—	2	—	1

One other rule is involved: the noncrossing rule, which forbids the crossing of potential energy curves of molecular states of the same electronic species (e.g., two Σ_g^+ states).

It is also possible to imagine that two nuclei of a molecule derive from the splitting of a single atom.[47] For small internuclear distances we again

TABLE LI. Correlation of Molecular Ω with Atomic J's for Like Atoms in the Same State[48]

Atomic J's	Molecular Ω								
	0_g^+	0_u^-	1_g	1_u	2_g	2_u	3_g	3_u	4_g
$0 + 0$	1	—	—	—	—	—	—	—	—
$\frac{1}{2} + \frac{1}{2}$	1	1	—	1	—	—	—	—	—
$1 + 1$	2	1	1	1	1	—	—	—	—
$1\frac{1}{2} + 1\frac{1}{2}$	2	2	1	2	1	1	—	1	—
$2 + 2$	3	3	2	2	2	1	1	1	1

have the effect of an electric field on an atom. If **L** and **S** are the orbital and spin angular momenta of the united atom, then

$$\Lambda = L, L-1, \ldots, 0$$

and **S** remains the same in the molecular and atomic states. The occurrence of Σ^+ or Σ^- states depends on whether $L + \sum l_i$ for the united atom is even or odd. For large spin–orbital coupling molecular states with

$$\Omega = J, J-1, \ldots, \tfrac{1}{2} \text{ or } 0$$

are obtained, where **J** is the total angular momentum vector of the united atom. For homonuclear diatomics the parity has to be conserved.

As examples, ^3D_g of Mg gives the states $^3\Delta$, $^3\Pi$, and $^3\Sigma$ of BeO and the states $^3\Delta_g$, $^3\Pi_g$, and $^3\Sigma_g^+$ of C_2.

APPENDIX D. Symbols Used in Sections 3–5

A_2, A_4	axial ligand field splitting parameters
B, C	Racah parameters of interelectronic repulsion
E_P, E_F	Energies of F and P terms (baricenters in the case of complexed ions)
F_2, F_4	Slater parameters of interelectronic repulsion; $B = F_2 - 5F_4$, $C = 35F_4$
f	oscillator strength
K_1, K_2, K_3	stability constants
β	nephelauxetic ratio, that of any interelectronic parameter in a complex to its free ion value
Δ	cubic crystal (ligand) field splitting parameter
Δ_1, Δ_2	square-planar crystal field splitting parameters
δ	half-width of absorption band
ε	molar absorptivity (extinction coefficient)
ζ	one-electron spin–orbital coupling constant
μ	magnetic moment
ν	wave number, 1 kK = 1 kilokayser = 1000 cm^{-1}

ACKNOWLEDGMENTS

We are grateful to the pioneers in this field, Dr. D. M. Gruen and Dr. G. P. Smith, for inspiration and useful discussions; to the authors and publishers of several books and articles for permission to reproduce figures; to the Science Research Council (Great Britain) and the National Research Council of Canada for support of our own researches, and to the University of Saskatchewan for a grant to cover publication expenses;

to Dr. W. D. Chandler, Dr. K. W. Fung, Dr. L. G. Boxall, and Dr. M. E. Stone for discursive and technical help; and to Miss Ruth Derges for typing the manuscript and associated secretarial work.

REFERENCES

1. H. A. Levy, P. A. Agron, M. A. Bredig, and M. D. Danford, *Ann. N. Y. Acad. Sci.* **79**:762 (1960).
2. S. C. Wait, Jr., A. T. Ward, and G. J. Janz, *J. Chem. Phys.* **45**:133 (1966).
3. S. Hafner and N. H. Nachtrieb, *J. Chem. Phys.* **40**:2891 (1964).
4. J. Brown, UCRL 9944 (UC-4 Chemistry), TID 4500 (16th ed.), December 13, 1961.
5. L. Yarmus, M. Kukk, and B. R. Sundheim, *J. Chem. Phys.* **40**:33 (1964).
6. T. B. Swanson, *J. Chem. Phys.* **45**:179 (1966).
7. N. H. Nachtrieb, *J. Phys. Chem.* **66**:1163 (1962).
8. A. Berlin and N. Nghi, *Compt. Rend.* (C) **262**:1421 (1966).
9. D. Inman, D. G. Lovering, and R. Narayan, *Trans. Faraday Soc.* **64**:2476 (1968).
10. S. H. White, D. Inman, and B. Jones, *Trans. Faraday Soc.* **64**:2841 (1968).
11. D. Inman, B. Jones, and S. H. White, *J. Inorg. Nucl. Chem.* **32**:927 (1970).
12. R. S. Juvet, Jr., V. R. Shaw, and M. A. Khan, *J. Am. Chem. Soc.* **91**:3788 (1969).
13. G. J. Janz, *Molten Salts Handbook*, Academic Press, New York (1967).
14. J. E. Ricci, in: *Molten Salt Chemistry* (M. Blander, ed.), Wiley—Interscience, New York (1964).
15. C. M. Cook and W. E. Dunn, Jr., *J. Phys. Chem.* **65**:1505 (1961).
16. K. F. Zmbov and J. L. Margrave, *J. Phys. Chem.* **70**:3379 (1966).
17. E. Iberson, R. Gut, and D. M. Gruen, *J. Phys. Chem.* **66**:65 (1962).
18. F. A. Cotton and G. Wilkinson, *Advanced Inorganic Chemistry*, 2nd ed., Wiley—Interscience, New York (1966).
19. M. M. Jones, *Elementary Coordination Chemistry*, Prentice-Hall, Englewood Cliffs, N. J. (1964).
20. W. P. Griffith, *The Chemistry of the Rarer Platinum Metals*, Wiley—Interscience, New York (1967).
21. H. J. Emeleus and A. G. Sharpe (eds.), *Advances in Inorganic Chemistry and Radiochemistry*, Academic Press, New York (one or two volumes per annum from 1959).
22. *Progress in Inorganic Chemistry* (F. A. Cotton, ed., to Vol. 10 and S. J. Lippard, ed., from Vol. 11), Wiley—Interscience, New York (one or two volumes per annum from 1959).
23. R. L. Carlin (ed.), *Transition Metal Chemistry*, Marcel Dekker, New York (from 1965).
24. E. U. Condon and G. H. Shortley, *The Theory of Atomic Spectra*, University Press, Cambridge (1935).
25. H. E. White, *Introduction to Atomic Spectra*, McGraw-Hill, New York (1934).
26. R. Mavrodineanu and H. Boiteux, *Flame Spectroscopy*, Wiley, New York (1965).
27. G. Bract, *Phys. Rev.* **28**:334 (1926).
28. H. N. Russell, *Phys. Rev.* **29**:782 (1927).
29. B. G. Wybourne, "Spectroscopic Properties of Rare Earths," Wiley—Interscience, New York (1965).

30. G. H. Shortley and B. Fried, *Phys. Rev.* **54**:749 (1938).
31. B. Edlen, in: *Handbuch der Physik*, Vol. XXVII, Springer-Verlag, Heidelberg (1964).
32. G. Racah, *Phys. Rev.* **85**:381 (1952).
33. J. C. Slater, *Quantum Theory of Atomic Sctructure*, McGraw-Hill, New York (1960).
34. E. U. Condon, *Phys. Rev.* **36**:1121 (1930).
35. B. O. Jordan and E. Wigner, *Z. Physik* **47**:631 (1928).
36. R. Stevenson, *Multiplet Structure of Atoms and Molecules*, Saunders, Philadelphia (1965).
37. G. Racah, *Phys. Rev.* **62**:438 (1942).
38. J. H. Van Vleck, *Phys. Rev.* **45**:405 (1934).
39. G. W. King, *Spectroscopy and Molecular Structure*, Holt, Rinehart and Winston, New York (1960).
40. M. Born and R. Oppenheimer, *Ann. Physik* **84**:457 (1927).
41. G. Herzberg, *Molecular Spectra and Molecular Structure, I. Diatomic Molecules*, Prentice-Hall, New York (1939).
42. R. S. Mulliken, *Rev. Mod. Phys.* **3**:89 (1931).
43. F. Hund, *Z. Physik* **36**:657 (1925).
44. R. S. Mulliken, *Rev. Mod. Phys.* **2**:60 (1930).
45. E. Wigner and E. E. Witmer, *Z. Physik* **51**:859 (1928).
46. R. S. Mulliken, *Rev. Mod. Phys.* **4**:1 (1932).
47. R. S. Mulliken, *Phys. Rev.* **36**:1440 (1930).
48. A. G. Gaydon, *Dissociation Energies and Spectra of Diatomic Molecules*, 3rd ed., Chapman and Hall, London (1968).
49. R. S. Mulliken, *Phys. Rev.* **32**:186 (1928).
50. R. S. Mulliken, *Int. J. Quant. Chem.* **1**:103 (1967).
51. W. Heitler and F. London, *Z. Physik* **44**:455 (1927).
52. J. C. Slater, *Phys. Rev.* **37**: 481 (1931).
53. J. C. Slater, *Phys. Rev.* **38**:1109 (1931).
54. L. Pauling, *J. Am. Chem. Soc.* **53**:1367, 3225 (1931).
55. G. Herzberg, *Molecular Spectra and Molecular Structure, III. Electronic Spectra and Electronic Structure of Polyatomic Molecule*, Van Nostrand, Princeton, N. J. (1966).
56. F. Hund, *Z. Physik* **51**:759 (1928).
57. F. Hund, *Z. Physik* **63**:719 (1930).
58. J. E. Lennard-Jones, *Trans. Faraday Soc.* **25**: 668 (1929).
59. J. E. Lennard-Jones, *Proc. Roy. Soc. London* **A198**:1, 14 (1949).
60. J. E. Lennard-Jones, J. A. Pople, and G. G. Hall, *Proc. Roy. Soc. London* **A202**:155, 166, 323 (1950).
61. D. R. Hartree, *Proc. Camb. Phil. Soc.* **24**:89 (1928).
62. V. Fock, *Z. Physik* **61**:126 (1930).
63. L. Brillouin, *Actualités Scientifiques* (Paris) IV (1934).
64. C. J. Roothaan, *Rev. Mod. Phys.* **23**:69 (1951).
65. C. J. Roothaan, *Rev. Mod. Phys.* **32**:179 (1960).
66. W. Moffitt, *Rept. Progr. Phys.* **17**:173 (1954).
67. P.-O. Löwdin, *Adv. Chem. Phys.* **2**:207 (1959).
68. P.-O. Löwdin, *Quantum Theory of Atoms, Molecules and Solids*, Academic Press, New York (1966).
69. K. D. Carlson and C. R. Claydon, in: *Advances in High-Temperature Chemistry*, Vol. 1, p. 43 (L. Eyring, ed.), Academic Press, New York (1967).

70. R. A. Berg and O. Sinanoglu, *J. Chem. Phys.* **32**:1082 (1960).
71. C. K. Jørgensen, *Mol. Phys.* **7**:417 (1964).
72. C. J. Cheetham and R. F. Barrow, in: *Advances in High-Temperature Chemistry*, Vol. 1, p. 7 (L. Eyring, ed.), Academic Press, New York (1967).
73. S. B. Schneiderman, *Int. J. Quant. Chem.* **2**:89 (1968).
74. J. S. Griffith, *The Theory of Transition-Metal Ions*, University Press, Cambridge (1961).
75. F. A. Cotton, *Chemical Applications of Group Theory*, Wiley—Interscience, New York (1963).
76. C. K. Jørgensen, *Absorption Spectra and Chemical Bonding in Complexes*, Addison-Wesley, Reading (1962).
77. C. K. Jørgensen, *Adv. Chem. Phys.* **5**:33 (1963).
78. Y. Tanabe and S. J. Sugano, *J. Phys. Soc. Japan* **9**:753, 766 (1954).
79. M. Rotenberg, R. Bivins, N. Metropolis, and J. K. Wooten, Jr., *The 3-j and 6-j Symbols*, MIT Press, Cambridge, Mass. (1959).
80. C. W. Nielson and G. F. Koster, *Spectroscopic Coefficients for p^n, d^n, and f^n Configurations*, MIT Press, Cambridge, Mass. (1963).
81. G. Racah, *Phys. Rev.* **61**:186 (1942).
82. G. Racah, *Phys. Rev.* **63**:367 (1943).
83. G. Racah, *Phys. Rev.* **76**:1352 (1949).
84. L. E. Biedenharn and H. Van Dam, *Quantum Theory of Angular Momentum*, Academic Press, New York (1965).
85. W. Moffitt, G. L. Goodman, M. Fred, and B. Weinstock, *Mol. Phys.* **2**:109 (1959).
86. A. D. Liehr, *J. Phys. Chem.* **64**:43 (1960).
87. K. W. Fung and K. E. Johnson, *J. Inorg. Nucl. Chem.* **33**:1407 (1971).
88. J. S. Griffith, *Disc. Faraday Soc.* **26**:173 (1958).
89. A. Abragam and M. H. L. Pryce, *Proc. Roy. Soc. London* **A205**:135 (1951).
90. R. F. Fenske, D. S. Martin, Jr., and K. Ruedenberg, *Inorg. Chem.* **1**:441 (1962).
91. D. S. Martin, Jr., M. A. Tucker, and A. J. Kassman, *Inorg. Chem.* **5**:1298 (1966).
92. H. B. Gray and C. J. Ballhausen, *J. Am. Chem. Soc.* **85**:260 (1963).
93. W. R. Mason, III and H. B. Gray, *J. Am. Chem. Soc.* **90**:5721 (1968).
94. H. B. Gray, *Transition Metal Chemistry* **1**:239 (1965).
95. J. Chatt, L. E. Orgel, and G. A. Gamlen, *J. Chem. Soc.* **1958**:486.
96. J. T. Hougen, G. E. Leroi, and T. C. James, *J. Chem. Phys.* **34**:1670 (1961).
97. C. W. DeKock and D. M. Gruen, *J. Chem. Phys.* **44**:4387 (1966).
98. C. W. DeKock and D. M. Gruen, *J. Chem. Phys.* **46**:1096 (1967).
99. H. B. Gray, in: *Structural Chemistry and Molecular Biology*, (A. Rich and N. Davidson, eds.), p. 783, Freeman, San Francisco (1968).
100. C. J. Ballhausen and H. B. Gray, *Inorg. Chem.* **1**:111 (1962).
101. P. T. Manoharan and H. B. Gray, *Inorg. Chem.* **5**:823 (1966).
102. C. J. Ballhausen and H. B. Gray, *Molecular Orbital Theory*, Benjamin, New York (1964).
103. H. Basch, Ph. D. thesis, Columbia Univ., New York (1966).
104. M. Wolfsberg and L. Helmholz, *J. Chem. Phys.* **20**:837 (1952).
105. H. Basch, A. Viste, and H. B. Gray, *Theoret. Chim. Acta* **3**:458 (1965).
106. H. Basch, A. Viste, and H. B. Gray, *J. Chem. Phys.* **44**:10 (1966).
107. S. Yamada and R. Tsuchida, *Bull. Chem. Soc. Japan* **26**:15 (1953).
108. R. F. Barrow and M. Senior, *Nature* **223**:1359 (1969).

109. W. J. M. Gissane and R. F. Barrow (1966); unpublished, cf. Ref. 72.
110. J. R. Marquart and J. Berkowitz, *J. Chem. Phys.* **39**:283 (1963).
111. P. Coppens, S. Smoes, and J. Drowart, *Trans. Faraday Soc.* **64**:630 (1968).
112. B. Rosen, *Données Spectroscopiques Concernant les Molécules Diatomiques*, Hermann, Paris (1951).
113. S. P. Reddy and P. T. Rao, *J. Mol. Spectry.* **4**:16 (1960).
114. S. P. Reddy, *J. Sci. Ind. Res. (India)* **18B**:188 (1959).
115. A. Kant and B. Strauss, *J. Chem. Phys.* **41**: 3806 (1964).
116. L. Kynning and H. Neuhaus, *Z. Naturforsch.* **18a**:1142 (1963).
117. S. V. K. Rao and P. T. Rao, *Indian J. Phys.* **35**:556 (1961).
118. S. V. K. Rao and P. T. Rao, *Indian J. Phys.* **36**:609 (1962).
119. A. Kant, *J. Chem. Phys.* **41**:1872 (1964).
120. N. Åslund, H. Neuhaus, A. Lagerqvist, and ⁻. Andersén, *Arkiv Fysik* **28**:271 (1964).
121. R. T. Grimley, R. P. Burns, and M. G. Inghram, *J. Chem. Phys.* **35**:551 (1961).
122. V. G. Krishnamurty, *Indian J. Phys.* **27**:354 (1953).
123. E. M. Bulewiez, L. F. Phillips, and T. M. Sugden, *Trans. Faraday Soc.* **57**:921 (1961).
124. S. P. Reddy and P. T. Rao, *Proc. Phys. Soc. (London)* **75**:275 (1960).
125. S. V. K. Rao, S. P. Reddy, and P. T. Rao, *Z. Physik* **166**:261 (1962).
126. A. Kant, *J. Chem. Phys.* **44**:2450 (1966).
127. N. S. McIntyre, A. V. Auwera-Mahieu, and J. Drowart, *Trans. Faraday Soc.* **64**: 3006 (1968).
128. R. Scullman, private communication.
129. V. Raziunas, G. Macur, and S. Katz, *J. Chem. Phys.* **43**:1010 (1965).
130. A. Lagerqvist, H. Neuhaus, and R. Scullman, *Z. Naturforsch.* **20a**:751 (1965).
131. A. Lagerqvist and R. Scullman, *Arkiv Fysik* **32**:479 (1966).
132. J. H. Norman, H. G. Staley, and W. E. Bell, *J. Phys. Chem.* **68**:662 (1964).
133. M. Ackerman, F. E. Stafford, and G. Verhaegen, *J. Chem. Phys.* **36**:1560 (1962).
134. C. Malmberg, R. Scullman, and P. Nylén, *Arkiv Fysik* **39**:495 (1969).
135. J. H. Norman, H. G. Staley, and W. E. Bell, *J. Phys. Chem.* **69**:1373 (1965).
136. A. Gatterer, J. Junkes, E. W. Salpeter, and B. Rosen, *Molecular Spectra of Metallic Oxides*, Specola Vaticana, Città del Vaticano (1957).
137. K. Jansson, R. Scullman, and B. Yttermo, *Chem. Phys. Letters* **4**:188 (1969).
138. J. H. Norman, H. G. Staley, and W. E. Bell, *J. Phys. Chem.* **42**:1123 (1965).
139. R. Scullman, *Arkiv Fysik* **28**:255 (1964).
140. V. A. Loginov, *Opt. Spectry. (USSR)* **20**:88 (1966).
141. R. Scullman and B. Yttermo, *Arkiv Fysik* **33**:231 (1966).
142. M. A. Catalan, F. Rohrlich, and A. G. Shenstone, *Proc. Roy. Soc. London* **A221**: 421 (1954).
143. R. F. Barrow, private communication.
144. C. B. Alcock and G. W. Hooper, *Proc. Roy. Soc. London* **A254**:551 (1960).
145. H. Schäfer and A. Tebben, *Z. Anorg. Allgem. Chem.* **304**:317 (1960).
146. A. Büchler and J. B. Berkowitz-Mattuck, in: "Advances in High Temperature Chemistry" (L. Eyring, ed.), Academic Press, New York (1967).
147. A. Büchler, J. L. Stauffer, and W. Klemperer, *J. Chem. Phys.* **40**:3471 (1964).
148. J. W. Hastie, R. H. Hauge, and J. L. Margrave, *High Temp. Sci.* **1**:76 (1969).
149. J. W. Hastie, R. H. Hauge, and J. L. Margrave, *J. Chem. Soc. D.* **1969**:1452.
150. J. W. Hastie, R. H. Hauge, and J. L. Margrave, *J. Chem. Phys.* **51**:2648 (1969).
151. D. E. Milligan, M. E. Jacose, and J. D. McKinley, *J. Chem. Phys.* **42**:902 (1965).

152. K. R. Thompson and K. D. Carlson, *J. Chem. Phys.* **49**:4379 (1968).
153. A. Trutia and M. Musa, *Spectrochim. Acta* **23**:1165 (1967).
154. D. M. Gruen and C. W. DeKock, *J. Chem. Phys.* **43**:3395 (1965).
155. C. W. DeKock and D. M. Gruen, *J. Chem. Phys.* **49**:4521 (1968).
156. K. F. Zmbov and J. L. Margrave, *J. Inorg. Nucl. Chem.* **29**:673 (1967).
157. R. C. Schoonmaker, A. H. Friedman, and R. F. Porter, *J. Chem. Phys.* **31**:1586 (1959).
158. N. W. Gregory and R. O. Macharen, *J. Phys. Chem.* **59**:110 (1955).
159. W. E. Bell, U. Merten, and M. Tagami, *J. Phys. Chem.* **65**:510 (1961).
160. H. Schäfer, U. Wiese, K. Rinke, and K. Brendel *Angew. Chem.* **6**:253 (1967).
161. F. A. Cotton and T. E. Haas, *Inorg. Chem.* **3**:10 (1964).
162. W. E. Bell, M. C. Garrison, and U. Merten, *J. Phys. Chem.* **65**:517 (1961).
163. V. S. Rao and P. Kusch, *J. Chem. Phys.* **34**:832 (1961).
164. J. A. Plambeck, *J. Chem. Eng. Data* **12**:77 (1967).
165. H. A. Laitinen and C. H. Liu, *J. Am. Chem. Soc.* **80**:1015 (1958).
166. H. A. Laitinen and J. W. Pankey, *J. Am. Chem. Soc.* **81**:1053 (1959).
167. H. A. Laitinen and J. A. Plambeck, *J. Am. Chem. Soc.* **87**:1202 (1965).
168. S. N. Flengas and T. R. Ingraham, *J. Electrochem. Soc.* **106**:714 (1959).
169. H. E. Bartlett and K. E. Johnson, *J. Electrochem. Soc.* **114**:457 (1967).
170. K. E. Johnson and H. A. Laitinen, *J. Electrochem. Soc.* **110**:314 (1963).
171. W. J. Hamer, M. S. Malmberg, and B. Rubin, *J. Electrochem. Soc.* **103**:8 (1956).
172. W. J. Hamer, M. S. Malmberg, and B. Rubin, *J. Electrochem. Soc.* **112**:750 (1965).
173. G. W. Mellors and S. Senderoff, *J. Electrochem. Soc.* **112**:642 (1965).
174. S. V. Winbush, E. Griswold, and J. Kleinberg, *J. Am. Chem. Soc.* **83**:3197 (1961).
175. W. L. Magnuson, E. Griswold, and J. Kleinberg, *Inorg. Chem.* **3**:88 (1964).
176. D. M. Gruen, *J. Inorg. Nucl. Chem.* **4**:74 (1957).
177. N. W. Silcox and H. M. Haendler, *J. Phys. Chem.* **64**:303 (1960).
178. G. Harrington and B. R. Sundheim, *Ann. N. Y. Acad. Sci.* **79**:950 (1960).
179. S. Balt, *Rec. Trav. Chim. des Pay-Bas* **86**:1025 (1967).
180. N. Islam, Ph. D. thesis, New York Univ. (1968).
181. B. Zaslow and R. E. Rundle, *J. Phys. Chem.* **61**:490 (1957).
182. S. Balt, *Mol. Phys.* **14**:233 (1968).
183. A. M. A. Verwey and S. Balt, private communication.
184. D. M. Gruen and R. L. McBeth, *Nature* **194**:468 (1962).
185. D. M. Gruen and R. L. McBeth, *Pure Appl. Chem.* **6**:23 (1963).
186. H. A. Øye and D. M. Gruen, *Inorg. Chem.* **3**:836 (1964).
187. J. P. Young, U. S. At. Energy Comm. ORNL-P-403 (1964).
188. J. P. Young, *Inorg. Chem.* **8**:825 (1969).
189. A. D. Liehr and C. J. Ballhausen, *J. Mol. Spectry.* **2**:342 (1958).
190. D. M. Gruen, *Nature* **178**:1181 (1956).
191. I. V. Tananaev and B. F. Dzhurinskii, *Dokl. Akad. Nauk SSSR* **134**:1374 (1960).
192. I. V. Tananaev and B. F. Dzhurinskii, *Dokl. Akad. Nauk SSSR* **135**:94 (1960).
193. I. V. Tananaev and B. F. Dzhurinskii, *Dokl. Akad. Nauk SSSR* **139**:120 (1961).
194. I. V. Tananaev and B. F. Dzhurinskii, *Dokl. Akad. Nauk SSSR* **140**:374 (1961).
195. K. W. Fung and K. E. Johnson, *Can. J. Chem.* **47**:4699 (1969).
196. J. R. Dickinson and K. E. Johnson, *J. Mol. Spectry.* **36**:1 (1970).
197. K. E. Johnson and T. S. Piper, *Disc. Faraday Soc.* **32**:32 (1962).
198. J. A. Duffy, F. P. Glasser, and M. D. Ingram, *J. Chem. Soc. (A)* **1968**:551.

199. J. A. Duffy and M. D. Ingram, *J. Chem. Soc. (A)* **1969**:2398.
200. R. A. Bailey, M. El Guindy, and J. A. Walden, *Inorg. Chem.* **8**:2526 (1969).
201. J. P. Young and J. C. White, *Anal. Chem.* **32**:799 (1960).
202. B. R. Sundheim and M. Kukk, *Disc. Faraday Soc.* **32**:49 (1962).
203. M. Kukk, Ph. D. thesis, New York Univ. (1964).
204. H. C. Egghart, *J. Phys. Chem.* **73**:4014 (1969).
205. H. A. Øye and D. M. Gruen, *Inorg. Chem.* **4**:1173 (1965).
206. K. W. Fung and K. E. Johnson, *Can. J. Chem.* **48**:3635 (1970).
207. C. A. Angell and D. M. Gruen, *J. Inorg. Nucl. Chem.* **29**:2243 (1967).
208. H. A. Øye and D. M. Gruen, in: *Selected Topics in High Temperature Chemistry*, (T. Førland, K. Grjotheim, K. Motzfeldt, and S. Urnes, eds.), University Press, Oslo (1966).
209. D. M. Gruen, in: *Fused Salts* (B. R. Sundheim, ed.), p. 322, McGraw-Hill, New York (1964).
210. J. R. Dickinson and K. E. Johnson, *J. Mol. Spectry.* **33**:414 (1970).
211. D. K. Straub, R. S. Drago, and J. T. Donoghue, *Inorg. Chem.* **1**:848 (1962).
212. F. A. Cotton and J. G. Bergman, *J. Am. Chem. Soc.* **86**:2941 (1964).
213. R. E. Isbell, E. W. Wilson, Jr., and D. F. Smith, *J. Phys. Chem.* **70**:2493 (1966).
214. E. W. Wilson, private communication.
215. R. Stahl-Breda and W. Low, *Phys. Rev.* **113**:775 (1959).
216. C. Simo, E. Banks, and S. Holt, *Inorg. Chem.* **8**:1446 (1969).
217. J. P. Young, private communication.
218. R. Pappalordo, *Spectrochim. Acta* **19**:2093 (1963).
219. A. D. Liehr and C. J. Ballhausen, *J. Mol. Spectry.* **4**:190 (1960).
220. D. M. Gruen, private communication.
221. M. Goffman, Ph. D. thesis, Temple Univ. (1966).
222. J. G. Bergman, Jr., and F. A. Cotton, *Inorg. Chem.* **5**:1420 (1966).
223. D. M. Gruen and R. L. McBeth, *J. Phys. Chem.* **63**:393 (1959).
224. C. H. Liu, J. Hasson, and G. P. Smith, *Inorg. Chem.* **7**:2244 (1968).
225. J. R. Dickinson, Ph. D. thesis, Univ. of Sask. (1969).
226. T. R. Griffiths, *Chem. Comm.* **1967**:1222.
227. G. P. Smith, C. H. Liu, and T. R. Griffiths, *J. Am. Chem. Soc.* **86**:4796 (1964).
228. G. P. Smith and C. R. Boston, *J. Chem. Phys.* **43**:4051 (1965).
229. C. R. Boston and G. P. Smith, *J. Am. Chem. Soc.* **85**:1006 (1963).
230. G. P. Smith, C. R. Boston, and J. Brynestad, *J. Chem. Phys.* **45**:829 (1966).
231. C. R. Boston, J. Brynestad, and G. P. Smith, *J. Chem. Phys.* **47**:3193 (1967).
232. J. Brynestad, C. R. Boston, and G. P. Smith, *J. Chem. Phys.* **47**:3179 (1967).
233. C. R. Boston and G. P. Smith, *J. Phys. Chem.* **62**:409 (1958).
234. B. R. Sundheim and G. Harrigton, *J. Chem. Phys.* **31**:700 (1959).
235. J. Brynestad, H. L. Yakel, and G. P. Smith, *J. Chem. Phys.* **45**:4652 (1966).
236. J. Brynestad and G. P. Smith, *J. Chem. Phys.* **47**:3190 (1967).
237. C. A. Angell and D. M. Gruen *J. Phys. Chem.* **70**:1601 (1966).
238. W. E. Smith, J. Brynestad, and G. P. Smith, *J. Am. Chem. Soc.* **89**:5983 (1967).
239. W. E. Smith, J. Brynestad, and G. P. Smith, *J. Chem. Phys.* **52**:3890 (1970).
240. G. P. Smith and S. Von Wimbush, *J. Am. Chem. Soc.* **88**:2127 (1966).
241. C. R. Boston, C. H. Liu, and G. P. Smith, *Inorg. Chem.* **7**:1938 (1968).
242. C. K. Jørgensen, *Acta Chem. Scand.* **9**:1362 (1955).

243. G. P. Smith, in: *Molten Salt Chemistry* (M. Blander, ed.), McGraw-Hill, New York (1964).
244. K. Knox, R. G. Shulman, and S. Sugano, *Phys. Rev.* **130**:512 (1963).
245. C. K. Jørgensen, *Mol. Phys.* **1**:410 (1958).
246. G. P. Smith and C. R. Boston, *J. Chem. Phys.* **46**:412 (1967).
247. J. Brynestad and G. P. Smith, *J. Am. Chem. Soc.* **92**:3198 (1970).
248. A. D. Liehr and C. J. Ballhausen, *Ann. Phys. (New York)* **6**:134 (1959).
249. W. Trzebiatowski and J. Mulak, *Bull. Acad. Sci. Pol.* **13**:759 (1965).
250. B. N. Figgis and J. Lewis, *Prog. Inorg. Chem.* **6**:37 (1964).
251. C. K. Jørgensen, *Acta Chem. Scand.* **10**:518 (1956).
252. C. K. Jørgensen, *Mol. Phys.* **2**:309 (1959).
253. J. R. Dickinson and K. E. Johnson, *Mol. Phys.* **19**:19 (1970).
254. R. B. Johannesen and G. A. Candela, *Inorg. Chem.* **2**:67 (1963).
255. E. von Blasius and W. Preetz, *Z. Anorg. Allgem. Chem.* **335**:1 (1965).
256. L. L. Larson and C. S. Garner, *J. Am. Chem. Soc.* **76**:2180 (1954).
257. P. B. Dorain, H. H. Patterson, and P. C. Jordon, *J. Chem. Phys.* **49**:3845 (1968).
258. P. C. Jordan, H. H. Patterson, and P. B. Dorain, *J. Chem. Phys.* **49**:3858 (1968).
259. C. K. Jørgensen, *Acta Chem. Scand.* **16**: 793 (1962).
260. R. Dingle, *J. Mol. Spectry.* **18**:276 (1965).
261. F. B. Ogilvie and O. G. Holmes, *Can. J. Chem.* **44**:447 (1966).
262. J. R. Dickinson and K. E. Johnson, *Can. J. Chem.* **45**:1631 (1967).
263. J. R. Dickinson and K. E. Johnson, *Can. J. Chem.* **45**:2457 (1967).
264. J. Chatt, L. E. Orgel, and G. A. Gamlen, *J. Chem. Soc.* **1958**:486.
265. C. K. Jørgensen, *Acta Chem. Scand.* **10**:500 (1956).
266. R. A. Bailey and J. A. McIntyre, *Inorg. Chem.* **5**:1824 (1966).
267. D. S. Martin and C. A. Lenhardt, *Inorg. Chem.* **3**:1368 (1964).
268. A. J. McCaffery, P. N. Schatz, and P. J. Stephens, *J. Am. Chem. Soc.* **90**:5730 (1968).
269. P. Day, M. J. Smith, and R. J. P. Williams, *J. Chem. Soc. A* **1968**:668.
270. H. B. Gray, *J. Chem. Ed.* **41**:2 (1964).
271. L. Sacconi, *Pure Appl. Chem.* **17**:95 (1968).
272. K. E. Johnson, *Electrochim. Acta* **11**:129 (1966).
273. C. K. Jørgensen, in: *Halogen Chemistry*, Vol. 1, p. 265, Academic Press, New York (1967).
274. B. N. Figgis and J. Lewis, in: *Technique of Inorganic Chemistry* (H. B. Jonassen and A. Weissberger, eds.), Vol. IV, Wiley—Interscience, New York (1965).
275. L. F. Audrieth and J. Kleinberg, *Nonaqueous Solvents*, Wiley, New York (1958).
276. L. F. Audrieth, A. Long, and R. E. Edwards, *J. Am. Chem. Soc.* **58**:428 (1936).
277. K. E. Johnson and M. E. Stone, *Can. J. Chem.* **49**:3836 (1971).
278. G. Maki, *J. Chem. Phys.* **29**:162 (1958).
279. D. W. Smith, *Inorg. Chim. Acta* **5**:231 (1971).
280. S. Suzuki and K. Tanaka, *Nippon Kinzoku Gakkaishi* **34**:461 (1970); *Chem. Abstr.* **73**:30356v (1970).
281. A. J. Barnes and H. E. Hallam, *Quart, Rev.* **23**:392 (1969).
282. T. Folkman and J. A. Plambeck, *Can. J. Chem.* **50**:3911 (1972).
283. J. R. Dickinson and M. E. Stone, *Can. J. Chem.* **50**:2946 (1972).
284. K. W. Fung and K. E. Johnson, *Inorg. Chem.* **10**:1347 (1971).

Chapter 4

ELECTROANALYTICAL CHEMISTRY IN MOLTEN SALTS—A REVIEW OF RECENT DEVELOPMENTS

K. W. Fung* and G. Mamantov

Department of Chemistry
University of Tennessee
Knoxville, Tennessee

1. INTRODUCTION

This chapter is concerned with electroanalytical chemistry in molten salts. Electroanalysis has been applied extensively for *in situ* determination and characterization of solutes in molten salt solvents. This field has been reviewed previously[1-6]† either alone or in conjunction with other topics. In this review we shall stress more recent developments (1965–1971) in several important solvent systems. We shall exclude the discussion of transport properties and related topics, the electrical double layer, and the rates of charge-transfer processes, since the reviews of these topics[6-8] are quite up to date.

---

* Present address: Department of Chemistry, University of Malaya, Kuala Lumpur, Malaysia.
† Also see Ref. 6 for a listing of other related reviews.

2. GENERAL REMARKS

2.1. Acid–Base Concepts

The acid–base concept of Lux and Flood,[9,10]

$$\text{acid} + O^{2-} = \text{base}$$

has been quite useful in oxide-containing melts, such as the carbonates and silicates. The concept, which is related to the Lewis acid–base concept, can be extended to molten halides, namely

$$\text{acid} + X^- = \text{base}$$

Some examples, such as

$$BeF_2 + 2F^- = BeF_4^{2-}$$

have been given by Bloom.[11] Tremillon and Letisse[12] have applied these ideas to aluminum chloride–alkali chloride melts. The acid–base properties of such melts near the composition $NaAlCl_4$ are controlled by the equilibrium[12–14]

$$2AlCl_4^- \rightleftarrows Al_2Cl_7^- + Cl^-$$

in which $Al_2Cl_7^-$ is a strong acid and Cl^- is a strong base. The basicity is usually expressed as pCl^-.

Thus, in the modified Lewis (or Lux–Flood) concept, pure alkali halides represent the highest degree of basicity; as the solvent composition changes from alkali halide-rich to alkali halide-deficient melts, the solvent becomes acidic. Acid–base properties of molten halides may be used to explain stabilization of unusually low (or high) oxidation states,[15–18] the differences in stability[19] of the same oxidation state in related melts, and the effects on coordination[20] observed spectrally for certain metal ions. Or, restating the idea in other terms, the redox potentials depend[19] on melt basicity. Thus, the systematic variation of melt composition is a useful technique in the arsenal of the molten salt electrochemist who is interested in the chemistry of solute species in molten salt solvents. In this respect, it is important to note that variation of temperature may be used to serve the same purpose; for example, it has been shown[13] that in neutral chloroaluminates pCl^- decreases with temperature.

2.2. Methodology

Much of the early electrochemical work in molten salts[3] was performed with solid working electrodes under poorly defined mass transfer conditions. The definition of current–potential curves and the resulting precision were relatively poor. A troublesome feature of slow-scan voltammetry at solid electrodes is the "increasing limiting current" effect[21] caused by the deposited metal accumulating on the electrode surface as dendritic growths which both increase the electrode area and protrude the electrode surface out into the diffusion or convection field. In order to minimize or eliminate such problems and still obtain well-defined ⌡⌐-shaped voltammograms, one may resort to the use of a rotating disc electrode,[22] apply pulse polarography,[23] or construct the voltammogram from current–time curves.[18,19,24,25] The second approach is probably the simplest to implement, assuming a pulse polarograph is available.

Much attention, particularly among the Russian workers, has been given to the applicability of the Heyrovsky–Ilkovic versus the Kolthoff–Lingane equation[3] for electrode reactions involving metal deposition. In practice, one plots the electrode potential versus $\log[(i_l - i)/i]$ or $\log(i_l - i)$ and examines the linearity of the plot and also the value of the slope (theoretically $2.3RT/nF$). It is expected[1,26] that the Heyrovsky–Ilkovic equation should be applicable when alloy formation with the electrode material takes place and the metal formed diffuses away from the surface so that its surface activity is a function of the current density. Alloying and diffusion in the electrode will be functions of the metal deposited, electrode material, temperature, and the rate of deposition (current density); therefore, comparison is difficult or meaningless if several of these variables are varied simultaneously.

Both linear sweep voltammetry (oscillographic polarography, stationary electrode polarography, chronoamperometry with linear sweep) and chronopotentiometry have been extensively applied for studies in molten salts. The advantages of linear sweep voltammetry include: (1) extensively developed theory* enabling the experimentalist to interpret the mechanisms of relatively complex electrode reactions; (2) well-defined mass transfer conditions, particularly when faster scan rates (~ 1 V/sec) are employed; (3) the decrease of the faradaic charge with the square root of the scan rate[28] and the resulting decrease of any modifications of the solid electrode caused by the faradaic process. Chronopotentiometry,[29] a related electroanalytical

---

* For recent reviews see Refs. 27a and 27b.

method, possesses some of the same advantages but suffers from the fact that the transition times obtained in molten salts are frequently not well defined, due to double layer charging; the measurements of such transition times usually involve semi-empirical procedures.[30]

Controlled-potential coulometry has been used[31,32] for high-precision (better than 1–2%) electroanalytical work in molten salts. Before embarking on controlled-potential coulometry, pertinent electrode potentials and current–potential characteristics must be known.

In addition to the more or less classical electroanalytical methods discussed above, the application of thin-layer electrochemical cells,[33] spectroscopic characterization coupled with electrochemical generation,[34] and the use of small digital computers coupled to electrochemical instrumentation[23,35] for molten salt studies have been reported.

2.3. Experimental Techniques

One cannot overemphasize the importance of using a high-purity molten salt solvent for electroanalytical measurements. The method of purification differs from one molten salt solvent system to another and will be discussed briefly for each melt system; however, removal of water without hydrolysis is commonly an important consideration. Clearly the exposure of the melt to the atmosphere must be avoided; in this respect the use of sealed cells is safer than the use of cells containing ball or standard taper joints, gaskets, or flanges. If a cover gas (commonly argon or helium) is to be used, it should be of the highest purity available.

The choice of the working electrode is governed by the operating temperature, inertness of the electrode material, and the degree of convenience desired. Dropping electrodes (mercury[36] for low-melting systems, bismuth or lead[37] for higher-melting systems, such as the LiCl–KCl eutectic) normally result in greater reproducibility, but are more complex to use than solid electrodes. Solid metal (platinum, gold, tungsten) or carbon (pyrolytic graphite,[38] vitreous carbon[39]) indicator electrodes are most commonly used in molten salts; the configuration is either planar, cylindrical, or spherical. For insulation, glass or quartz are used wherever possible; thermal coefficients of expansion have to be considered. As in all solid electrode work,[40] the accurate measurement of electrode area is a definite problem. The change in the nature and the area of electrode surface due to metal deposition during electrolysis should be avoided as much as possible. The use of several materials (for example, platinum, tungsten, and pyrolytic graphite) as working electrodes in the same cell may be advantageous if interaction with the electrode is unavoidable. For potentiometric measurements in molten nitrates a metal–ion sensitive electrode system, Pd/PdO, CdO/Cd^{2+}, (an electrode of the third kind)* was employed by Inman[41] and Braunstein et al.[42]

The subject of reference electrodes in molten salts has been treated extensively by Laity[43] and Alabyshev et al.[44] The applicability of a reference electrode is determined by the nature of the molten salt solvent and the operating temperature. It has been stated[1] that, "It is unlikely that a universal reference electrode of thermodynamic significance can be devised for general use in different molten solvents at varied temperature." Metal–metal ion systems (for example, Pt^{2+}/Pt,[45] Ag$^+$/Ag,[46] Ni^{2+}/Ni,[47] Al^{3+}/Al[12,48]) constitute most frequently used reference electrodes in molten solvents. Separation of the reference electrode from the bulk of the melt is an important consideration; glass frits,[45] membranes,[1] asbestos fiber,[42] other porous insulators,[47,49] and ionic crystals,[50] have been used for this purpose.

Quite frequently a large noble metal electrode[51] is used as a quasi-reference electrode (QRE) in electroanalytical work. Such an electrode is equivalent to the mercury pool electrode in conventional polarography. Since the QRE is normally used in a three-electrode system, the current passing through the reference electrode is extremely small ($\leq 10^{-8}$ A). Therefore, the potential of the QRE remains quite constant (± 10 mV for prolonged periods of time) provided oxidants or reductants are not added to the melt. In that case the potential shifts in the direction expected from the Nernst equation.

2.4. EMF Series

Electromotive force series based on experimentally determined potentials or on potentials calculated from appropriate free energies have been developed for several molten solvents. Plambeck[52] in 1966 critically reviewed and tabulated values for the LiCl–KCl eutectic (450°C), equimolar NaCl–KCl (700–900°C), NaF–KF eutectic (850°C), MgCl$_2$–NaCl–KCl eutectic (475°C), AlCl$_3$–NaCl–KCl (66–20–14 mole %) (218°C), Li$_2$SO$_4$–K$_2$SO$_4$ eutectic (625°C), and Li$_2$SO$_4$–Na$_2$SO$_4$–K$_2$SO$_4$ eutectic (575°C). By far the most extensive series is that for the LiCl–KCl eutectic, developed

---

* It should be noted that formulas such as Cd^{2+} and Hg$_2^{2+}$ are only intended to specify the oxidation states and the number of atoms per unit. No implications should be drawn regarding coordination between central ion and surrounding ions.

primarily by Laitinen and co-workers.[45,53] Electrode potentials of several redox couples in molten $LiF-BeF_2-ZrF_4$ (65.6–29.4–5.0 mole %) and LiF–NaF–KF (46.5–11.5–42.0 mole %) at 500°C have been measured.[49,54] A tabulation of electrode potentials in molten $LiF-BeF_2$ (66–34 mole %) is available.[55] Boxall and Johnson,[56] based on their measurements of intersolvent and thermoelectric potentials, have expressed the potentials of several electrode couples in several molten solvents with respect to the standard aqueous hydrogen electrode at 25°C.

3. PRESENTLY IMPORTANT SOLVENT SYSTEMS

For each solvent system the material is presented in roughly the following order: (1) purification of the melt, (2) work that is primarily methodologically oriented (application of new methods to molten salts, new electrodes, etc.), (3) studies involving transition metal elements, (4) studies of main group elements, (5) a tabular summary of solutes investigated and the method employed.

It is clear, from the perusal of the material below, that in a number of cases conflicting results have been obtained by different workers. In these cases a critical evaluation must await further work.

3.1. LiCl–KCl Eutectic (58.8–41.2 mole %)

Molten LiCl eutectic has been widely used as a solvent for electrochemical studies, probably because for an alkali chloride melt it has a relatively low liquidus temperature (361°C) and it may be easily studied in Pyrex glass. However, if the moisture is not completely removed, hydrolysis occurs upon fusion, resulting in HCl and OH^- ions. The OH^- ions precipitate solute metal ions, attack glass, and diminish the available potential span.[57] Laitinen et al.[58] have shown that a pure melt can be obtained by bubbling dry HCl followed by nitrogen to remove HCl through a vacuum-dried salt mixture. Hill et al.[59] modified the method by treating the melt with magnesium powder after the HCl treatment (to replace any metal ions with Mg^{2+} ions which are reduced at very cathodic potentials). Maricle and Hume[57] treated the melt with chlorine instead of using vacuum and HCl. Such a treatment also gives a very pure melt.

Naryshkin et al.[60] have studied the reduction of Ag^+, Bi^{3+}, Cd^{2+}, Co^{2+}, Cr^{2+}, Cr^{3+}, Cu^+, Mn^{2+}, Ni^{2+}, Pb^{2+}, and Tl^+ ions at a platinum electrode by linear sweep voltammetry. In all cases a linear relationship between the peak current and the concentration of the metal ion was observed. All the processes except the reduction of Ni^{2+} (>1 V/sec), Co^{2+}, and Mn^{2+} (>2 V/sec) were found to be reversible up to ~ 8 V/sec. Behl[61] has studied the reduction of Cd^{2+}, Co^{2+}, Ni^{2+}, and Pb^{2+} at a glassy carbon electrode by the same technique. Linear dependence of peak current on metal ion concentration and the square root of scan rate were observed up to 1 V/sec. The diffusion coefficients for the ions reported are in agreement with those obtained by Naryshkin et al.[60] Fondanaiche et al.[62] examined the reduction of Ag^+ and Cd^{2+} by linear sweep voltammetry and chronopotentiometry. They also found that the peak currents were directly proportional to concentration. Tungsten electrode was found to be more suitable than platinum for the study of those ions. Hladik et al.[63] used linear sweep voltammetry to study the systems Cd^{2+}/Cd, Co^{2+}/Co, Ni^{2+}/Ni, and Pb^{2+}/Pb with a gold electrode in the molten and solid-state (~ 300–$350°C$) eutectics. The reduction of the ion and the oxidation of a thin electrodeposited metallic layer were studied. Two peaks were found for the system Cd^{2+}/Cd; this result was interpreted in terms of Cd_2^{2+}, formed from cadmium and $CdCl_2$.

The use of ac polarography at a solid electrode for analytical purposes was demonstrated by Delimarskii et al.[64,65] They found that the peak height was linearly related to concentration for the reduction of Ag^+, Cd^{2+}, Co^{2+}, Cu^{2+}, and Tl^+. Simultaneous determination of two or three metallic ions in the melt was also studied.

Panchenko[66] used a dropping bismuth electrode to study the reduction of Ag^+; he found that the limiting current was proportional to concentration of Ag^+ in the melt and that the Kolthoff–Lingane equation was valid. Naryshkin et al.[67] obtained a linear plot of E versus $\log[(i_d - i)/i]$ for the reduction of Zn^{2+} at a dropping lead electrode.

The electrochemistry of VO^{2+} (as $VOSO_4$) was studied by Scrosati and Laitinen.[68] Three reduction steps were observed in the chronopotentiogram. The first step was well defined and attributed to $VO^{2+} + e \rightarrow VO^+$. The product undergoes a rapid, irreversible chemical reaction to give V_2O_3 and V^{3+}. This mechanism was supported by the results of controlled-potential electrolysis. The ill-defined second step was attributed to the reduction of the remaining VO^+ or some other intermediate species. The third step was attributed to the reduction of SO_4^{2-}. By measuring electrode potentials, Suzuki[69] concluded that V^{2+} is the species in equilibrium with the metal.

Laitinen and co-workers[70–73] examined the reduction of the chromate ion in this melt. A three-electron reduction of CrO_4^{2-} to CrO_4^{5-} was observed. The product subsequently decomposed into CrO_3^{3-} and O^{2-} ions. At high surface concentration of CrO_4^{5-} a solid deposit of Li_5CrO_4 was obtained on the platinum electrode, which proved to be very stable. In the presence of Ni^{2+}, Mg^{2+}, Zn^{2+}, and Co^{2+} ions very stable compounds involving CrO_4^{5-}, Li^+, and the second cation were obtained.

Alabyshev et al.[74] examined the equilibrium between manganese and its ions in LiCl–KCl (50–50 mole %) melt potentiometrically. The average n value for Mn ions changed from 1.92 at 427°C to 1.70 at 927°C. Alabyshev et al. assumed Mn^{2+} and Mn^+ to be present and determined the equilibrium potentials of Mn^{2+}/Mn, Mn^+/Mn, and Mn^{2+}/Mn^+ couples. Spectroscopic studies will be required to prove the presence of Mn in the $+1$ state.

The electrochemical behavior of zirconium and hafnium was studied by Baboian et al.[75] potentiometrically and voltammetrically. They showed that Zr^{4+} was the stable species produced in the melt by anodic dissolution of zirconium at lower temperatures. At higher temperatures (550°C) Zr^{2+} was the predominant species produced by anodic dissolution; it disproportionates, however, into Zr and Zr^{4+}. The reduction of Zr^{4+} occurred in two steps at high temperatures (500–550°C); it was a single four-electron process at 450°C. The reduction of Hf^{4+} was found to be one four-electron step at all temperatures. The disproportionation of $ZrCl_2$ was supported by voltammetric studies at various temperatures.[76] No oxidation wave was observed by adding $ZrCl_2$ to the melt up to 550°C due to its insolubility, whereas above this temperature an unstable oxidation wave attributed to the formation of Zr^{4+} was observed. The electrochemical properties of zirconium were further investigated by Eon et al.[77] using linear sweep voltammetry and coulometry. The results of coulometric studies are similar to those reported by Baboian et al.;[75] namely, up to 400°C zirconium was oxidized to Zr^{4+}, whereas at 510°C and higher temperatures only Zr^{2+} was formed. Eon et al. also reported that the reduction of Zr^{4+} at high temperatures proceeds in two steps:

$$Zr^{4+} + 2e \rightleftharpoons Zr^{2+} \quad \text{and} \quad Zr^{2+} + 2e \rightleftharpoons Zr$$

At lower temperatures (400°C) only one reduction step ($Zr^{4+} + 2e \rightleftharpoons Zr^{2+}$) was observed, due to the precipitation of $ZrCl_2$.

Johnson and co-workers[78] reported $NbCl_3$ to be sparingly soluble in the melt. At low current densities the niobium anode dissolved with 100% current efficiency to give Nb^{3+}, whereas at current densities exceeding about 5A/cm², a black deposit of $NbCl_{2.95}$ was obtained on the anode and the "n" value increased gradually to four and then rapidly to higher values at higher current densities. Caton and Freund[32] carried out coulometric and voltammetric studies on niobium. Niobium ions were generated by controlled-potential electrolysis with an average n value of 3.64. They suggested that the solution contained Nb^{4+} and Nb^{3+} in a concentration ratio of 3:1. Nb^{4+} was reduced reversibly to Nb^{3+}; however, the mechanism of further reduction was not clearly established. Attempts to reduce the species completely to Nb^{3+} at a controlled potential were complicated by the formation of a deposit. Nb^{5+} was unstable in the melt at 450°C. Saeki and co-workers[79,80] studied the equilibrium between niobium and its subchlorides by potentiometric measurements. The apparent valence of niobium ions formed immediately after anodic dissolution of niobium was about 3.1. After 9 hr the equilibrium potentials were stable and the composition of the subchloride was Nb_3Cl_8. Pimenov[81] also found the average valence for niobium ions in equilibrium with the metal to be less than three. However, he assumed that this result was due to the presence of Nb^{2+} and Nb^{3+} ions in equilibrium. Based on this assumption, he obtained the standard potentials for Nb^{2+}/Nb, Nb^{3+}/Nb, and Nb^{3+}/Nb^{2+} couples and the equilibrium constant for the reaction $2Nb^{3+} + Nb \rightleftharpoons 3Nb^{2+}$. Later, by studying the anodic polarization curves,[82] he concluded that Nb^{3+} is the predominant species formed by anodic dissolution of niobium metal. Recently Inman et al.[83] reinvestigated the niobium system by chronopotentiometry. Niobium ions were prepared in situ by anodic dissolution of the metal. No reduction wave could be detected for Nb^{3+} that was formed at low current densities. Nb^{4+}, produced at high current densities (>50 mA/cm^2), was reduced reversibly to Nb^{3+}.

Senderoff and Mellors[84] studied the reduction of Mo^{3+} (added as K_3MoCl_6) chronopotentiometrically. The reduction was found to be a single three-electron irreversible process at 600°C. It became more reversible with increasing temperature. The source of the irreversibility was not unambiguously established; however, based on the lack of stability of the solute at lower concentrations, Senderoff and Mellors suggested that molybdenum is present as a stable binuclear (or polymeric) ion in equilibrium with a less stable mononuclear ion which disproportionates into molybdenum metal and $MoCl_5$. The electrode reaction involves the reduction of the mononuclear species. Suzuki[69] investigated the equilibrium between molybdenum and its subchloride. The metal ions were introduced by the anodization of molybdenum metal. Stable electrode potentials were obtained up to 675°C. No disproportionation of the generated species (Mo^{3+}) was indicated. On the other hand, Ryzhik and Smirnov,[85] by measuring the equilibrium potential, suggested that Mo, Mo^{2+}, and Mo^{3+} are in equilibrium: $2Mo^{3+} + Mo \rightleftharpoons 3Mo^{2+}$. The equilibrium potentials for Mo^{3+}/Mo, Mo^{2+}/Mo, Mo^{3+}/Mo^{2+} couples were reported.

Suzuki[69] concluded from potentiometric measurements that Ta^{4+} is the species in equilibrium with the metal in the melt. He also studied the electrolytic reduction and oxidation of $TaCl_4$ by chronopotentiometric and coulometric measurements.[86] The reduction occurred in two steps: a diffusion-controlled, reversible two-electron reaction from Ta^{4+} to Ta^{2+}; and an irreversible reaction from Ta^{2+} to the metal. Ta^{4+} was oxidized to Ta^{5+} reversibly.

Johnson and Mackenzie[87] reported the formal potential of the couple W^{2+}/W in this eutectic. The W^{2+} species was produced by anodic dissolution of a tungsten electrode in the presence of Yb^{2+} (reducing agent). Attempts to directly generate tungsten ions in the melt were unsuccessful; instead chlorine evolution was observed. The passivity was attributed to the adherent film of the insoluble polymeric chlorides formed by anodic polarization.

Scrosati[88] determined the solubility of PdO in the melt by chronopotentiometry. The results, which are in good agreement with those obtained by potentiometry, indicated that PdO is completely dissociated in the melt. The diffusion coefficient for Pd^{2+} was reported. Schmidt et al.[89] studied the reduction of Bi^{3+}, Pd^{2+}, Pt^{2+}, and Sb^{3+} voltammetrically with an intermittenly polarized platinum electrode. The shape of the current–voltage curves for Pd^{2+} and Pt^{2+} was described by the Kolthoff–Lingane equation, whereas for Bi^{3+} and Sb^{3+} the Heyrovsky–Ilkovic equation was valid. The diffusion coefficients for the species were reported.

Johnson and Mackenzie[90] measured the standard potentials for Eu^{3+}/Eu^{2+}, Sm^{3+}/Sm^{2+}, and Yb^{3+}/Yb^{2+} couples. They observed that europium, samarium, and ytterbium metals displace alkali metal ions in the melt. No evidence for the disproportionation of the divalent ions was found but scrupulous precautions were needed to keep the melt free from impurities which readily react with the highly reducing ions. Martinot, Duyckaerts, and Caligara[91-93] examined the systems Np^{4+}/Np^{3+}, Np^{3+}/Np, U^{4+}/U^{3+}, U^{3+}/U, and UO_2^{2+}/UO_2^+ potentiometrically and chronopotentiometrically in the temperature range 400–550°C. The results showed that a slow rearrangement of the shell of complexing Cl^- ions follows the electron transfer step. They discussed the discrepancies between the earlier results on uranium in this melt. Martinot and Duyckaerts[94] also studied the electrochemical behavior of NpO_2^{2+} and NpO_2^+ by using similar techniques. They also investigated the electrochemistry of Pu^{3+} by chronopotentiometry.[95] The reversibility of the electrode process Pu^{4+}/Pu^{3+} increased with temperature. The diffusion coefficients for the neptunium, plutonium, and uranium species were reported. The voltammetric behavior of lanthanide elements was studied by Panchenko and co-workers.[96,97] The diffusion current was found to be directly proportional to the concentration. The half-wave potentials and diffusion coefficients were determined. The results may be of interest from the standpoint of the electrolytic separation of these elements. The standard electrode potentials of La^{3+}/La and Y^{3+}/Y couples were measured by Plambeck and co-workers.[98,99]

The electrochemical behavior of sulfur, selenium, and tellurium was studied by Bodewig and Plambeck[100] potentiometrically and voltammetrically. The standard potentials for S/S^{2-}, Se/Se^{2-}, and Te^{2+}/Te couples were reported. The formation of Te^{2-} was not observed. The anodic and cathodic waves in different systems were ascribed to $2S + 2Cl^- \rightarrow S_2Cl_2 + 2e$, $S + 2e \rightarrow S^{2-}$; $2Se + 2Cl^- \rightarrow Se_2Cl_2(g) + 2e$, $Se + 2e \rightarrow Se^{2-}$, respectively. The observed blue color of $S-S^{2-}$ solution was ascribed to polysulfide ions. Shafir and Plambeck[101] reexamined the standard potentials of Ga^{3+}/Ga, In^{3+}/In^+, and In^+/In couples. The results were compared with those reported previously.

Wrench and Inman[102] carried out a potentiometric investigation of the oxygen electrode in this melt. The potentials of gold and platinum electrodes varied with the oxide ion concentration in the melt and with the changes in the partial pressure of O_2 above the melt. The potential-determining reaction was suggested to be $2O^{2-} + 2e \rightarrow O_2^{2-}$ (see the discussion below under nitrate melts). The voltammetric and chronopotentiometric studies of Na_2O_2 were made by Mignonsin et al.[103]

The electrochemical investigation of trimetaphosphate as well as other phosphates has been reported by Laitinen and Lucas.[104] The reduction of trimetaphosphate was found to proceed in two steps at 450°C and in three steps above 450°C. Voltammetric results showed that a film is formed in the reduction. The reduction process was observed to be very complex because of a prior chemical reaction. Franks and Inman[105] reported similar observations for the reduction of orthophosphate and pyrophosphate in this solvent.

Wrench and Inman[106] reinvestigated the reduction of sulfate ions in this melt. They concluded that the sulfate ion is not electroactive. The appearance of reduction waves reported previously may possibly be due to the catalytic effect of traces of water or hydroxide in the melt.

Hladik and Morand[107] examined the systems NO_3^-/NO_2^- and NO_2^-/NO voltammetrically in molten and solid eutectics with different oxide concentrations. The processes were found to be irreversible.

The results of electroanalytical studies in the molten LiCl–KCl eutectic are summarized in Table I.

TABLE I. Results of Electroanalytical Studies in Molten LiCl-KCl Eutectic

Experimental technique	Temperature, °C	Solutes investigated	Ref.
Voltammetry			
Solid metal electrodes	400–550	Ag^+, Al^{3+}, Bi^{3+}, Cd^{2+}, Co^{2+}, Cr^{2+}, Cr^{3+}, Cu^+, Cu^{2+}, Fe^{2+}, Fe^{3+}, Ga^{3+}, Hg^{2+}, In^{3+}, Ni^{2+}, Pb^{2+}, Sn^{2+}, Th^{4+}, Ti^{3+}, Ti^{4+}, Tl^+, U^{3+}, U^{4+}, UO_2^{2+}, V^{2+}, V^{3+}, Zn^{2+}, Bi_2O_3, CdO, CuO, MgO, NiO, V_2O_5, OH^-, O^{2-}, CrO_4^{2-}	1
Intermittently polarized Pt electrode	450	Bi^{3+}, Pd^{2+}, Pt^{2+}, Sb^{3+}	89
Dropping Bi electrode	420	Ag^+	66
Dropping Pb electrode	450	Zn^{2+}	67
Pt electrode	450	Nb^{4+}, Nb^{3+}	32
Au electrode	220–450	NO_3^-, NO_2^-	107
Rotating Pt electrode	400	Ce^{3+}, Dy^{3+}, Er^{3+}, Eu^{3+}, Gd^{3+}, Ho^{3+}, La^{3+}, Lu^{3+}, Nd^{3+}, Pr^{3+}, Sm^{3+}, Tb^{3+}, Y^{3+}, Yb^{3+}, Tm^{3+}	96, 97
Pt electrode	450–550	Hf^{4+}, Zr^{2+}, Zr^{4+}	75
Pt electrode	420–700	Zr^{2+}	76
Pt electrode	585	Na_2O_2	103
Pt electrode	450	Metaphosphates	104
Graphite electrode	400 or 420	S, Se, Te	100
Ac voltammetry			
Pt electrode	500–600	Cd^{2+}, Hg^{2+}	64
Pt electrode	400	Ag^+, Cd^{2+}, Co^{2+}, Cu^{2+}, Tl^+	65
Linear sweep voltammetry			
Pt electrode	450	Ag^+, Bi^{3+}, Cd^{2+}, Co^{2+}, Cr^{2+}, Cr^{3+}, Cu^+, Mn^{2+}, Ni^{2+}, Pb^{2+}, Tl^+	60
Glassy carbon electrode	450	Cd^{2+}, Co^{2+}, Ni^{2+}, Pb^{2+}	61
Au electrode	300–420	Cd^{2+}, Co^{2+}, Ni^{2+}, Pb^{2+}	63
W electrode	400–700	Zr^{2+}, Zr^{4+}	77

W and Pt electrodes	450	Ag^+, Cd^{2+}	62
Chronopotentiometry			
Solid metal electrodes	400–450	Ag^+, Bi^{3+}, Cd^{2+}, Co^{2+}, Cr^{2+}, Cr^{3+}, Cu^+, Li^+, Pb^{2+}, Pt^{2+}, Tl^+, U^{4+}, V_2O_5, Zn^{2+}	1, 2
Pt and W electrodes	450	Ag^+, Cd^{2+}	62
Pt electrode	450	$VOSO_4$	68
Pt electrode	600–800	Mo^{3+}	84
Pt electrode	720	Nb^{4+}	83
Pt electrode	485	PdO, Pd^{2+}	88
Pt electrode	500–675	Ta^{4+}	86
Pt and pyrolytic graphite electrodes	400–550	UO_2^{2+}, U^{4+}, U^{3+}, NpO_2^{2+}, NpO_2^+, Np^{4+}, Np^{3+}	92–94
Pt electrode	400–650	Pu^{3+}	95
Pt electrode	585	Na_2O_2	103
Pt electrode	450–500	CrO_4^{2-}	70–73
Pt, W, and graphite electrodes	450–800	Phosphates	104, 105
Pt electrode	—	SO_4^{2-}	106
Potentiometry	427–927 (50–50 mole %)	Mn^+/Mn, Mn^{2+}/Mn, Mn^{2+}/Mn^+	74
	500–675	Mo^{3+}/Mo, V^{2+}/V, Ta^{4+}/Ta	69
	500–860	Mo^{2+}, Mo^{3+}	85
	500	Nb^{3+}, Nb_3Cl_8	79, 80
	400–700	Nb^{2+}/Nb, Nb^{3+}/Nb, Nb^{3+}/Nb^{2+}	81
	450–500	Hf^{4+}/Hf, Zr^{2+}/Zr, Zr^{4+}/Zr	75
	450	W^{2+}/W	87
	450	Eu^{3+}/Eu, Eu^{3+}/Eu^{2+}, Sm^{3+}/Sm, Sm^{2+}, Yb^{3+}/Yb, Yb^{3+}/Yb^{2+}	90
	400–550	NpO_2^{2+}/NpO_2^+, NpO_2, NpO_2^+/NpO_2, Np^{4+}/Np^{3+}, Np^{3+}/Np, U^{4+}/U^{3+}, U^{3+}/U	91, 94
	450–600	La^{3+}/La, Y^{3+}/Y	98, 99
	400 or 420	S/S^{2-}, S/Se^{2-}, Te^{2+}/Te	100
	450	Ga^{3+}/Ga, In^{3+}/In^+, In^{++}/In	101
	450	Oxygen electrode	102

3.2. NaCl–KCl (50–50 mole %)

An important drawback of LiCl–KCl eutectic as a solvent is the hygroscopic nature of LiCl. In the case of molten equimolar NaCl–KCl both components are relatively nonhygroscopic; however, the liquidus temperature of this mixture is ~300°C higher than that of LiCl–KCl eutectic. The solvent can be prepared by mixing reagent-grade salts and then dried under vacuum at 400–500°C for several hours. The melts can be further purified by preelectrolysis[78] or by the same treatment as used for the LiCl–KCl eutectic.[57] Another method involves the use of highly purified NaCl and KCl (vacuum sublimation at 750°C).[83]

A standard potential series has been established primarily by Flengas and co-workers in this solvent over the temperature range 700–900°C.[108] Combes et al.[109] reinvestigated the redox couples Ag^+/Ag, Au^+/Au, Co^{2+}/Co, Cu^+/Cu, Ni^{2+}/Ni, Pb^{2+}/Pb, Pd^{2+}/Pd, Pt^{2+}/Pt, Sn^{2+}/Sn, and Zn^{2+}/Zn. The electroactive species were introduced by coulometric oxidation of the metal rather than by adding the anhydrous chlorides. The results are in substantial agreement with those obtained by Flengas and co-workers.

The reduction mechanism of Cr^{3+} has been studied by Cho and Kuroda.[110] The chronopotentiograms showed a two-step reduction apparently involving a one-electron step followed by a two-electron step. Only the first step was found to be a diffusion-controlled process. The diffusion coefficient of Cr^{3+} was reported.

The diffusion coefficient for Fe^{2+} was reported by Kornilov et al.[111] A well-defined one-step chronopotentiogram was obtained by adding $FeCl_2$ to the melt. Dubinin et al.[112] showed that Fe^{2+} was generated by anodic dissolution of iron in this solvent. Recently Smirnov et al.[113] have carried out emf measurements on the iron system in NaCl, NaCl–KCl, and KCl. They suggested the existence of Fe^+. The standard potentials of Fe^+/Fe, Fe^{2+}/Fe, and Fe^{2+}/Fe^+ couples were reported. However, the evidence for the presence of Fe^+ in NaCl–KCl is very meager.

Kazantsev and co-workers[114,115] studied the anodic polarization of nickel and manganese electrodes in the melt at different temperatures. The ionic species in the melt were suggested to be Ni^{2+}, Ni^+, and Mn^{2+}. The fraction of Ni^{2+} ions increased with increasing current density and with decreasing temperature. The claim for the presence of Ni^+ ions is not in agreement with the emf measurements on the Ni^{2+}/Ni couple[109] and the voltammetric results.[116]

Recently Hoff[117] has developed a dropping silver electrode. He studied the polarographic reduction of Cu^+ at this electrode and showed that the Ilkovic equation, corrected for spherical diffusion, is in good agreement with the observed currents. A maximum of the first kind was observed, but its influence on the limiting current and half-wave potential could be eliminated by the analysis of the current–time curves obtained on a single drop. The diffusion coefficient for Cu^+ is much higher than that reported previously.[118]

Sakakura[119] studied the electrodeposition of zirconium by chronopotentiometry. A well-defined one-step chronopotentiogram corresponding to $Zr^{2+} + 2e \rightleftharpoons Zr$ was obtained for $ZrCl_2$. No interpretation of the chronopotentiograms of $ZrCl_4$ was attempted. The diffusion coefficients for Zr^{2+} and Zr^{4+} were reported.

Sakawa and Kuroda[120] studied the reduction of NbF_7^{2-} by voltammetry, chronopotentiometry, and constant-current and controlled-potential coulometry at a molybdenum electrode. They concluded that the reduction proceeds in two steps. The first step is a reversible two-electron process forming a soluble reduced species and the second step is an irreversible process with metallic niobium as a product. Inman et al.[83] investigated the electrochemistry of niobium using a different approach. The niobium ions were prepared *in situ* by anodic dissolution of the metal.[78] Nb^{3+} was found to form at low current densities; however, no reduction wave could be detected. The use of high current densities (>50 mA/cm^2) produced Nb^{4+} ions which were electroactive. The chronopotentiometric results indicated that the reduction of Nb^{4+} is a one-electron reversible process producing a soluble product. Inman et al.[83] concluded that the reduction of Nb^{3+} occurs only by a secondary process following the deposition of alkali metals. Saeki and Suzuki[121] obtained evidence for a polynuclear ion (Nb_3^{8+}) in the melt by studying the equilibrium between niobium and its subchlorides; the results were similar to those obtained in LiCl–KCl eutectic.[79,80] The average valence of niobium after anodic dissolution was about 3.1. The existence of polynuclear niobium ions in other media is well known.[122] On the other hand, Pimenov and Baimakov[123] reported the diffusion coefficients for Nb^{3+} and Nb^{2+}. Nb^{3+} was produced by anodization and Nb^{2+} was formed by keeping Nb^{3+} in contact with niobium for 15–20 hr. Recently Pimenov[82] studied the anodic polarization of niobium; the n value, from Tafel slopes, was less than three; he concluded that Nb^{2+} and Nb^{3+} were present in the melt. More work is needed to resolve the contradictory results.

Inman et.al.[83] found that Mo^{3+} is significantly volatile in equimolar NaCl–KCl melt. However, the volatility is minimized in the NaCl–KCl

(20–80 mole %); therefore a study of the reduction of Mo^{3+} was carried out in this melt at 700°C. The results obtained in this work provided support for the earlier observation of Senderoff and Mellors[84] in LiCl–KCl eutectic, namely that the reduction of monomeric Mo^{3+} is preceded by a slow dissociation of a dimeric species.

Naryshkin et al.[124] studied the reduction of Ag^+, Cd^{2+}, and Pb^{2+} by linear sweep voltammetry. They found that all three electrode processes are reversible at a platinum electrode. An insoluble product (Ag metal) was formed in the reduction of Ag^+; soluble products (alloys) were formed for Cd^{2+} and Pb^{2+}. The peak currents were directly proportional to concentration. The diffusion coefficients were also reported. The value for Cd^{2+} agrees with that obtained by chronopotentiometry;[125] it is four times smaller than that for Pb^{2+}. Goto and Maeda[126] reexamined the voltammetric behavior of Ag^+, Cd^{2+}, and Pb^{2+} by using a tungsten electrode. Similar results were obtained. The polarograms for Cd^{2+} and Pb^{2+} were found to obey the Heyrovsky–Ilkovic equation with $n = 2$. For the reduction of Ag^+, the Kolthoff–Lingane equation was found to be valid. The wave heights were proportional to the concentration of the reducible species. Delimarskii and Kuzmovich[127] used a dropping bismuth electrode to study the polarographic reduction of Cd^{2+} and reported that the Heyrovsky–Ilkovic equation describes the Cd^{2+} reduction wave. However, the reported diffusion coefficients for Cd^{2+}, Pb^{2+}, and Tl^+ were three times larger than those obtained by other methods. They suggested that the discrepancy may be due to convection caused by the extreme mobility of the bismuth drop. The limiting currents were shown to be proportional to concentrations of reducible ions.

The chronopotentiometric studies of $HfCl_4$ were reported by Sakakura and Kirihara.[128] At mole fractions less than 10^{-3} the reduction proceeds in three steps at 700–750°C: $Hf^{4+} + e \rightleftharpoons Hf^{3+}$; $Hf^{3+} + e \rightleftharpoons Hf^{2+}$; $Hf^{2+} + 2e \rightleftharpoons Hf$. At 850–900°C the reduction occurs in two steps: $Hf^{4+} + 2e \rightleftharpoons Hf^{2+}$; $Hf^{2+} + 2e \rightleftharpoons Hf$. The diffusion coefficient of Hf^{4+} was estimated. Smirnov et al.[129] reported the presence of only Hf^{2+} and Hf^{4+} in equilibrium with Hf. The standard potentials of Hf^{2+}/Hf, Hf^{4+}/Hf^{2+}, and Hf^{4+}/Hf couples in molten NaCl, NaCl–KCl, and KCl were reported.

Hladik et al.[130] studied the oxidation of various metallic electrodes (platinum, gold, and silver) in molten and solid NaCl–KCl by linear sweep voltammetry. They also investigated the effects observed upon the addition of oxide to the system.

Val'tsev and Didora[131] extended the voltammetric studies to rare earth elements. Two-step reductions were observed for La^{3+}, Nd^{3+}, and

Pr^{3+}. They assigned the reduction steps to M^{3+} + e $\rightleftharpoons$ M^{2+} and M^{2+} + 2e $\rightleftharpoons$ M; however, the n values were poorly defined. The wave heights were not proportional to concentrations of the solute at high concentrations, because of maxima formation. The electrochemical behavior of thorium in molten NaCl, NaCl–KCl, and KCl was studied by Smirnov et al.[132] Their results indicated that thorium is in equilibrium with a mixture of Th^{2+} and Th^{4+}. The standard electrode potentials for Th^{2+}/Th and Th^{4+}/Th couples were reported. The equilibrium constant for Th^{4+} + Th $\rightleftharpoons$ 2Th^{2+} was obtained.

Smirnov and Podlesnyak[133] carried out emf measurements to study the solubility of sodium and potassium in the melt. The redox potentials of alkali metal solutions varied linearly with the concentrations of metals at mole fractions below 2×10^{-3}. At higher concentrations the redox potential decreased rapidly; this was attributed to the increase of activity coefficients. The results are claimed to be in good agreement with data obtained by other methods.

The concentration and temperature dependences of the equilibrium potentials of tin and its ions in this melt were measured by Karzhavin et al.[134] The average valence of tin was found to be ~1.6. This was taken as evidence that there exists monovalent as well as divalent tin in the melt. The standard potentials of Sn^{+}/Sn, Sn^{2+}/Sn, and Sn^{2+}/Sn^{+} and the equilibrium constant for the reaction Sn^{2+} + Sn $\rightleftharpoons$ 2Sn^{+} were reported.

The electroreduction of phosphates in this melt by voltammetry and chronopotentiometry was studied by Franks and Inman.[105] They reported that the reduction of PO$_4^{3-}$ is a reversible two-electron process. The final product of the reaction is phosphorus that is produced through subsequent chemical reactions involving soluble reduced species (PO$_2^{-}$). Two steps were observed in the reduction of P$_2$O$_7^{4-}$. The results indicated that both PO$_4^{3-}$ and P$_2$O$_7^{2-}$ are present in solution. PO$_4^{3-}$ and P$_2$O$_7^{2-}$ were also present in the melt upon addition of P$_3$O$_9^{3-}$ or P$_3$O$_{10}^{5-}$.

Wrench and Inman[106] reexamined the reduction of SO$_4^{2-}$ in solution. The sulfate ion is not directly reduced to lower-valent sulfur compounds. However, upon the addition of sodium metaphosphate, cathodic chronopotentiometric waves were observed which were not produced by the metaphosphate alone. This was attributed to the reduction of SO$_3$ produced by the reaction of SO$_4^{2-}$ + 2PO$_3^{-}$ $\rightarrow$ SO$_3$ + P$_2$O$_7^{4-}$. The final product of the reaction is sulfide that is produced by a subsequent chemical reaction involving SO$_3^{-}$.

The results obtained in this melt are summarized in Table II.

TABLE II. Results of Electroanalytical Studies in Molten Equimolar NaCl-KCl

Experimental technique	Temperature, °C	Solutes investigated	Ref.
Voltammetry			
Pt, W, and graphite electrodes	660–740	Ag^+, Cd^{2+}, Co^{2+}, Cu^+, Cu^{2+}, Fe^{2+}, Fe^{3+}, Mn^{2+}, Ni^{2+}, Pb^{2+}, S^{2-}, Sn^{2+}, Tl^+, Zn^{2+}, Zr^{2+}, Zr^{4+}	1, 118
W electrode	710	Ag^+, Cd^{2+}, Pb^{2+}	126
Dropping Bi electrode	700–800	Cd^{2+}, Pb^{2+}, Tl^+	127
Dropping Ag electrode	1000	Cu^+	117
Graphite electrode	650	Nb^{3+}, Nb^{4+}	78
Mo electrode	750	Nb^{5+}	120
Pt electrode	700	La^{3+}, Nd^{3+}, Pr^{3+}	131
Graphite and W electrodes	400–800	Phosphates	105
Linear sweep voltammetry			
Pt electrode	700–900	Ag^+, Cd^{2+}, Pb^{2+}	124
Ag, Au, and Pt electrodes	540–750	Ag, Au, Pt species	130
Chronopotentiometry			
Pt, W, and graphite electrodes	710–740	Ag^+, Cd^{2+}, Ni^{2+}, Pb^{2+}, UO_2^+, Zn^{2+}, Zr^{2+}, Zr^{4+}	118
Pt electrode	680–760	Cd^{2+}, Pb^{2+}, Zn^{2+}	125

Electroanalytical Chemistry in Molten Salts—A Review of Recent Developments

Pt electrode	700–800	Cr^{3+}	110
Mo electrode	691–909	Fe^{2+}	111
Pt electrode	760	Nb^{3+}, Nb^{4+}	83
Mo electrode	800	Nb^{2+}, Nb^{3+}	123
Mo electrode	750	Nb^{5+}	120
Pt electrode (20 mole % NaCl)	760	Mo^{3+}	83
W electrode	700–900	Zr^{2+}, Zr^{4+}	119
W electrode	700–900	Hf^{4+}	128
Au and Pt electrodes	750	SO_4^{2-}	106
Graphite and W electrodes	400–800	Phosphates	105
Potentiometry	727	$Ag^+, Au^+, Co^{2+}, Cu^+, Ni^{2+}, Pb^{2+}, Pd^{2+}, Pt^{2+}, Sn^{2+}, Zn^{2+}$	109
	700–900	Fe^+, Fe^{2+}	113
	675	Nb_3^{8+}	121
		Hf^{2+}, Hf^{4+}	129
	680–900	Th^{2+}, Th^{4+}	132
	690–900	Na, K	133
	700–850	Sn^+, Sn^{2+}	134
Coulometry	700–800	Fe^{2+}	112
	700–900	Ni^+, Ni^{2+}	114
	700–900	Mn^{2+}	115
	700–900	Nb^{2+}, Nb^{3+}	82

3.3. Chloroaluminates (AlCl$_3$–Alkali Chlorides)

Molten chloroaluminates have recently gained popularity as solvents for electrochemical and spectral studies. In considering the solvent properties, it is helpful to apply acid–base concepts. "Acids" are defined as Cl$^-$ acceptors and "bases" as Cl$^-$ donors. Tremillon and Letisse[12] have considered the equilibrium $2\text{AlCl}_4^- \rightleftharpoons \text{Al}_2\text{Cl}_7^- + \text{Cl}^-$ in AlCl$_3$–NaCl systems at 175°C. They determined the equilibrium constant in the composition region near 50–50 mole % AlCl$_3$–NaCl by potentiometric measurements. Torsi and Mamantov[13,14] extended the studies to a wide temperature range (175–400°C), to different melt-compositions (melt saturated with NaCl to that saturated with AlCl$_3$), and to other alkali chlorides.* According to Tremillon and Letisse, melts with $p\text{Cl}^- > 2.7$ are "acidic" and melts with $p\text{Cl}^- < 2.7$ are "basic." The acidity of the melt increases with increasing content of AlCl$_3$ and decreasing temperature. In general the low-valence states of metals are less acidic than the higher-valence states and therefore they are stabilized in strongly acidic solvents. Also, as pointed out by Corbett,[16] the stability of the low-oxidation-state cations is affected by the size of the anion that it is combined with. Acidic chloroaluminates provide a medium with a very low concentration of Cl$^-$ and a high concentration of big anions (AlCl$_4^-$ and Al$_2$Cl$_7^-$); thus they are good media for studying unusually low oxidation states. The existence of Cd$_2^{2+}$, Pb$^+$, Sn$_2^+$,[135] lower oxidation states of bismuth (Bi$^+$, Bi$_3^+$, Bi$_5^{3+}$, Bi$_5^{2+}$, and Bi$_8^{2+}$),[136] lower oxidation states of sulfur,[137] selenium,[137,138] tellurium,[139] and Hg$_3^{2+}$[18] in these melts is well established. In addition, the chloroaluminates have low liquidus temperatures (as low as 61°C).† The difficulties involved in the preparation and handling and high vapor pressure of highly acidic melts are the main disadvantages of chloroaluminates as solvents.

Two different methods have been used to prepare AlCl$_3$–alkali chloride mixtures. In one method the highest-purity aluminum metal and HCl are used to prepare AlCl$_3$ which is then mixed with purified alkali chlorides.[15] By this method clear melts, exhibiting very low voltammetric background currents,[24] are obtained. In the other method analytical-grade alkali chlorides are mixed with purified commercial AlCl$_3$; then the less active ions are removed using aluminum metal or by electrolysis.[19,140] The melts obtained are frequently light-brown, although the voltammetric background currents are still quite low.[19] The solvent reacts vigorously

---

* See Refs. 270 and 271 for related work.
† See Ref. 272.

with moisture to form insoluble Al_2O_3 and small amounts of soluble oxychlorides.[141]

The deposition potentials of metals in the $AlCl_3$–NaCl–KCl (66–20–14 mole %) melt[48,142,143] and in $NaAlCl_4$[144] have been studied by using current–voltage curves. The measurements were carried out using three-electrode systems; therefore the potentials may be compared to potentials obtained by equilibrium measurements. With both tungsten and molybdenum[142] similar values were obtained for deposition potentials from solutions containing the metals in the trivalent and in the hexavalent states. Marshall and Yntema[142] suggested that the lower oxidation states are in equilibrium with the metal in this melt. Recently Plambeck and co-workers[140,145] have measured the standard potentials for the metals of Groups IB, IIB, and IIIA in the $AlCl_3$–NaCl–KCl (66–20–14 mole %) melt.

Delimarskii et al.[146] used the voltammetric technique to measure the decomposition potentials of several compounds in $NaAlCl_4$. Two waves were observed for the reduction of Cd^{2+}, Co^{2+}, and Ni^{2+}. The possibility of the presence of lower oxidation states of these ions in the melt was suggested.

Fremont et al.[147] investigated the polarographic reduction of Co^{2+}, Ni^{2+}, and other ions at a dropping mercury electrode in the $AlCl_3$–NaCl–KCl eutectic (60–26–14 mole %). Only a single reduction step was observed for both Co^{2+} and Ni^{2+}. These results indicate that the Co(<II) and Ni(<II) species are not formed in this melt. Fremont et al. report half-wave potentials for the reduction of several metal ions. According to these workers the reduction of Ag^+, Cu^+, Pb^{2+}, and Sn^{2+} is reversible; the reduction of Co^{2+}, Fe^{2+}, and Ni^{2+} is irreversible.

Saito et al.[148] studied the polarographic behavior of Cd^{2+}, Pb^{2+}, and Zn^{2+} in molten $AlCl_3$–LiCl–KCl (59.2–21.1–19.7 mole %). The processes were found to be reversible and the plots of E versus $\log[(i_d - i)/i]$ were straight lines with slopes corresponding to $n = 2$. The diffusion currents for the reduction of Cd^{2+} and Zn^{2+} were found to be proportional to their concentrations. Francini et al.[149] employed linear sweep voltammetry to examine the electroreduction of Cd^{2+}, Cu^+, Pb^{2+}, and Sn^{2+} at a dropping mercury electrode in a slightly acidic $AlCl_3$–NaCl mixture (52–54 mole % $AlCl_3$). The reduction was found to proceed in one step and to be reversible up to sweep rates of 10 V/sec. The diffusion coefficients for the ions were determined (of the order of 10^{-6} cm^2/sec). A linear relationship between peak current and concentration was observed at concentrations as low as 10^{-4} M. The results do not support the presence of a low oxidation state of cadmium in the melt.

On the other hand, Munday and Corbett[135] carried out emf measurements (on the cell: Ag | Ag$^+$ in NaAlCl$_4$ | glass | M^{2+}, M$_m^{m(2-n)+}$ in NaAlCl$_4$ | Ta) to study the lower oxidation states of cadmium, lead, and tin in a slightly acidic AlCl$_3$–NaCl (51.3 mole % of AlCl$_3$) melt. The ions were added as M(AlCl$_4$)$_2$ and the reduced ions were produced coulometrically using a third electrode separated by a glass frit. The results showed that the products were Cd$_2^{2+}$, Pb$^+$, and Sn$_2^+$ (tentatively). Hames and Plambeck[140] repeated the emf measurements on the cadmium system in the more acidic AlCl$_3$–NaCl–KCl eutectic (66–20–14 mole %). They reported that Cd^{2+} is the only species produced by anodization of the metal and no stable emf values were obtained by reducing Cd^{2+} at platinum and tungsten electrodes. However, they observed two waves in the voltammetric reduction of Cd^{2+}. The first was small and ill defined and may correspond to the first step in the reduction Cd^{2+} → Cd$_2^{2+}$ → Cd; the second was well defined and corresponded to the reduction to the metal. Chronopotentiometric results were also obtained for CdCl$_2$ dissolved in the eutectic. Two well-defined steps were observed with a ratio of τ_2/τ_1 expected for the Cd^{2+} → Cd$_2^{2+}$ → Cd system. The existence of Cd$_2^{2+}$ in chloroaluminates is substantiated by spectral studies.[150]

The systems Pb^{2+}/Pb and Fe^{3+}/Fe^{2+} were studied by Delimarskii and Tumanova[151] in a very acidic melt: 2AlCl$_3$–NaCl. The processes were found to be reversible and a linear relationship was observed between the potential and log[$(i_d - i)/i$]. No subvalent lead ion was observed by emf measurements, in agreement with the polarographic results reported previously.[147,148] Recently Chovnyk and Myshalov[152] have studied the voltammetric reduction of Fe^{3+} to Fe^{2+} at a rotating platinum disk electrode in NaAlCl$_4$. Well-defined voltammograms were obtained with the limiting currents directly proportional to the concentration of Fe^{3+} in solution. The process was found to be reversible. The diffusion coefficient for Fe^{3+} was determined as 1.2×10^{-5} cm^2/sec.

Fung and Mamantov[19] studied the electrochemical oxidation of Ti^{2+} [added as Ti(AlCl$_4$)$_2$] in AlCl$_3$–NaCl (50–65 mole % AlCl$_3$) melts by linear sweep voltammetry, voltammetry, and chronopotentiometry at solid electrodes. Ti^{2+} is oxidized stepwise to Ti^{3+} and Ti^{4+}; both processes are electrochemically reversible at a platinum electrode. The lower oxidation states were found to be more stable at lower temperatures and in more acidic melts. The diffusion coefficient for Ti^{2+} was reported.

Gut[153] studied voltammetrically the reduction of Nb^{5+} and Ta^{5+} in the AlCl$_3$–NaCl–KCl (60–26–14 mole %) eutectic. The voltammograms for Nb^{5+} reduction showed two waves corresponding to Nb^{5+} → Nb^{4+}

→ Nb^{3+}. They were well separated in the less acidic melt. In the more acidic melt the two steps almost overlap. Gut interpreted the wave as being due to a one-step two-electron reduction. However, a nonlinear plot for E versus $\log[(i_d - i)/i]$ was obtained. The reduction of Ta^{5+} was found to be a two-electron process in two different melt compositions. The $E_{1/2}$ values for both cases shift to less cathodic values with increasing acidity of the melt.

Voltammetric studies of Hg^{2+} in molten $AlCl_3$–NaCl–KCl (50–25–25 mole %) and of a mixture of Cu^{2+} and Hg^{2+} in molten $AlCl_3$–NaCl–KCl (60–26–14 mole %) eutectic were made by Gut.[153] For the case of Cu^{2+} a two-step reduction was observed voltammetrically. The voltammogram for the reduction of Hg^{2+} in the less acidic melt showed two waves corresponding to $Hg^{2+} \to Hg_2^{2+} \to Hg$ with an ill-defined first wave. In acidic melts the two reduction steps were well defined, and an additional ill-defined wave was shown in the voltammogram; however, no attempt was made to explain this additional wave for the reduction of Hg_2^{2+} to Hg. The polarogram of Hg_2^{2+} in $AlCl_3$–NaCl–KCl eutectic[147] (60–26–14 mole %) showed only one well-defined wave ($Hg_2^{2+} \to Hg$). Hames and Plambeck[140] investigated the electrochemistry of mercury in the $AlCl_3$–NaCl–KCl melt (66–20–14 mole %) by emf measurements, voltammetry, and chronopotentiometry. The results of voltammetric studies showed that Hg^{2+} is reduced in a conventional manner to Hg_2^{2+} and Hg; this was supported by emf measurements. The chronopotentiogram for the reduction of Hg^{2+}, however, showed a long sloping region between the plateaus of the first and second steps; the expected ratio of τ_2/τ_1 was not obtained. Recently we have reinvestigated the system in $AlCl_3$–NaCl melts.[18] In the melt saturated with NaCl (basic) only two voltammetric waves of equal height were obtained. In more acidic melts and at low temperatures the voltammogram indicated an intermediate oxidation state, Hg_3^{2+}, between Hg_2^{2+} and Hg. The new ion has been well characterized by absorption spectra and stoichiometric measurements as well as by electrochemistry. Raman spectra[18] of solid $Hg_3(AlCl_4)_2$ and single-crystal X-ray diffraction data[154] have also been obtained. The equilibrium constants for $Hg^{2+} + Hg_3^{2+} \rightleftharpoons 2Hg_2^{2+}$, $Hg_2^{2+} + Hg \rightleftharpoons Hg_3^{2+}$, and $Hg^{2+} + Hg \rightleftharpoons Hg_2^{2+}$ in the melts were determined.

Delimarskii et al.[155] examined the anodic dissolution of aluminum metal in molten $2AlCl_3$–NaCl. The anodic current efficiency was found to be more than 100% at low current densities. This result was explained by the formation of Al^+. During the polarization of the aluminum anode a layer of dark powder on its surface was observed. They suggested that the powder is finely dispersed metallic aluminum formed by the disproportiona-

TABLE III. Results of Electroanalytical Studies in AlCl$_3$–Alkali Chlorides

Solvent composition	Temperature, °C	Experimental technique	Solutes investigated	Ref.
AlCl$_3$–NaCl–KCl melt (66–20–14 mole %)	156 or 218	Current-voltage curve to determine the deposition potential (Pt electrode)	Ag$^+$, Al^{3+}, As^{3+}, Bi^{3+}, Cd^{2+}, Co^{2+}, Cr^{2+}, Cu$^+$, Fe^{2+}, Ga^{3+}, H$_2$O, Hg$^+$, Mo^{3+}, Mn^{2+}, Ni^{2+}, Sb^{3+}, Sn^{2+}, Te^{2+}, W^{3+}, Zn^{2+}	48, 142, 143
NaAlCl$_4$	500 and 700	Current-voltage curve to determine the deposition potential (Pt electrode)	Ag$^+$, Bi^{3+}, Cd^{2+}, Co^{2+}, Cu$^+$, Tl$^+$, Mn^{2+}, Ni^{2+}, Pb^{2+}, Sb^{3+}, Sn^{2+}, Zn^{2+}	144
AlCl$_3$–NaCl (~51.3 mole % AlCl$_3$)	277	Potentiometry	Cd, Pb, and Sn species	135
AlCl$_3$–NaCl–KCl melt (66–20–14 mole %)	135 or 150	Potentiometry	Ag, Au, Cd, Cu, Ga, Hg, In, Tl, Zn species	140, 145
AlCl$_3$–NaCl–KCl eutectic (60–26–14 mole %)	140	Polarography (dropping Hg electrode)	Ag$^+$, Co^{2+}, Cu$^+$, Cu^{2+}, Fe^{2+}, Fe^{3+}, Hg$_2^{2+}$, Ni^{2+}, Pb^{2+}, Sn^{2+}	147
AlCl$_3$–LiCl–KCl (59.2–21.1–19.7 mole %)	160	Polarography (dropping Hg electrode)	Cd^{2+}, Pb^{2+}, Zn^{2+}	148
NaAlCl$_4$	300	Voltammetry (Pt electrode)	Ag$^+$, Al^{3+}, Bi^{3+}, Cd^{2+}, Co^{2+}, Cu$^+$, Fe^{3+}, Hg^{2+}, Mn^{2+}, Ni^{2+}, Pb^{2+}, Sb^{3+}, Sn^{2+}, Zn^{2+}	146
2AlCl$_3$–NaCl	160	Voltammetry (Pt electrode)	Fe^{2+}, Fe^{3+}, Pb^{2+}, H$^+$	151
NaAlCl$_4$	200–300	Voltammetry (rotating Pt electrode)	Fe^{3+}	152

Melt	Temperature (°C)	Technique	Species	Ref.
AlCl$_3$–NaCl–KCl (60–26–14 mole % and 50–25–25 mole %)	120 or 200	Voltammetry (Pt electrode)	Cu^{2+}, Hg^{2+}, Nb^{5+}, Ta^{5+}	153
AlCl$_3$–NaCl (52–54 mole % AlCl$_3$)	150–230	Linear sweep voltammetry (dropping Hg electrode)	Cd^{2+}, Cu$^+$, Pb^{2+}, Sn^{2+}	149
AlCl$_3$–NaCl–KCl (50–36–14 mole %)	150	Linear sweep voltammetry (W electrode)	Aromatic hydrocarbons	157
AlCl$_3$–NaCl–KCl melt (66–20–14 mole %)	136 or 150	Voltammetry and chronopotentiometry (W electrode)	Cd, Ga, Hg, Zn species	140, 145
2AlCl$_3$–NaCl	160	Chronopotentiometry	Pb^{2+}, Sn^{2+}	158
AlCl$_3$–NaCl (melt saturated with NaCl to saturated with AlCl$_3$)	125–250	Voltammetry, linear sweep voltammetry, and chronopotentiometry (Au, Pt, and pyrolytic graphite electrodes)	Hg^{2+}, Hg$_2^{2+}$, Hg$_3^{2+}$	18
AlCl$_3$–NaCl (50–65 mole % AlCl$_3$)	150–330	Voltammetry, linear sweep voltammetry, and chronopotentiometry (Au and Pt electrodes)	Ti^{2+}, Ti^{3+}, Ti^{4+}	19
AlCl$_3$–NaCl (52–63 mole % AlCl$_3$)	125–300	Voltammetry, linear sweep voltammetry, and chronopotentiometry (Au, Pt, and pyrolytic graphite electrodes)	Bi^{3+}, Bi$^+$	24
2AlCl$_3$–NaCl	160	Coulometry	Al$^+$	155
AlCl$_3$–NaCl (50–50 mole %)	175	Cyclic voltammetry (W electrode)	Aromatic amines	157a
AlCl$_3$–NaCl (63–37 to 50–50 mole %)	140–165	Linear sweep voltammetry (Pt and W electrodes)	Sulfur heterocycles	157b

tion of Al$^+$. In studies of the reaction of aluminum with molten chloroaluminates the present authors[156] have also obtained evidence for the existence of an unstable subvalent aluminum ion in these melts.*

The electrochemical reduction of Bi^{3+} in AlCl$_3$–NaCl melts (52–63 mole % AlCl$_3$) has been studied by voltammetry, linear sweep voltammetry, and chronopotentiometry at solid electrodes. Torsi and Mamantov[24] found that Bi^{3+} is reduced reversibly to Bi$^+$; further reduction of Bi$^+$ to Bi is complex and involves intermediate oxidation states which were not characterized electrochemically. Strong adsorption of Bi$^+$ at gold and platinum electrodes was observed. As in other systems,[18,19] the lower oxidation states were found to be more stable at lower temperatures and in more acidic melts.

Fleischmann and Pletcher[157] have reported the cyclic voltammetric data for a series of aromatic hydrocarbons in the AlCl$_3$–NaCl–KCl (50–36–14 mole %) melt. Large polynuclear hydrocarbons showed several electron transfers, while simple hydrocarbons showed a single wave. The electrode reactions were found to be almost reversible. Jones *et al.*[157a] who obtained firm voltammetric evidence for the stabilization of radical cations of triphenylamine, diphenylamine, and N,N-dimethylaniline in chloroaluminate melts, were, however, unable to reproduce the results of Fleischmann and Pletcher. Quite recently Fung *et al.*[157b] generated voltammetrically the radical cation and the dication of tetrathioethylene; they also studied the electrochemical oxidation of thianthrene in acidic molten chloroaluminates. It was found that the dication of tetrathioethylene is considerably more stable in molten chloroaluminates at 150°C than in dry acetonitrile at room temperature.

Table III summarizes the results of electrochemical studies in chloroaluminate melts.

3.4. Other Chlorides

Molten MgCl$_2$–alkali chlorides have been used as solvents for electrochemical studies by Gaur and co-workers. Their earlier studies have been summarized in previous reviews.[1,4,6] Recently Gaur and Jindal[159] measured the standard electrode potentials of transition and post-transition metals in MgCl$_2$–NaCl–KCl eutectic (50–30–20 mole %) and MgCl$_2$–KCl eutectic (32.5–67.5 mole %). The potentials of M^{n+}/M (see Table IV) in these solvents and those in molten LiCl–KCl eutectic and equimolar

---

* Also see Ref. 272 for related work.

NaCl–KCl differed from the emf's calculated for the formation cells involving the solutes; the difference was attributed to the polarizing power of the cations of the solvents. Gaur and Jindal[160] also examined the voltammetric behavior of the metal ions at platinum and tungsten electrodes in the molten $MgCl_2$–KCl eutectic. They again concluded that (a) in the absence of any interaction with the electrode material the activities of the deposited metals are invariable and the voltammetric wave is described by the Kolthoff–Lingane equation; this applies to the reduction of metal ions at a tungsten electrode (no alloy formation) and to the metal deposited as a solid at the working temperature on platinum; (b) in the case of alloy formation the Heyrovsky–Ilkovic equation describes the wave. The reduction of Cd^{2+}, Sn^{2+}, and Tl^+ at a platinum electrode followed the latter behavior. Gaur and Jindal also studied the reduction of Zn^{2+} at a platinum electrode in the $MgCl_2$–NaCl–KCl eutectic at 410 and 475°C;[160] at the lower temperature the metal deposits as a solid, while at the higher temperature it is found as a liquid. They found that the Kolthoff–Lingane equation is valid at 410°C but the Heyrovsky–Ilkovic equation is valid at 475°C for the above process. A similar conclusion was reached for the voltammetric reduction of Pb^{2+} in the $MgCl_2$–KCl eutectic at platinum and tungsten electrodes.[161]

Grebenik and Grachev[162] studied the electrochemical behavior of Fe^{2+} and Fe^{3+} in NaCl–$CaCl_2$ (58–42 mole %) melt. The voltammograms were well defined and obeyed the Heyrovsky–Ilkovic equation. The reduction of Fe^{3+} and Fe^{2+} was found to proceed directly to metal. On the other hand, Pokrovskii et al.[163] measured the equilibrium potentials of iron and its ions as a function of temperature and concentration in molten CsCl. They found that the valence of iron was less than two and not an integral value as was found in molten NaCl, KCl, and NaCl–KCl.[113] The presence of Fe^+ and Fe^{2+} in equilibrium with iron was suggested. The temperature dependence of apparent standard potentials of Fe^+/Fe and Fe^{2+}/Fe was reported. Using the same technique Kochergin et al.[164] also reported the presence of Fe^+ in the molten $ZnCl_2$–NaCl eutectic. The apparent standard potentials of Fe^+/Fe, Fe^{2+}/Fe^+, and Fe^{2+}/Fe and the equilibrium constant for the reaction $Fe^{2+} + Fe \rightleftharpoons 2Fe^+$ were reported. Further work is needed to characterize the Fe^+ ion.

The influence of the solvent cations on metal–ion equilibria was studied by Smirnov and Ryzhik.[165] They measured the equilibrium potentials of molybdenum and its ions in molten LiCl, KCl, and CsCl. The results showed that the increase of the radius of the alkali cation causes a shift of the potentials Mo^{2+}/Mo, Mo^{3+}/Mo^{2+}, and Mo^{3+}/Mo toward more negative values. Simultaneously the equilibrium of the reaction $2Mo^{3+} + Mo \rightleftharpoons 3Mo^{2+}$ shifts to the left. A linear relationship between potential and the reciprocal of the radius of the solvent cation was obtained. The existence of Mo^{2+} was supported by the anodic dissolution of molybdenum in these melts.[166] The average valence of molybdenum in the melts was found to increase from 2.12 to 3.00 with current density.

Smirnov et al.[167–171] carried out equilibrium potential measurements on the systems $Ti/(Ti^{2+}, Ti^{3+}, Ti^{4+})$, $Zr/(Zr^{2+}, Zr^{4+})$, $Hf/(Hf^{2+}, Hf^{4+})$, and $Th/(Th^{2+}, Th^{4+})$ in alkali (LiCl, NaCl, KCl, CsCl) and alkaline earth ($MgCl_2$, $CaCl_2$, $BaCl_2$, $SrCl_2$) chlorides. The results supported the earlier conclusion, namely that the lower oxidation state is more stable at higher temperatures and in solvents containing the more polarizing cation. For the extreme case of hafnium, only Hf^{2+} is present in molten LiCl, whereas in CsCl both Hf^{2+} and Hf^{4+} are present. Increase in the polarizing power of the cation is equivalent to increasing the acidity of the solvent (the opposite behavior has been observed[14] in molten chloroaluminates). The entropy for the reaction of $M^{n+} + M \rightleftharpoons 2M^{m+}$ (where $n > m$) is generally positive.[172] The increase in temperature shifts the equilibrium to the right side. It should be noted that in chloroaluminates at moderate temperatures the change in acid–base properties of the solvent with temperature is more important than the above temperature effect. Thus opposite behavior was observed[18,19] in chloroaluminates.

Molten $NaCl-ZnCl_2-PbCl_2$ (47.4–46.2–6.4 mole %)[173] and $ZnCl_2$–InCl[174] (47.5–52.5 mole %) have been used as solvents by Pogorelyi and co-workers to measure the standard electrode potentials of some transition and post-transition metals in these melts. Kolotii et al.[175] have studied the behavior of the Ag^+/Ag couple in $SrCl_2$–NaCl–KCl eutectic by emf measurements.

It is well known that many metals will dissolve to some extent in their molten chloride salts. Chronopotentiometry, emf measurements, and voltammetry have been employed to study such systems in an effort to determine the form of the dissolved metal in the melt. The earlier studies have been discussed in previous reviews.[176,177] Vetyukov and Fransis[178] reexamined $Pb-PbCl_2$ and $Zn-ZnCl_2$ systems. They found that the metals interact with their chlorides by forming a subchloride (Pb_2Cl_2 and Zn_2Cl_2). The results agree with those reported by Van Norman et al.[179] The system $Cd-CdCl_2$ has been reinvestigated by Crawford and Tomlinson[180] by emf measurements. The strong positive departures from ideality in this system were attributed to the formation of a subchloride Cd_2Cl_2 as indicated by previous results.[177] Komura et al.[181] measured the emf of magnesium concentration cells in molten $MgCl_2$. Their results indicated that Mg_2^{2+} is present in the melt, supporting the results obtained by anodic chronopotentiometry.[177] Corbett et al.[182] extended the emf measurements in the $Bi-BiCl_3$ system to higher metal concentrations and to lower temperatures. They suggested that the lower-oxidation-state species are Bi^+ and Bi_3^+ (rather than Bi_4^{4+} as suggested by Topol et al.[177]). Bronstein[183] reported emf measurements on the systems $Ce-CeCl_3$, $Pr-PrCl_3$, and $Nd-NdCl_3$ using a metal cell to contain the rare earth metal–chloride solutions. The results showed the presence of a subvalent ion M^{2+} from the reaction $2M^{3+} + M \rightleftharpoons 3M^{2+}$ in all three cases. The concentrations of metals in the melt at saturation, determined by this technique, are in agreement with those obtained by other methods where the solutions were not in contact with ceramic materials.

Ohmae and Kuroda[184] studied voltammetrically the electrochemical properties of $BeCl_2$ in LiCl–NaCl–KCl (43–33–24 mole %). A single wave was obtained corresponding to the reduction $Be^{2+} + 2e \rightleftharpoons Be$. However, when appreciable quantities of beryllium metal were present an additional wave was observed. The formation of Be^+ by the reduction of Be^{2+} with the beryllium was suggested.

The diffusion coefficient for dissolved chlorine was determined chronopotentiometrically in molten NaCl, KCl, and CsCl.[185] The D values were $\sim 1-6 \times 10^{-4}$ cm²/sec; the diffusion coefficient increased going from CsCl to NaCl. The unusually high values were attributed to the formation of Cl_3^-. The ionic transport was believed to proceed in chain fashion with the formation and decomposition of the Cl_3^- complex.

The voltammetric determination of moisture in molten $NaCl-CaCl_2$ (58–42 mole %) was investigated by Grachev and Grebenik.[186] A cathodic wave was obtained in a nonhydrated melt. The height of this wave increased when superheated steam was passed through the melt; it was not affected by adding NaOH to the melt, so that the reduction was not due to the hydronium ion. The wave height increased greatly initially after treatment with steam and then leveled off. Thus the results indicated that the moisture content of the melt tends to an equilibrium value. The electrochemical process $H_2O + e \rightarrow OH^- + \frac{1}{2}H_2$ was found to be irreversible. Grachev and Grebenik also determined O^{2-} and CO_3^{2-} in this melt.[187] An anodic wave described by the Heyrovsky–Ilkovic equation was obtained for CaO at a platinum electrode. The wave height was found to be directly proportional to CaO content at low concentrations. A cathodic wave was found for CO_3^{2-}. The results indicated that this technique can be applied to the determination of CO_3^{2-} in the melt.

Texier et al.[188,189] measured the standard electrode potentials of

TABLE IV. Results of Electroanalytical Studies in Molten Chlorides Other Than Those Given in Tables I–III

Solvent composition	Temperature, °C	Experimental technique	Solutes investigated	Ref.
CsCl	750–1000	Potentiometry	Fe^+, Fe^{2+}	163
Individual alkali chlorides	750–1000	Potentiometry	Mo^{2+}, Mo^{3+}	165
Individual alkali and alkaline earth chlorides	750–1000	Potentiometry	Hf^{2+}, Hf^{4+}, Th^{2+}, Th^{4+}, Ti^{2+}, Ti^{3+}, Ti^{4+}, Zr^{2+}, Zr^{4+}	167–171
Individual alkali chlorides (LiCl, KCl, CsCl)	750–1000	Coulometry	Mo^{2+}, Mo^{3+}	166
Individual alkali chlorides (NaCl, KCl, CsCl)	800–900	Chronopotentiometry (C electrode)	Cl_2	185
LiCl–NaCl–KCl eutectic (43–33–24 mole %)	460	Voltammetry (Pt electrode)	Be^+, Be^{2+}	184
MgCl$_2$–KCl eutectic (32.5–67.5 mole %)	475	Potentiometry and voltammetry (Pt and W electrodes)	Ag^+, Au^+, Cd^{2+}, Co^{2+}, Cr^{2+}, Cu^+, Fe^{2+}, Mn^{2+}, Ni^{2+}, Pd^{2+}, Pt^{2+}, Sb^{3+}, Sn^{2+}, Tl^+, U^{3+}, Zn^{2+}	1, 2, 159–161
MgCl$_2$–NaCl–KCl eutectic (50–30–20 mole %)	475			
NaCl–CaCl$_2$ (58 mole % NaCl)	600	Voltammetry (Pt electrode)	CaO, CO_3^{2-}, Fe^{2+}, Fe^{3+}, H_2O	162, 186, 187
NaCl–ZnCl$_2$ eutectic	400–500	Potentiometry	Fe^+, Fe^{2+}, Fe^{3+}	164

NaCl–ZnCl$_2$–PbCl$_2$ (47.4–46.2–6.4 mole %)	360–450	Potentiometry	Ag, Bi, Cu, Tl species	173
ZnCl$_2$–InCl (47.5 mole % ZnCl$_2$)	200–250	Potentiometry	Cu, Fe, Ni, Pb, Sn species	174
NaCl–KCl–SrCl$_2$ eutectic	600–650	Potentiometry	Ag$_2^+$, Ag$^+$	175
BiCl$_3$	240	Potentiometry	Bi$^+$, Bi$_3^+$	177, 182
CdCl$_2$	600–700	Potentiometry	Cd$_2^{2+}$, Cd^{2+}	177, 180
MgCl$_2$	750	Spectrophotometry and chronopotentiometry (Ta electrode)	Mg$_2^{2+}$, Mg^{2+}	177, 181
PbCl$_2$	518–600	Spectrophotometry and chronopotentiometry (Pt electrode)	Pb$_2^{2+}$	178, 179
ZnCl$_2$	500–600	Potentiometry and chronopotentiometry (Pt electrode)	Zn$_2^{2+}$	178, 179
CeCl$_3$	853	Potentiometry	Ce^{2+}, Ce^{3+}	183
NdCl$_3$	800	Potentiometry	Nd^{2+}, Nd^{3+}	183
PrCl$_3$	855	Potentiometry	Pr^{2+}, Pr^{3+}	183
SbCl$_3$	99	Potentiometry	Cu$^+$, Fe^{3+}, Hg^{2+}, Sb^{3+}, Sb^{5+}	188, 189
SbCl$_3$	100	Voltammetry (rotating C electrode)	Aromatic amines	190

$FeCl_4^-/FeCl_2(s)$, Sb^{3+}/Sb, Sb^{5+}/Sb^{3+}, Hg^{2+}/Hg, and Cu^+/Cu couples in molten $SbCl_3$. The potentials were found to be linearly dependent on pCl^-. The reduction of Fe^{3+} to $FeCl_2(s)$ was also studied by voltammetry. Molten $SbCl_3$ has also been used to study the electrochemical oxidation of aromatic amines;[190] organic cation radicals are apparently more stable in this melt than in organic solvents. Similar results have been obtained in molten chloroaluminates.[157a] The oxidation of amines proceeds by a one-electron reversible step; they are harder to oxidize in this solvent than in organic non-aqueous solvents. This was ascribed to stronger solvation in $SbCl_3$ and also to complexation with $SbCl_2^+$. Two types of electrochemical behavior were found for amines: For some amines the half-wave potential increases with increasing pCl^-, while for others it s independent of pCl^-.

The results are summarized in Table IV.

3.5. Other Halides

The electrochemical studies in molten fluorides have been recently reviewed.[191] Studies reported since that time include the use of a LaF_3 conducting membrane for a Ni^{2+}/Ni reference electrode,[50,192] studies of the titanium system in molten LiF–NaF–KF (46.5–11.5–42.0 mole %), $LiF-BeF_2-ZrF_4$ (65.6–29.4–5.0 mole %), and $NaBF_4$[192] as well as of uranium and thorium systems in LiF–NaF–KF.[192]

The results of electroanalytical studies in molten fluorides are summarized in Table V.

Little work has been reported for molten bromide and iodide systems.

3.6. Nitrates

Molten nitrates have been widely used as solvents for electrochemical studies. Advantages of low liquidus temperatures and relatively non-hygroscopic nature of molten nitrates have outweighed the disadvantages of these melts, i.e., limited potential span and oxidizing properties. A fairly pure melt can be obtained by simply melting the dried recrystalized salts under vacuum and continuing the evacuation for several hours.

A comprehensive study of "oxide chemistry" in molten $NaNO_3$–KNO_3 eutectic was made recently by Zambonin and Jordan.[211] In this study the results were critically dependent on the moisture in the melt and the container material. Zambonin and Jordan used a rotating electrode in a specially designed cell made from platinum; the moisture level was decreased to less than 10^{-6} m to avoid the side reactions of oxide ions with

TABLE V. Results of Electroanalytical Studies in Molten Fluorides

Solvent composition	Temperature, °C	Experimental technique	Solutes investigated	Ref.
LiF–NaF–KF eutectic (46.5–11.5–42.0 mole %)	500	Potentiometry	Ni^{2+}/Ni	47
	500	Potentiometry	Fe^{2+}/Fe, Fe^{3+}/Fe^{2+}	49
	502	Potentiometry	Ni^{2+}/Ni	192
	470–545	Voltammetry (Pt electrode)	Fe^{2+}	193
	498	Voltammetry (Pt electrode)	Zr^{4+}	194
	500–600	Voltammetry (pyrolytic graphite electrode)	Ni^{2+}	195
	500	Linear sweep voltammetry (Ag, Pt, W, and pyrolytic graphite electrodes)	Fe^{2+}, Ni^{2+}	196
	500	Linear sweep voltammetry (Ni, Pt, and W electrodes)	Th^{4+}	192
	500	Linear sweep voltammetry, chronopotentiometry, chronoamperometry, and voltammetry (Pt and pyrolytic graphite electrodes)	Ti^{4+}, U^{4+}	192
	500	Linear sweep voltammetry and chronopotentiometry (pyrolytic graphite electrodes)	Fe^{2+}	197
	750	Chronopotentiometry (Ta and Pt electrodes)	Ta^{5+}	198
	750 or 770	Chronopotentiometry (graphite electrode)	Ta^{5+}, Nb^{5+}	199
	500–750	Chronopotentiometry (Ta and Pt electrodes)	Zr^{4+}	200

TABLE V (*continued*)

Solvent composition	Temperature, °C	Experimental technique	Solutes investigated	Ref.
LiF–NaF–KF eutectic	650–800	Chronopotentiometry (Pt electrode)	Nb^{5+}	201
	600–800	Chronopotentiometry (Pt electrode)	Mo^{3+} (Mo^{3+} produced by Mo^{6+} and Mo)	202
	600		W species (oxidation number not certain; species produced by the reaction of W^{6+} and W)	
NaF–KF eutectic (40–60 mole %)	850	Potentiometry	Ag^+/Ag, Co^{2+}/Co, Cr^{3+}/Cr, Cu^+/Cu, Fe^{3+}/Fe, Mn^{2+}/Mn, Ni^{2+}/Ni	203
LiF–BeF$_2$–ZrF$_4$ (65.6–29.4–5.0 mole %)	500	Potentiometry	Ni^{2+}/Ni	47
	500	Potentiometry	Be^{2+}/Be, Cr^{3+}/Cr^{2+}, Cr^{2+}/Cr, Fe^{3+}/Fe^{2+}, Fe^{2+}/Fe, Zr^{4+}/Zr	49
	500	Potentiometry	U^{4+}/U^{3+}	204
	505	Potentiometry	Fe^{2+}/Fe, Ni^{2+}/Ni	192
	500	Linear sweep voltammetry (pyrolytic graphite and glassy carbon electrodes)	Cr^{2+}	205
	500	Linear sweep voltammetry and chronopotentiometry (pyrolytic graphite, Pt, Pt–10% Rh electrodes)	U^{4+}	38

Medium	Temp (°C)	Technique	Species	Ref.
LiF–BeF$_2$–ZrF$_4$ (65.6–29.4–5.0 mole %)	404–500	Linear sweep voltammetry and chronopotentiometry (Pt, Pt–10% Rh, and pyrolytic graphite electrodes)	U^{4+}	206
	480–620	Linear sweep voltammetry, chronopotentiometry, and chronoamperometry (Pt, Mo, Ta, W, and graphite electrodes)	U^{4+}	51
	500	Linear sweep voltammetry, chronopotentiometry, and voltammetry (glassy carbon electrode)	Ti^{3+}	192
LiF–BeF$_2$				
(67–33 mole %)	600	Potentiometry	Ni^{2+}/Ni	50
	507–707	Potentiometry	NiO/Ni	207
(66–34 mole %)	500	Linear sweep voltammetry (pyrolytic graphite electrode)	Fe^{2+}, Ni^{2+}	196
	507	Linear sweep voltammetry and chronopotentiometry (pyrolytic graphite electrode)	Fe^{2+}	197
(30–90 mole % BeF$_2$)	500–900	Potentiometry	$Be^{2+}/Be, HF/H_2$	208
(27–70 mole % BeF$_2$)	500 or 610	Potentiometry	Be^{2+}/Be	209
(12–58 mole % BeF$_2$)	455–545	Potentiometry	Be^{2+}/Be	210
NaBF$_4$	420	Potentiometry	Fe^{2+}/Fe	192
	420	Linear sweep voltammetry, chronopotentiometry, and voltammetry (Pt electrode)	Ti^{4+}	192

silica and H_2O. Oxide ions added to the melt as Na_2O were almost completely converted to peroxide and superoxide ions in accordance with the equations $NO_3^- + O^{2-} \rightleftharpoons O_2^{2-} + NO_2^-$ and $O_2^{2-} + 2NO_3^- \rightleftharpoons 2O_2^- + 2NO_2^-$. The presence of O_2^{2-}, O_2^-, and NO_2^- was deduced from the comparison of voltammetric results with the voltammograms obtained by dissolving Na_2O_2, KO_2, and $NaNO_2$, respectively, in the pure melt. The peroxide ions could be converted almost completely to superoxide ions by bubbling oxygen through the melt: $O_2^{2-} + O_2 \rightarrow 2O_2^-$. In view of the work by Zambonin and Jordan, previous and some recent[212-214] results pertaining to the behavior of oxide ions in molten nitrates are apparently incorrect.

Zambonin[215] carried on further studies in pure molten $NaNO_3$–KNO_3 eutectic and in the melt containing species produced *in situ* by massive electrolysis. He concluded that the electrochemical reduction of NO_3^- at a platinum electrode can be represented by

$$NO_3^- + 2e \rightleftharpoons NO_2^- + O^{2-}$$

followed by

$$O^{2-} + 2Na^+ \rightleftharpoons Na_2O \quad \text{(solid)}$$
$$O^{2-} + NO_3^- \rightleftharpoons NO_2^- + O_2^{2-} \quad \text{(fast)}$$
$$O_2^{2-} + 2NO_3^- \rightleftharpoons 2NO_2^- + 2O_2^- \quad \text{(slow)}$$

The solid Na_2O formed on the electrode surface can be oxidized according to

$$2Na_2O(\text{solid}) \rightleftharpoons 4Na^+ + O_2^- + 3e, \quad O_2^- \rightleftharpoons O_2 + e$$

He suggested that in previous investigations[1] which conflict with this model the side reactions of the oxide ions with H_2O and silica were studied.

The solubility and detection of water in molten $NaNO_3$–KNO_3 eutectic were studied by Zambonin and co-workers[216] using a rotating platinum electrode. Zambonin[217] also reported the voltammetric study of the interaction between O_2^- and H_2O ($4O_2^- + 2H_2O \rightleftharpoons 4OH^- + 3O_2$) in the same solvent. The possibility of determining the four species in the above equilibrium was demonstrated. The presence of NO_2^+ in the melt,[218] formed by self-dissociation $NO_3^- \rightleftharpoons NO_2^+ + O^{2-}$ which cannot be detected by direct measurements because of its extremely low concentration, was also shown indirectly by Zambonin by measuring the concentration of OH^- obtained from O^{2-}.

The nature of the oxygen electrode (O_2/O_n^{m-}) in molten $NaNO_3$–KNO_3 eutectic at platinum and gold electrodes has been investigated by potentiometry and voltammetry by Zambonin.[219] In complete absence of water the oxygen electrode behaves as a superoxide electrode ($O_2 + e$

$\rightleftharpoons O_2^-$), while in the presence of an excess of water the oxygen electrode is actually a hydroxide electrode ($O_2 + 2H_2O + 4e \rightleftharpoons 4OH^-$). Zambonin suggested that the results for the oxygen electrode obtained by adding metal oxide to molten KNO_3,[220] LiCl–KCl eutectic,[102] and Li_2SO_4–K_2SO_4 eutectic[221] ($n = 1$ from the Nernst slope) may also correspond to the O_2/O_2^- couple.

The question of the nature of the oxygen electrode in molten nitrates has been quite recently reopened by Temple and co-workers,[221a] who state that potentiometric results at platinized electrodes are in agreement with the reaction $\frac{1}{2}O_2 + 2e \rightleftharpoons O^{2-}$. These results imply a greater stability for the oxide ions in molten nitrates than that arising from the studies of Zambonin and Jordan.[211] It appears that further work will be needed to resolve this controversy. One reason[221a] for the discrepancy between potentiometric and voltammetric (chronopotentiometric) results in nitrates may be the different electrochemical behavior of platinized (used for potentiometric work) and bright (used for voltammetric studies) platinum.

Voltammetric and controlled-potential coulometric investigations on the CO_2–O_2–CO_3^{2-} system in molten $NaNO_3$–KNO_3 eutectic were made by Zambonin.[222] The possibility of employing the voltammetric technique in the analytical determination of CO_2, O_2, and CO_3^{2-} in the melt was demonstrated.

Inman et al.[223] reinvestigated the reduction of Cd^{2+}, Co^{2+}, Ni^{2+}, and Pb^{2+} in $LiNO_3$–$NaNO_3$–KNO_3 eutectic by conventional polarography. The reduction of Cd^{2+}, Ni^{2+}, and Pb^{2+}, respectively, was a two-electron reversible process described by the Heyrovsky–Ilkovic equation, whereas the results for Co^{2+} were complex. The half-wave potentials and diffusion coefficients for these ions were reported. The results were compared with those reported previously. The discrepancies were attributed to the effect of impurities (i.e., H_2O, Cl^-) in the melt.

Francini and Martini[36,224] examined the electrochemical behavior of Cd^{2+}, Ni^{2+}, Pb^{2+}, Tl^+, and Zn^{2+} in $LiNO_3$–$NaNO_3$–KNO_3 eutectic and Cd^{2+}, Pb^{2+}, and Tl^+ in $NaNO_3$–KNO_3 eutectic by conventional and linear sweep polarography. The results in the ternary eutectic showed that the electrochemical processes for Cd^{2+}, Pb^{2+}, and Tl^+ were reversible up to 10 V/sec; however, the reduction of Ni^{2+} and Zn^{2+} was somewhat irreversible. The order of reversibility was $Tl^+ > Cd^{2+} > Pb^{2+} > Zn^{2+} > Ni^{2+}$. In $NaNO_3$–KNO_3 eutectic only Tl^+ was reduced reversibly.

The polarographic behavior of Cd^{2+}, Ni^{2+}, Pb^{2+}, Tl^+, and Zn^{2+} in $NaNO_3$–KNO_3 eutectic was studied by Tridot et al.[225] They found that the reduction of Cd^{2+} and Tl^+ was reversible, whereas that of Ni^{2+}, Pb^{2+}, and Zn^{2+} was irreversible. Co^{2+}, Cu^{2+}, and Fe^{3+} precipitated as oxides in the melt.

Kalvoda and Tumanova[226] used linear sweep voltammetry to analyze several transition metal ions at a platinum electrode in molten $NaNO_3$–KNO_3 eutectic. In most cases the peak current was proportional to the concentration of the electroactive ions. Mamantov et al.[227] used the same technique and found that the reduction of several metal ions at a platinum electrode was more complicated than the same processes studied by others at a dropping mercury electrode. The reduction of Ag^+ was reversible whereas the reduction of Cd^{2+} and Pb^{2+} was complicated by the oxidation of the reduction product by NO_3^-. The behavior for the reduction of Cu^{2+} could be explained by multistep charge transfer with catalytic regeneration of the reactant. The results for Ag^+ and Cd^{2+} were confirmed by Casadio et al.[228] using chronopotentiometry and linear sweep voltammetry.

The diffusion coefficient of Ag^+ over a wide composition range of $NaNO_3$–KNO_3 melt was studied by linear sweep voltammetry and chronopotentiometry[229] and also by chronoamperometry.[230] The values obtained by the latter method were always lower than by the former; both studies showed that the value decreased with increasing KNO_3 content. Sternberg and Herdlicka[231] determined the diffusion coefficient of Ag^+ in molten $LiNO_3$ and KNO_3 chronopotentiometrically.

Behl and Gaur[232] studied the electrochemical behavior of Ag^+ and Pb^{2+} in molten $NaNO_3$–$Ba(NO_3)_2$ eutectic. The results showed that the reduction of Ag^+ occurred reversibly and uncomplicated by any reaction of the deposited metal with either the platinum electrode or the nitrate ion. However, for Pb^{2+} the deposited liquid metal reacted with the nitrate ion to form oxide. The reaction was negligible at fast scan rates (20–130 V/sec).

Shilina and Delimarskii[233] studied the voltammetric behavior of Bi^{3+}, Cr^{3+}, and Cu^{2+} in molten $Ca(NO_3)_2 \cdot 4H_2O$, which is a low-temperature ionic melt (m.p. 42.7°C). The diffusion coefficients for the ions were reported to be of the order of 10^{-8} cm²/sec at 60°C. Moynihan and Angell[234] examined the reduction of Ag^+, Cd^{2+}, and Tl^+ chronopotentiometrically in the same solvent. The diffusion coefficients for the ions were in the range 10^{-8} (15°C) to 10^{-6} cm²/sec (60°C). Non-Arrhenius temperature dependence of the diffusion coefficients was observed in all cases. The diffusion coefficients for Ag^+ and Tl^+ were about twice as large as for Cd^{2+}, which was attributed to the fact that Cd^{2+} was hydrated, while Ag^+ and Tl^+ were not. The value for Cd^{2+} is in good agreement with that obtained by Braunstein et al.[235] polarographically.

The voltammetric behavior of Ce^{3+}, Nd^{3+}, and Pr^{3+} at a rotating platinum electrode in molten $NaNO_3$-KNO_3 (40–60 mole %) was studied by Shilina and Delimarskii.[236] Only the reduction of Pr^{3+} was reversible. The diffusion coefficients for the ions were reported. Vol'tsev and Didora[131] investigated the reduction of Er^{3+}, Nd^{3+}, Pr^{3+}, and Sm^{3+} in molten KNO_3. Only one reduction wave was observed for Er^{3+}, Pr^{3+}, and Sm^{3+}, whereas a three-step reduction was observed for Nd^{3+}. The limiting currents were proportional to concentrations of the ions. The electroreduction of Eu^{3+} in $LiNO_3$-$NaNO_3$-KNO_3 eutectic was studied by Francini and Martini[237] using conventional and linear sweep polarography. The reduction was a one-electron reversible process for sweep rates up to 1 V/sec. A very low diffusion coefficient for Eu^{3+} ($\sim 10^{-7}$ cm^2/sec) was reported.

Swofford and Holifield[238] employed conventional polarography to study the depolarization of mercury by halide ions in molten $NaNO_3$-KNO_3 eutectic. They attributed the anodic waves observed to two-electron irreversible processes involving formation of Hg_2X_2. Francini et al.[239] carried out similar studies using linear sweep voltammetry. The electrode reactions attributed to the oxidation of halide ions were found to be reversible and involved the transfer of one electron only. No reliable results were obtained when $LiNO_3$ was added to the solvent. Hills and Power[240] also observed abnormal behavior in the oxidation of Cl^- in the ternary eutectic, attributing it to adsorption effects. Recently O'Deen and Osteryoung[23] reinvestigated the halide depolarization of mercury in molten $NaNO_3$-KNO_3 eutectic by pulse polarography. Well-defined waves were obtained due to the limited charge involved in this technique. The results obtained were in essential agreement with those of Francini et al.[239] The different results reported by Swofford and Holifield were apparently due to the ill-defined conventional polarograms obtained for these processes.

Swofford and co-workers[241,242] examined the electrochemical oxidation of Br^- and I^- at a platinum electrode in molten $NaNO_3$-KNO_3 eutectic. The oxidation of Br^- to bromine was a one-electron reversible process. The I^- ion was oxidized in two one-electron steps to an unstable positive univalent state of iodine. The first step was a reversible process (I^- to iodine). The univalent iodine formed as the product in the second step was unstable in the melt and was reduced back to iodine by the reaction with NO_3^-. Triaca et al.[243] reinvestigated the oxidation of I^- in the same solvent at a rotating platinum electrode. They found that both steps were reversible and suggested the following scheme: $2I^- \rightleftharpoons I_2 + 2e$; $I_2 \rightleftharpoons I_2^+ + e$ ($I_2^+ + I^- \rightleftharpoons \frac{3}{2}I_2$). A fairly high diffusion coefficient for the second electroactive species was reported (10^{-4} cm^2/sec). The standard potentials of Br_2/Br^- and I_2/I^- in molten $LiNO_3$-KNO_3 (45–55 mole %) were determined by

TABLE VI. Results of Electroanalytical Studies in Molten Nitrates

Solvent composition	Temperature, °C	Experimental technique	Solutes investigated	Ref.
$LiNO_3$–$NaNO_3$–KNO_3 eutectic (30–17–53 mole %)	145	Polarography (dropping Hg electrode)	Cd^{2+}, Co^{2+}, Ni^{2+}, Pb^{2+}	223
	150–200	Dropping Hg electrode	Cd^{2+}, Ni^{2+}, Pb^{2+}, Tl^{+}, Zn^{2+}	224
	150	Dropping Hg electrode	Eu^{3+}	237
	150–200	Linear sweep voltammetry (dropping Hg electrode)	Cd^{2+}, Ni^{2+}, Pb^{2+}, Tl^{+}, Zn^{2+}	224
	150	Dropping Hg electrode	Eu^{3+}	237
	150	Dropping Hg electrode	Cl^{-}, Br^{-}, I^{-}	239
	150	Dropping Hg and Pt electrodes	Ag^{+}, Pb^{2+}, Cl^{-}	240
	150–230	Voltammetry (rotating Pt electrode)	Cl_2, Br_2, I_2	246
	150	Chronopotentiometry (Pt electrode)	I^{-}, NO_2^{-}, Tl^{+}	249
$LiNO_3$–KNO_3 (34 mole % $LiNO_3$)	150	Linear sweep voltammetry (dropping Hg and Pt electrodes)	Ag^{+}, Pb^{2+}, Cl^{-}	240
$LiNO_3$–KNO_3 (43 mole % $LiNO_3$)	147–217	Potentiometry	Br_2/Br^{-}, I_2/I^{-}	244, 245
$LiNO_3$–KNO_3 (43 mole % $LiNO_3$)	147–207	Potentiometry	Cd^{2+}/Cd, Hg_2^{2+}/Hg, Hg^{2+}/Hg	248
$LiNO_3$–KNO_3	340	Chronopotentiometry (Pt electrode)	Ag^{+}	231
$NaNO_3$–KNO_3 eutectic (50–50 mole %)	250	Linear sweep voltammetry (dropping Hg electrode)	Cd^{2+}, Pb^{2+}, Tl^{+}, Cl^{-}, Br^{-}, I^{-}	36
	240	Polarography (dropping Hg electrode)	Cd^{2+}, Ni^{2+}, Pb^{2+}, Tl^{+}, Zn^{2+}	225
	250	Dropping Hg electrode	Cl^{-}, Br^{-}, I^{-}	238
	240	Pulse polarography (dropping Hg electrode)	Cl^{-}, Br^{-}, I^{-}	23

Electroanalytical Chemistry in Molten Salts—A Review of Recent Developments

Medium	Temp.	Technique	Species	Ref.
$NaNO_3-KNO_3$ eutectic (50–50 mole %)	250	Linear sweep voltammetry (dropping Hg electrode)	Cd^{2+}, Pb^{2+}, Tl^+	36
	250	Dropping Hg electrode	Cl^-, Br^-, I^-	239
	230–280	Dropping Hg and Pt electrodes	O_2, O^{2-}, OH^-	212
	250	Dropping Hg and Pt electrodes	Ag^+, Pb^{2+}, Cl^-	240
	250	Pt electrode	$Cd^{2+}, Cu^{2+}, Ni^{2+}, Pb^{2+}, Tl^+$	226
	245	Pt electrode	$Ag^+, Cd^{2+}, Cu^{2+}, In^{3+}, Pb^{2+}$	227
	300–325	Pt electrode	Ag^+, Cd^{2+}	228
	265	Voltammetry (rotating Pt electrode)	NO_2^-, O_2^{2-}, O_2^-	211
	229	Rotating Pt electrode	NO_3^-, Na_2O	215
	230–342	Rotating Pt electrode	H_2O	216
	227	Rotating Pt electrode	O_2, O_2^-, H_2O, OH^-	217
	227–327	Rotating Pt electrode	CO_2, O_2, CO_3^{2-}	222
	220–420	Pt electrode	$Ag^+, Cd^{2+}, Co^{2+}, Cu^{2+}, Ni^{2+}, Pb^{2+}, Tl^+, Zn^{2+}, Br^-, I^-, NO_2^-, NO_3^-$	1
	250	Rotating Pt electrode	Br^-, I^-	241, 242
	234–314	Rotating Pt electrode	I^-	243
	300–325	Pt electrode	$C_2O_4^{2-}$	247
	240–350	Chronopotentiometry (Pt electrode)	O^{2-}	214
$NaNO_3-KNO_3$ eutectic	250–400	Pt and Hg electrodes	$Ag^+, Cd^{2+}, Pb^{2+}, NO_3^-$	1
	300–325	Pt electrode	Ag^+, Cd^{2+}	228
	250	Pt electrode	I^-	241, 242
	250	Pt electrode	NO_2^-, O^{2-}	213
	250	Pt electrode	$Cd^{2+}, Ce^{3+}, Pr^{3+}$	249
	230	Potentiometry	NO_2^+	218
	230		O_2/O_2^-	219

TABLE VI (continued)

Solvent composition	Temperature, °C	Experimental technique	Solutes investigated	Ref.
$NaNO_3$–KNO_3 (40–60 mole %)	230–340	Voltammetry (rotating Pt electrode)	Ce^{3+}, Pr^{3+}, Nd^{3+}	236
$NaNO_3$–KNO_3 (0–100 mole % $NaNO_3$)	315–360	Linear sweep voltammetry (Pt electrode)	Ag^+	229
	315–360	Chronopotentiometry (Pt electrode)	Ag^+	229
	240–350	Chronoamperometry (Pt electrode)	Ag^+	230
KNO_3	360	Voltammetry (Pt electrode)	Ag^+, Cd^{2+}, Pb^{2+}, CrO_4^{2-}	1
	360	Pt electrode	Er^{3+}, Nd^{3+}, Pr^{3+}, Sm^{3+}	131
$NaNO_3$–$Ba(NO_3)_2$ eutectic (94.2–5.8 mole %)	350	Linear sweep voltammetry (Pt electrode)	Ag^+, Pb^{2+}	232
$Ca(NO_3)_2 \cdot 4H_2O$	60	Voltammetry (rotating Pt electrode)	Bi^{3+}, Cr^{3+}, Cu^{2+}	233
	15–60	Chronopotentiometry (Pt electrode)	Ag^+, Cd^{2+}, Tl^+	234
	50	Polarography (dropping Hg electrode)	Cd^{2+}	235

Bombi and co-workers.[244,245] The standard potential of Cl_2/Cl^-, which could not be determined experimentally due to the oxidation of NO_3^- by chlorine, was estimated. On the other hand, Delimarskii and Shilina[246] claimed that the halide ions were oxidized to halogens by $LiNO_3$ in the ternary eutectic. They studied the voltammetric behavior of the dissolved halogen in the melt. Irreversible cathodic waves corresponding to $X_2 + 2e \rightleftharpoons 2X^-$ were observed, but no anodic waves. The diffusion coefficients of Cl_2, Br_2, and I_2 were reported; the order observed was $D_{Cl_2} > D_{Br_2} > D_{I_2}$.

The stability of oxalate ions in $NaNO_3$–KNO_3 eutectic was studied by El Hosary and Shams El Din[247] voltammetrically. The decomposition reaction $C_2O_4^{2-} + NO_3^- \rightleftharpoons CO_3^{2-} + NO_2^- + CO_2$ was suggested by comparing the voltammogram to those of CO_3^{2-} and NO_2^- in the melt. Oxalate ion itself was electroactive and was oxidized to CO_2. The results agree with those obtained by the titrimetric method.

The results obtained in molten nitrates are summarized in Table VI.

3.7. Other Melts

Most of the earlier studies were carried out in molten nitrates and chlorides. Recently some studies have been made in more basic solvents: molten alkali acetates, thiocyanates, hydroxides, and carbonates. Molten LiOAc–NaOAc–KOAc (20–35–45 mole %) (liquidus temperature ~187°C) was employed as a solvent by Bombi et al.[250] to study the polarographic reduction of Cd^{2+}, Pb^{2+}, and Zn^{2+}. The polarograms were well defined and were described by the Heyrovsky–Ilkovic equation. The diffusion currents for the reduction of Cd^{2+} and Pb^{2+} were found to be directly proportional to concentration. Giess et al.[251] examined the reduction of several metal ions in the same solvent by conventional and linear sweep polarographic techniques. The Tl^+ ion showed reversible behavior up to 29 V/sec, while for Cd^{2+}, Pb^{2+}, and Zn^{2+} a negative deviation from linearity for the plot of peak current against square root of scan rate at rates higher than 6 V/sec was observed. The degree of irreversibility was in the following order: $Pb^{2+} > Zn^{2+} > Cd^{2+}$. The diffusion currents and peak currents were found to be proportional to concentration. The diffusion coefficients reported were of the order of 10^{-7} cm²/sec. Marassi et al.[252,253] investigated several systems in a slightly different acetate melt composition (20–30–50 mole %). They measured the standard electrode potentials of Cd^{2+}/Cd, In^{3+}/In, Pb^{2+}/Pb, Tl^+/Tl, and Zn^{2+}/Zn couples with a dropping amalgam electrode. They also carried out polarographic studies of anodic oxidation of the metal amalgams. All the couples, except Tl^+/Tl, exhibited a certain degree of irreversibility. The results agree with those reported by Giess et al.[251]

Francini et al.[254] studied the reduction of Cd^{2+}, Cu^+, Pb^{2+}, Tl^+, and Zn^{2+} in the molten NaSCN–KSCN eutectic by conventional and linear sweep polarography. Tl^+ showed reversible behavior at sweep rates up to ~12 V/sec, whereas Cu^+, Cd^{2+}, Zn^{2+}, and Pb^{2+} showed increasing departures from reversible behavior at sweep rates >1.5 V/sec. The diffusion or peak currents were found to be directly proportional to concentration. The standard electrode potentials of Ag^+/Ag, Cd^{2+}/Cd, Co^{2+}/Co, In^{3+}/In, Tl^+/Tl, and Zn^{2+}/Zn couples in molten NaSCN–KSCN (26.3–73.7 mole %) were reported by Cescon et al.[255] The electrochemical behavior of Cd^{2+}, Pb^{2+}, Zn^{2+}, and UO_2^{2+} in molten KSCN was studied polarographically and chronopotentiometrically by Yanagi and co-workers.[256,257] Well-defined and reproducible chronopotentiograms for the reduction of Cd^{2+}, Pb^{2+}, and Zn^{2+} were obtained; the product of current density and the square root of the transition time was proportional to concentration for each system. The reduction of Pb^{2+} and Zn^{2+} was found to be reversible. The diffusion coefficients reported were of the order of 10^{-6} cm²/sec. Two reduction steps were observed for UO_2^{2+} polarographically and chronopotentiometrically. The processes are diffusion controlled and were proposed to be $UO_2^{2+} + e \rightleftharpoons UO_2^+$; $UO_2^+ + e \rightleftharpoons UO_2$. The diffusion coefficient of UO_2^{2+} is ~1/3 that of Pb^{2+}.

Voltammetric and chronopotentiometric studies of PbO, Sb_2O_3, and Bi_2O_3 in molten NaOH were made by Delimarskii et al.[258] Three waves were observed in the conventional voltammogram of PbO. The first wave was ascribed to the reversible discharge of simple Pb^{2+}: $Pb^{2+} + 2e \rightleftharpoons Pb$. The second wave (not observed at fast scans) was ascribed to the reduction of some complex lead ions. The third wave was attributed to the decomposition of water that was formed by the reaction $2NaOH + PbO \rightarrow Na_2PbO_2 + H_2O$, or present as an impurity in the melt. Similar results were obtained by chronopotentiometric measurements. Sb_2O_3, which was unstable in the melt (due to oxidation by air to a higher oxidation state), was reduced to the metal. Two-step reduction of Bi_2O_3 was observed. The mechanism was postulated to be: (a) $Bi^{3+} + 2e \rightleftharpoons Bi^+$ (first wave), (b) $Bi^+ + Bi^{3+} \rightleftharpoons 2Bi^{2+}$, (c) $Bi^{2+} + 2e \rightleftharpoons Bi$ (second wave). The diffusion coefficient for Bi^{3+} was 7×10^{-6} cm²/sec. Tremillon and co-workers[259,260] studied the electrochemical and chemical properties of the ion, oxide, and oxygenated complex ion of a large number of elements in the molten NaOH–KOH eutectic. The stability of different species in acidic (hydrated), neutral (dehydrated), and basic (in the presence of O^{2-}) melts was investigated. For example, they found that HgO, insoluble in very strong acidic media, dissolves as Hg^{2+} in acidic media and HgO_2^{2-} in neutral and basic media. The potential of the Hg^{2+}/Hg couple in acid media and the ratio Hg^{2+}/HgO_2^{2-} as a function of $p(H_2O)$ were reported. The voltammograms of Hg^{2+} in acidic melts and HgO_2^{2-} in basic melts at a rotating platinum electrode were obtained. The results showed that the reduction is a two-electron step and that the mercurous ion is unstable in the melt. Electrochemical properties of heptavalent neptunium as NpO_5^{3-}, prepared by oxidizing NpO_2 by bubbling O_2, in the molten NaOH–KOH eutectic were examined by Martinot and Duyckaerts.[261] The ill-defined chronopotentiograms showed a one-step reduction at 240°C and two at 290°C. The processes were suggested as $NpO_5^{3-} + 3e \rightleftharpoons NpO_2 + 3O^{2-}$ at 240°C, and $NpO_5^{3-} + 2e \rightleftharpoons NpO_2^+ + 3O^{2-}$ and $NpO_2^+ + e \rightleftharpoons NpO_2$ at 290°C. The diffusion coefficient for NpO_5^{3-} was reported as 10^{-6} cm^2/sec.

A copper(I)–copper reference electrode for use in molten hydroxides has been described quite recently.[261a]

Linear sweep voltammetric studies of Bi_2O_3, CuO, Fe_2O_3, PbO, and $K_2Cr_2O_7$ in molten equimolar Li_2CO_3–K_2CO_3 were made by Delimarskii and Tumanova.[262] The large separation between the anodic and cathodic peaks for Bi_2O_3, Fe_2O_3, and PbO was attributed to irreversibility. An anodic but no cathodic peak was obtained for CuO; this result was attributed to the oxidation of Cu^+ formed by the reaction of Cu^{2+} and the deposited copper. The method was found to be suitable for the determination of $K_2Cr_2O_7$ in the melt, since the first reduction wave (two reduction waves were observed) was proportional to concentration.

Well-defined voltammograms for Sb_2O_3, SnO_2, and V_2O_5 in molten equimolar Li_2CO_3–K_2CO_3 were reported by Kuzmovich and Tumanova.[263] The results showed that these processes were also irreversible, which was confirmed by a linear sweep voltammetric investigation. The wave height for the reduction of SnO_2 was directly proportional to concentration up to 0.1 mole %. The electrochemical reduction of O_2 in Li_2CO_3 and Li_2CO_3–Na_2CO_3–K_2CO_3 eutectic was studied by Lorenz and Janz[264] voltammetrically. The reduction occurred via two reactions:

$$\tfrac{1}{2}O_2 + CO_2 + 2e \rightleftharpoons CO_3^{2-} \quad \text{and} \quad \tfrac{1}{2}O_2 + 2e \rightleftharpoons O^{2-}$$

A note dealing with the significance of the peroxide and superoxide ions in the oxygen reduction in carbonate melts has appeared quite recently.[264a]

Baak and Frederick[265] obtained voltammograms for the oxides of Co, Cr, Cu, Fe, Mn, Ni, Ti, and V dissolved in sodium disilicate at a rotating platinum electrode. Linear E versus $\log[(i_d - i)/i]$ plots were obtained for

TABLE VII. Results of Electroanalytical Studies in Molten Salts (Other Than Halides or Nitrates)

Solvent composition	Temperature, °C	Experimental technique	Solutes investigated	Ref.
LiOAc–NaOAc–KOAc (20–35–45 mole %)	197	Polarography (dropping Hg electrode)	Cd^{2+}, Pb^{2+}, Zn^{2+}	250
LiOAc–NaOAc–KOAc (20–35–45 mole %)	200–260	Polarography and linear sweep voltammetry (dropping Hg electrode)	Cd^{2+}, Pb^{2+}, Tl^{+}, Zn^{2+}	251
LiOAc–NaOAc–KOAc (20–30–50 mole %)	197–247	Potentiometry and polarography (dropping Hg amalgam electrode)	Cd^{2+}, In^{3+}, Pb^{2+}, Tl^{+}, Zn^{2+}	252, 253
NaSCN–KSCN eutectic	137–210	Polarography and linear sweep voltammetry (dropping Hg electrode)	Cd^{2+}, Cu^{+}, Pb^{2+}, Tl^{+}, Zn^{2+}	254
NaSCN–KSCN (73.7 mole % KSCN)	142–197	Potentiometry	Ag^{+}, Cd^{2+}, Co^{2+}, In^{3+}, Tl^{+}, Zn^{2+}	255
KSCN	185	Polarography and chronopotentiometry (dropping Hg electrode)	Cd^{2+}, Pb^{2+}, Zn^{2+}, UO_2^{2+}	256, 257
NaOH	340–440	Voltammetry and chronopotentiometry (Pt electrode)	Bi_2O_3, Sb_2O_3, PbO	258
NaOH–KOH eutectic (51–49 mole %)	227	Potentiometry and voltammetry (rotating metal electrodes)	Ag, Au, Ba, Be, Ca, Cu, Hg, Li, Mg, Mn, Ni, Pb, Sr, Pt species	259, 260
NaOH–KOH (61 mole % NaOH)	240 and 290	Chronopotentiometry (pyrolytic graphite electrode)	NpO_5^{3-}	261

Medium	Temp (°C)	Method	Species	Ref.
Li_2CO_3–K_2CO_3 (equimolar)	640	Linear sweep voltammetry (Pt electrode)	Bi_2O_3, CuO, Fe_2O_3, PbO, $K_2Cr_2O_7$	262
Li_2CO_3–K_2CO_3 (equimolar)	600	Voltammetry and linear sweep voltammetry (Pt electrode)	Sb_2O_3, SnO_2, V_2O_5	263
Li_2CO_3 and Li_2CO_3–Na_2CO_3–K_2CO_3 eutectic (43.5–31.5–25.0 mole %)	440–840	Voltammetry (Au electrode)	O_2	264
$Na_2Si_2O_5$	1000	Voltammetry (rotating Pt electrode)	CoO, Cr_2O_3, CuO, Fe_2O_3, NiO, TiO_2, V_2O_5, Mn_2O_3	265
$Na_2B_4O_7$	800–1000	Voltammetry (Pt and Sb electrodes)	Sb_2O_3	266
$Na_2B_4O_7$	820–920	Voltammetry (Pt electrode)	Bi_2O_3, CdO, CoO, Cr_2O_3, CuO, Fe_2O_3, GeO_2, MoO_3, NiO, Sb_2O_3, SnO, SnO_2, WO_3, ZnO	1
$NaPO_3$	740	Voltammetry (Pt electrode)	Bi_2O_3, CuO, Sb_2O_3, TiO_2	1
Li_2SO_4–Na_2SO_4–K_2SO_4	550	Voltammetry (Pt and W electrodes)	Co^{2+}, CrO_4^{2-}, Ni^{2+}, Pb^{2+}, V_2O_5	1
$(CH_3)_2NH_2Cl$	170–197	Potentiometry	Ag^+, Bi^{3+}, Cd^{2+}, Co^{2+}, Cu^{2+}, Cu^+, H^+, Hg^{2+}, Ni^{2+}, Pb^{2+}, Pt^{2+}, Sn^{2+}	267
$(C_2H_5)NH_3Cl$	127	Potentiometry and chronopotentiometry (Hg electrode)	Ag, Au, Bi, Cu, Fe, H, Hg, Sn, V species	268
$(CH_3)_2SO$ (1 m $LiClO_4$)	127	Potentiometry and voltammetry (rotating Pt electrode)	Halogen and halide ions (Cl_2, Br_2, I_2)	269

all cases, suggesting that the deposited metals formed an alloy with platinum. The reduction of Co(II), Fe(III), and Ni(II) oxides corresponded to a six-electron process; formation of cobaltite, ferrite, and nickelite was suggested. The limiting currents were found to be proportional to concentrations of the electroactive species. Voltammetric studies of Sb_2O_3 in molten borax were reported by Colom and de la Cruz.[266] A cathodic wave at a platinum electrode, which disappeared at high concentrations ($>1.5 \times 10^{-2}$ mole fraction), was observed. The reduction waves were best described by the Heyrovsky–Ilkovic equation at a platinum electrode and by the Kolthoff–Lingane equation at an antimony electrode.

Not much work has been reported on electrochemical studies in molten organic compounds as yet. One example of an organic melt is a molten alkylammonium chloride. Kisza[267] reported the electromotive series of several M^{n+}/M couples in molten dimethylammonium chloride. The same order as in the molten LiCl–KCl eutectic was observed. Picard and Vedel[268] measured the standard electrode potentials of several metals in molten ethylammonium chloride. The systems Fe^{3+}/Fe^{2+} and Sn^{4+}/Sn^{2+} were also studied by polarographic and chronopotentiometric techniques. The diffusion coefficients for Fe^{3+} and Sn^{4+} were of the order of 10^{-6} cm^2/sec. Molten dimethylsulfone is another example of a molten organic medium. It is a nonelectrolyte; thus its potential span depends on the supporting electrolyte. Bry and Tremillon[269] studied the electrochemical behavior of halogens (I_2, Br_2, Cl_2) and halide ions in this solvent voltammetrically and potentiometrically. Stable trihalide ions were found. The formal potentials of X^-/X_3^-, X_3^-/X_2, and X^-/X_2 couples were reported.

The results discussed in this section are summarized in Table VII.

ACKNOWLEDGMENTS

We would like to thank Dr. J. Braunstein and Dr. C. E. Bamberger of the Oak Ridge National Laboratory for valuable comments on the manuscript and K. A. Bowman, D. L. Brotherton, Dr. B. Gilbert, and G. Ting for checking a number of references. This work was supported by the United States Atomic Energy Commission and the National Science Foundation.

REFERENCES

1. C. H. Liu, K. E. Johnson, and H. A. Laitinen, in: *Molten Salt Chemistry* (M. Blander, ed.), pp. 681–733, Interscience, New York (1964).

2. H. A. Laitinen and R. A. Osteryoung, in: *Fused Salts* (B. R. Sundheim, ed.), pp. 255-300, McGraw-Hill, New York (1964).
3. Yu. K. Delimarskii and B. F. Markov, *Electrochemistry of Fused Salts*, Sigma Press, Washington, D.C. (1961).
4. H. C. Gaur and B. B. Bhatia, *J. Sci. and Ind. Res.* **21A**:16 (1962).
5. T. B. Reddy, *Electrochem. Technol.* **1**:325 (1963).
6. D. Inman, A. D. Graves, and R. S. Sethi, in: *Electrochemistry*, Vol. 1, A Specialist Periodical Report, The Chemical Society, London (1970); D. Inman, A. D. Graves, and A. A. Nobile in: *Electrochemistry*, Vol. 2, A Specialist Periodical Report, The Chemical Society, London (1972).
7. A. D. Graves, G. J. Hills, and D. Inman, in: *Advances in Electrochemistry and Electrochemical Engineering* (P. Delahay, ed.), Vol. 4, pp. 117-183, Interscience, New York (1966).
8. G. J. Janz and R. D. Reeves, in: *Advances in Electrochemistry and Electrochemical Engineering* (C. W. Tobias, ed.), Vol. 5, pp. 137-204, Interscience, New York (1967).
9. H. Lux, *Z. Elektrochem.* **45**:303 (1939).
10. H. Flood and T. Förland, *Acta Chem. Scand.* **1**:592 (1947).
11. H. Bloom, *The Chemistry of Molten Salts*, p. 158, Benjamin, New York (1967).
12. B. Tremillon and G. Letisse, *J. Electroanal. Chem.* **17**:371 (1968).
13. G. Torsi and G. Mamantov, *Inorg. Chem.* **10**:1900 (1971).
14. G. Torsi and G. Mamantov, *Inorg. Chem.* **11**:1439 (1972).
15. N. J. Bjerrum, C. R. Boston, and G. P. Smith, *Inorg. Chem.* **6**:1162 (1967); N. J. Bjerrum and G. P. Smith, *Inorg. Chem.* **6**:1968 (1967).
16. J. D. Corbett, in: *Preparative Inorganic Reactions* (W. L. Jolly, ed.), Vol. 3, pp. 1-33, Interscience, New York (1966).
17. R. A. Potts, R. D. Barnes, and J. D. Corbett, *Inorg. Chem.* **7**:2558 (1968).
18. G. Torsi, K. W. Fung, G. M. Begun, and G. Mamantov, *Inorg. Chem.* **10**:2285 (1971).
19. K. W. Fung and G. Mamantov, *J. Electroanal. Chem.* **35**:27 (1972).
20. G. P. Smith, J. Brynestad, C. R. Boston, and W. E. Smith, in: *Molten Salts: Characterization and Analysis* (G. Mamantov, ed.), pp. 143-168, Marcel Dekker, New York (1969).
21. H. A. Laitinen, C. H. Liu, and W. S. Ferguson, *Anal. Chem.* **30**:1266 (1958).
22. P. G. Zambonin, *Anal. Chem.* **41**:868 (1969).
23. W. O'Deen and R. A. Osteryoung, *Anal. Chem.* **43**:1879 (1971).
24. G. Torsi and G. Mamantov, *J. Electroanal. Chem.* **30**:193 (1971).
25. G. Mamantov, P. Papoff, and P. Delahay, *J. Am. Chem. Soc.* **79**:4034 (1957).
26. H. A. Laitinen, *Pure Appl. Chem.* **15**:227 (1967).
27a. E. R. Brown and R. F. Large, in: *Techniques of Chemistry* (A. Weissberger and B. W. Rossiter, eds.), Vol. I, Part IIA, pp. 423-530, Wiley—Interscience, New York (1971).
27b. S. Piekarski and R. N. Adams, in: *Techniques of Chemistry* (A. Weissberger and B. W. Rossiter, eds.), Vol. I, Part IIA, pp. 531-589, Wiley–Interscience, New York (1971).
28. R. A. Osteryoung, G. Lauer, and F. C. Anson, *J. Electrochem. Soc.* **110**:926 (1963).
29. P. J. Lingane, *Critical Revs. Anal. Chem.* **1**:587 (1971).

30. M. L. Olmstead and R. S. Nicholson, *J. Phys. Chem.* **72**:1650 (1968); *Anal. Chem.* **42**:796 (1970).
31. J. D. Van Norman, *Anal. Chem.* **34**:594 (1962).
32. R. D. Caton, Jr. and H. Freund, *Anal. Chem.* **36**:15 (1964).
33. A. T. Hubbard and L. P. Zajicek, Abstract ANAL 32, 161st National Meeting of the American Chemical Society, Los Angeles, Calif., 1971; L. P. Zajicek, *J. Electrochem. Soc.* **116**:80c (1969).
34. J. P. Young, G. Mamantov, and F. L. Whiting, *J. Phys. Chem.* **71**:782 (1967).
35. R. A. Osteryoung, Abstract CHED 22, 162nd National Meeting of the American Chemical Society, Washington, D. C., 1971.
36. M. Francini and S. Martini, *Electrochim. Metal.* **3**:136 (1968).
37. R. J. Heus, T. Tidwell, and J. J. Egan, in: *Molten Salts:Characterization and Analysis* (G. Mamantov, ed.), pp. 499–508, Marcel Dekker, New York (1969).
38. D. L. Manning and G. Mamantov, *J. Electroanal. Chem.* **18**:137 (1968).
39. J. D. Van Norman and R. J. Tiver, in: *Molten Salts: Characterization and Analysis* (G. Mamantov, ed.), pp. 509–527, Marcel Dekker, New York (1969).
40. R. N. Adams, *Electrochemistry at Solid Electrodes*, Marcel Dekker, New York (1969).
41. D. Inman, *Electrochim. Acta* **10**:11 (1965).
42. H. Braunstein, J. Braunstein, and D. Inman, *J. Phys. Chem.* **70**:2726 (1966).
43. R. W. Laity, in: *Reference Electrodes* (D. J. G. Ives and G. J. Janz, eds.), pp. 524–606, Academic Press, New York (1961).
44. A. F. Alabyshev, M. F. Lantratov, and A. G. Morachevskii, *Reference Electrodes for Fused Salts*, Sigma Press, Washington, D. C. (1965).
45. H. A. Laitinen and C. H. Liu, *J. Am. Chem. Soc.* **80**:1015 (1958).
46. S. N. Flengas and T. R. Ingraham, *Can. J. Chem.* **35**:1139 (1957).
47. H. W. Jenkins, G. Mamantov, and D. L. Manning, *J. Electroanal. Chem.* **19**:385 (1968).
48. R. G. Verdieck and L. F. Yntema, *J. Phys. Chem.* **46**:344 (1942).
49. H. W. Jenkins, G. Mamantov, and D. L. Manning, *J. Electrochem. Soc.* **117**:183 (1970).
50. H. R. Bronstein and D. L. Manning, *J. Electrochem. Soc.* **119**:125 (1972).
51. G. Mamantov and D. L. Manning, *Anal. Chem.* **38**:1494 (1966).
52. J. A. Plambeck, *J. Chem. Eng. Data* **12**:77 (1967).
53. H. A. Laitinen and J. W. Pankey, *J. Am. Chem. Soc.* **81**:1053 (1959); H. A. Laitinen and J. A. Plambeck, *J. Am. Chem. Soc.* **87**:1202 (1965).
54. F. R. Clayton, Jr., G. Mamantov, and D. L. Manning, paper in preparation.
55. C. F. Baes, Jr., in: *Nuclear Metallurgy*, Vol. 15, AIME Symposium on Reprocessing of Nuclear Fuels (P. Chiotti, ed.), pp. 617–644 (1969).
56. L. G. Boxall and K. E. Johnson, *J. Electroanal. Chem.* **30**:25 (1971).
57. D. L. Maricle and D. N. Hume, *J. Electrochem. Soc.* **107**:354 (1960).
58. H. A. Laitinen, W. S. Ferguson, and R. A. Osteryoung, *J. Electrochem. Soc.* **104**:516 (1957).
59. D. L. Hill, J. Perano, and R. A. Osteryoung, *J. Electrochem. Soc.* **107**:698 (1960).
60. I. I. Naryshkin, V. P. Yurkinskii, and P. T. Stangrit, *Soviet Electrochem.* **5**:981, 1401 (1969).
61. W. K. Behl, *J. Electrochem. Soc.* **118**:889 (1971).

62. J. C. Fondanaiche, Y. Antuszewicz, R. Cartier, and A. Leseur, *Bull. Soc. Chim. Fr.* **1970**:1689.
63. J. Hladik, Y. Pointud, and G. Morand, *J. Chim. Phys.* **66**:113 (1969).
64. Yu. K. Delimarskii, N. Kh. Tumanova, and M. U. Prikhodko, *Soviet Electrochem.* **6**:543, 1211 (1970).
65. Yu. K. Delimarskii, N. Kh. Tumanova, and M. U. Prikhodko, *Soviet Electrochem.* **7**:339 (1971).
66. I. D. Panchenko, *J. Phys. Chem. USSR* **39**:277 (1965).
67. I. I. Naryshkin, V. P. Yurkinskii, and B. S. Yavich, *Izv. Vyssh. Ucheb. Zaved. Khim. i Khim. Tekhnol.* **11**(1):23 (1968).
68. B. Scrosati and H. A. Laitinen, *Anal. Chem.* **38**:1894 (1966).
69. T. Suzuki, *Electrochim. Acta* **15**:127 (1970).
70. H. A. Laitinen and R. D. Bankert, *Anal. Chem.* **39**:1790 (1967).
71. J. H. Propp and H. A. Laitinen, *Anal. Chem.* **41**:644 (1969).
72. B. Popov and H. A. Laitinen, *J. Electrochem. Soc.* **117**:482 (1970).
73. K. W. Hanck and H. A. Laitinen, *J. Electrochem. Soc.* **118**:1123 (1971).
74. A. F. Alabyshev, M. V. Kamenetskii, A. G. Morachevskii, and V. A. Petrov, *J. Appl. Chem. USSR* **43**:1843 (1970).
75. R. Baboian, D. L. Hill, and R. A. Bailey, *J. Electrochem. Soc.* **112**:1221 (1965).
76. V. T. Barchuk and I. N. Sheiko, *Soviet Progr. Chem.* **33**(3):18 (1967).
77. C. Eon, C. Pommier, J. C. Fondanaiche, and H. Fould, *Bull. Soc. Chim. Fr.* **1969**: 2574.
78. J. Dartnell, K. E. Johnson, and L. L. Shreir, *J. Less-Common Metals* **6**:85 (1964).
79. Y. Saeki and T. Suzuki, *J. Less-Common Metals* **9**:362 (1965).
80. Y. Saeki and K. Funaki, *Bull. Tokyo Inst. Technol.* **86**:1 (1968).
81. V. F. Pimenov, *Izv. Vyssh. Ucheb. Zaved., Tsvet. Met.*, **11**(5):64 (1968).
82. V. F. Pimenov, *Izv. Vyssh. Ucheb. Zaved., Tsvet. Met.*, **12**(2):90 (1969).
83. D. Inman, R. S. Sethi, and R. Spencer, *J. Electroanal. Chem.* **29**:137 (1971).
84. S. Senderoff and G. W. Mellors, *J. Electrochem. Soc.* **114**:556 (1967).
85. O. A. Ryzhik and M. V. Smirnov, *Electrochem. Molten and Solid Electrolytes* **5**:41 (1967).
86. T. Suzuki, *Electrochim. Acta* **15**:303 (1970).
87. K. E. Johnson and J. R. Mackenzie, *Anal. Chem.* **41**:1483 (1969).
88. B. Scrosati, *Anal. Chem.* **38**:1588 (1966); *Ric. Sci.* **37**:829 (1967).
89. E. Schmidt, H. Pfander, and H. Siegenthaler, *Electrochim. Acta* **10**:429 (1965).
90. K. E. Johnson and J. R. Mackenzie, *J. Electrochem. Soc.* **116**:1697 (1969).
91. L. Martinot, G. Duyckaerts, and F. Caligara, *Bull. Soc. Chim. Belges* **76**:5, 15, 211 (1967); **77**:77 (1968); **78**:495 (1969).
92. F. Caligara, L. Martinot, and G. Duyckaerts, *J. Chim. Phys.* **64**:1740 (1967); *J. Electroanal. Chem.* **16**:335 (1968).
93. L. Martinot and G. Duyckaerts, *Anal. Lett.* **1**:669 (1968).
94. L. Martinot and G. Duyckaerts, *Inorg. Nucl. Chem. Lett.* **5**:909 (1969).
95. L. Martinot and G. Duyckaerts, *Anal. Lett.* **4**:1 (1971).
96. I. D. Panchenko, I. I. Penkalo, and Yu. K. Delimarskii, *Soviet Electrochem.* **2**:485 (1966).
97. I. D. Panchenko and I. I. Penkalo, *Ukr. Khim. Zh.* **34**:42 (1968).
98. J. B. P. F. Lesourd and J. A. Plambeck, *Can. J. Chem.* **47**:3387 (1969).
99. Y. Hoshino and J. A. Plambeck, *Can. J. Chem.* **48**:685 (1970).

100. F. G. Bodewig and J. A. Plambeck, *J. Electrochem. Soc.* **116**:607 (1969); **117**:618 (1970).
101. J. M. Shafir and J. A. Plambeck, *Can. J. Chem.* **48**:2131 (1970).
102. N. S. Wrench and D. Inman, *J. Electroanal. Chem.* **17**:319 (1968).
103. E. P. Mignonsin, L. Martinot, and G. Duyckaerts, *Inorg. Nucl. Chem. Lett.* **3**:511 (1967).
104. H. A. Laitinen and K. R. Lucas, *J. Electroanal. Chem.* **12**:553 (1966).
105. E. Franks and D. Inman, *J. Electroanal. Chem.* **26**:13 (1970).
106. D. M. Wrench and D. Inman, *Electrochim. Acta* **12**:1601 (1967).
107. J. Hladik and G. Morand, *Bull. Soc. Chim. Fr.* **1965**:828.
108. S. N. Flengas and T. R. Ingraham, *J. Electrochem. Soc.* **106**:714 (1959).
109. R. Combes, J. Vedel, and B. Tremillon, *J. Electroanal. Chem.* **27**: 174 (1970).
110. K. Cho and T. Kuroda, *Denki Kagaku* **39**:206 (1971).
111. N. I. Kornilov, V. E. Solomatin, and N. G. Ilyushchenko, *Soviet Electrochem.* **5**:885 (1969).
112. M. V. Dubinin, I. F. Nichkov, and S. P. Raspopin, *Izv. Vyssh. Ucheb. Zaved., Tsvet. Met.* **9**(3):73 (1966).
113. M. V. Smirnov, A. V. Pokrovskii, and N. A. Loginov, *Tr. Inst. Elektrochim. Akad. Nauk SSSR, Ural. Filial* **14**:47 (1970); *Chem. Abstr.* **74**:9001m (1971).
114. G. N. Kazantsev, Yu. P. Kanashin, I. F. Nichkov, S. P. Raspopin, and V. V. Chernyshev, *Izv. Vyssh. Ucheb. Zaved., Tsvet. Met.* **9**(3):39 (1966).
115. G. N. Kazantsev, I. F. Nichkov, and S. P. Raspopin, *Soviet Electrochem.* **2**:466 (1966).
116. Yu. K. Delimarskii and V. V. Kuzmovich, *J. Inorg. Chem. USSR* **4**:1263 (1959).
117. H. Hoff, *Electrochim. Acta* **16**:1059 (1971).
118. G. J. Janz, *Molten Salts Handbook*, Academic Press, New York (1967).
119. T. Sakakura, *J. Electrochem. Soc. Japan* **35**:75 (1967).
120. M. Sakawa and T. Kuroda, *Denki Kagaku* **36**:653 (1968); **37**:99 (1969).
121. Y. Saeki and T. Suzuki, *Denki Kagaku* **34**:691 (1966).
122. J. E. Fergusson, in: *Preparative Inorganic Reactions* (W. L. Jolly, ed.), Vol. 7, pp. 93–163. Interscience Publishers, New York (1971).
123. V. F. Pimenov and Yu. V. Baimakov, *Soviet Electrochem.* **4**:1220 (1968).
124. I. I. Naryshkin, V. P. Yukinskii, and B. S. Yavich, *Soviet Electrochem.* **2**:807 (1966).
125. T. Yanagi, T. Ikeda, and M. Shinagawa, *Nippon Kagaku Zasshi* **86**:898 (1965).
126. H. Goto and F. Maeda, *Nippon Kagaku Zasshi* **90**:787 (1969).
127. Yu. K. Delimarskii and V. V. Kuzmovich, *J. Appl. Chem. USSR* **37**:1484 (1964).
128. T. Sakakura and T. Kirihara, *Denki Kagaku* **37**:107 (1969).
129. M. V. Smirnov, T. A. Puzanova, N. A. Loginov, and V. A. Panishev, *Tr. Inst. Elektrokhim. Akad. Nauk SSSR, Ural. Filial* **14**:38 (1970); *Chem Abstr.* **74**:8999f (1971).
130. J. Hladik, Y. Pointud, M. C. Bellissent, and G. Morand, *Electrochim. Acta* **15**:405 (1970).
131. V. K. Val'tsev and N. F. Didora, *Izv. Sib. Otd. Akad. Nauk SSSR,* Ser. Khim. Nauk **26** (1968).
132. M. V. Smirnov, V. Ya. Kudyakov, Yu. V. Poskhin, and Yu. N. Krasnov, *Soviet Atomic Energy* **28**:530 (1970).
133. M. V. Smirnov and N. P. Podlesnyak, *J. Appl. Chem. USSR* **42**:2313 (1969).

134. S. V. Karzhavin, I. F. Nichkov, S. P. Raspopin, and L. B. Kirvonosov, *Izv. Vyssh. Ucheb. Zaved., Tsvet. Met.* **9**(4):60 (1966).
135. T. C. F. Munday and J. D. Corbett, *Inorg. Chem.* **5**:1263 (1966).
136. C. R. Boston, in: *Advances in Molten Salt Chemistry* (J. Braunstein, G. Mamantov, and G. P. Smith, eds.), Vol. 1, pp. 145–146, Plenum Press, New York (1971).
137. N. J. Bjerrum, private communication.
138. R. K. McMullan, D. J. Prince, and J. D. Corbett, *Inorg. Chem.* **10**:1749 (1971).
139. N. J. Bjerrum, *Inorg. Chem.* **9**:1965 (1970); **10**:2578 (1971); **11**:2648 (1972).
140. D. A. Hames and J. A. Plambeck, *Can. J. Chem.* **46**:1727 (1968).
141. G. Letisse and B. Tremillon, *J. Electroanal. Chem.* **17**:387 (1968).
142. E. E. Marshall and L. F. Yntema, *J. Phys. Chem.* **46**:353 (1942).
143. R. G. Verdieck and L. F. Yntema, *J. Phys. Chem.* **48**:268 (1944).
144. Yu. K. Delimarskii, L. S. Berenblyum, and I. N. Sheiko, *Zh. Fiz. Khim.* **25**:398 (1951).
145. U. Anders and J. A. Plambeck, *Can. J. Chem.* **47**:3055 (1969).
146. Yu. K. Delimarskii, E. M. Skobets, and L. S. Berenblyum, *Zh. Fiz. Khim.* **22**:1108 (1948).
147. R. M. de Fremont, R. Rosset, and M. Leroy, *Bull. Soc. Chim. Fr.* **1964**:706.
148. M. Saito, S. Suzuki, and H. Goto, *Nippon Kogaku Zasshi* **83**:883 (1962).
149. M. Francini, S. Martini, and C. Monfrini, *Electrochim. Metal.* **2**:3 (1967).
150. J. D. Corbett, *Inorg. Chem.* **1**:700 (1962).
151. Yu. K. Delimarskii and N. Kh. Tumanova, *Ukr. Khim. Zh.* **35**:133 (1969).
152. N. G. Chovnyk and M. V. Myshalov, *Soviet Electrochem.* **6**:1583 (1970).
153. R. Gut, *Helv. Chim. Acta* **43**:830 (1960).
154. R. D. Ellison, H. A. Levy, and K. W. Fung, *Inorg. Chem.* **11**:833 (1972).
155. Yu. K. Delimarskii, V. F. Makogon, and A. Ya. Zhigailo, *Soviet Electrochem.* **5**:98 (1969).
156. K. W. Fung, G. Mamantov, and J. P. Young, *Inorg. Nucl. Chem. Lett.* **8**:219 (1972).
157. M. Fleischmann and D. Pletcher, *J. Electroanal. Chem.* **25**:449 (1970).
157a. H. L. Jones, L. G. Boxall, and R. A. Osteryoung, *J. Electroanal. Chem.* **38**:476 (1972).
157b. K. W. Fung, J. Q. Chambers, and G. Mamantov, *J. Electroanal. Chem.* (in press).
158. Yu. K. Delimarskii and V. F. Makogon, *Zashch. Metal.* **5**:128 (1969); *Chem. Abstr.* **70**:73501u (1969).
159. H. C. Gaur and H. L. Jindal, *Electrochim. Acta* **13**:835 (1968); **15**:1113, 1127 (1970).
160. H. C. Gaur and H. L. Jindal, *J. Electroanal. Chem.* **16**:437 (1968); **23**:289 (1969).
161. H. C. Gaur, H. L. Jindal, and R. S. Sethi, *Electrochim. Acta* **15**:845 (1970).
162. V. Z. Grebenik and K. Ya. Grachev, *Soviet Electrochem.* **6**:1558 (1970).
163. A. V. Pokrovskii, M. V. Smirnov, and N. A. Loginov, *Tr. Inst. Elektrokhim. Akad. Nauk SSSR Ural. Filial* **15**:89 (1970); *Chem. Abstr.* **75**:44213z (1971).
164. V. P. Kochergin, R. I. Ufimtseva, and N. N. Bochkareva, *Izv. Vyssh. Ucheb. Zaved., Khim. i Khim. Tekhnol.* **14**(1):61 (1971).
165. M. V. Smirnov and O. A. Ryzhik, *Electrochem. Molten and Solid Electrolytes* **4**:27 (1967).
166. O. A. Ryzhik and Yu. P. Savochkin, *Tr. Ural. Politekh. Inst.* **148**: 49 (1966); *Chem. Abstr.* **77**:96154k (1967).
167. N. A. Loginov, M. V. Smirnov, and B. G. Rossokhin, *Electrochem. Molten and Solid Electrolytes* **4**:7, 17 (1967); *J. Appl. Chem. USSR* **42**:2105 (1969).

168. B. G. Rossokhin, M. V. Smirnov, and N. A. Loginov, *Electrochem. Molten and Solid Electrolytes* **5**:11, 27 (1967).
169. M. V. Smirnov, T. A. Puzanova, and N. A. Loginov, *Soviet Electrochem.* **7**:349 (1971).
170. M. V. Smirnov and V. Ya. Kudyakov, *Izv. Vyssh. Ucheb. Zaved., Tsvet. Met.* **9**(2):71 (1966).
171. M. V. Smirnov and V. Ya. Kudyakov, *Tr. Inst. Elektrokhim. Akad. Nauk SSSR, Ural. Filial* **12**:55 (1969); *Chem. Abstr.* **74**:150229y (1971).
172. *Handbook of Chemistry and Physics*, 47th ed., The Chemical Rubber Co., pp. D38–49 (1966).
173. I. V. Aleksanyants and A. D. Pogorelyi, *Izv. Vyssh. Ucheb. Zaved., Tsvet. Met.* **13**(6):136 (1970).
174. A. N. Kokoev and A. D. Pogorelyi, *Izv. Vyssh. Ucheb. Zaved., Tsvet. Met.* **13**(6):138 (1970).
175. A. A. Kolotii, Yu. K. Delimarskii, and G. V. Chernetskaya, *Soviet Electrochem.* **7**:476 (1971).
176. M. A. Bredig, in: *Molten Salt Chemistry* (M. Blander, ed.), pp. 367–425, Interscience, New York (1964).
177. J. D. Corbett, in: *Fused Salts* (B. R. Sundheim, ed.), pp. 341–407, McGraw-Hill, New York (1964).
178. M. M. Vetyukov and A. Fransis, *Soviet Electrochem.* **5**:331 (1969).
179. J. D. Van Norman, J. S. Bookless, and J. J. Egan, *J. Phys. Chem.* **70**:1276 (1966).
180. G. A. Crawford and J. W. Tomlinson, *Trans. Faraday Soc.* **62**:3046 (1966).
181. A. Komura, H. Imanaga, N. Watanabe, and K. Nakanishi, *Kogyo Kagaku Zasshi* **71**:1976 (1968).
182. J. D. Corbett, F. C. Alberts, and R. A. Sallach, *Inorg. Chim. Acta* **2**:22 (1968).
183. H. R. Bronstein, *J. Phys. Chem.* **73**:1320 (1969).
184. K. Ohmae and T. Kuroda, *J. Electrochem. Soc. Japan* **36**:163 (1968).
185. L. S. Leonova, Yu. M. Ryabukhin, and E. A. Ukshe, *Soviet Electrochem.* **5**:190 (1969).
186. K. Ya. Grachev and V. Z. Grebenik, *J. Appl. Chem. USSR* **39**: 495 (1966).
187. V. Z. Grebenik and K. Ya. Grachev, *J. Anal. Chem. USSR* **23**:147, 1398 (1968).
188. J. Badoz-Lambling, D. Bauer, and P. Texier, *Anal. Lett.* **2**:411 (1969).
189. P. Texier, *J. Electroanal. Chem.* **29**:343 (1971).
190. D. Bauer and J. P. Beck, *Coll. Czech. Chem. Commun.* **36**:323 (1971).
191. G. Mamantov, in: *Molten Salts: Characterization and Analysis* (G. Mamantov, ed.), pp. 529–561, Marcel Dekker, New York (1969).
192. F. R. Clayton, Jr., Ph. D. Dissertation, Univ. of Tennessee, Knoxville, Tennessee, 1971.
193. D. L. Manning, *J. Electroanal. Chem.* **6**:227 (1963).
194. D. L. Manning and G. Mamantov, *J. Electroanal. Chem.* **6**:328 (1963).
195. D. L. Manning, *J. Electroanal. Chem.* **7**:302 (1964).
196. D. L. Manning, J. M. Dale, and G. Mamantov, in: *Polarography, 1964*, (G. J. Hills, ed.), Vol. 2, pp. 1143–1151, Interscience, New York (1966).
197. D. L. Manning and G. Mamantov, *J. Electroanal. Chem.* **7**:102 (1964).
198. S. Senderoff, G. W. Mellors, and W. J. Reinhart, *J. Electrochem. Soc.* **112**:840 (1965).
199. G. W. Mellors and S. Senderoff, *J. Electrochem. Soc.* **112**:642 (1965).

200. G. W. Mellors and S. Senderoff, *J. Electrochem. Soc.* **113**:60 (1966).
201. S. Senderoff and G. W. Mellors, *J. Electrochem. Soc.* **113**:66 (1966).
202. S. Senderoff and G. W. Mellors, *J. Electrochem. Soc.* **114**:586 (1967).
203. K. Grjotheim, *Z. Phys. Chem., N. F.* **11**:150 (1957).
204. H. W. Jenkins, G. Mamantov, D. L. Manning, and J. P. Young, *J. Electrochem. Soc.* **116**:1712 (1969).
205. D. L. Manning and J. M. Dale, in: *Molten Salts: Characterization and Analysis* (G. Mamantov, ed.), pp. 563–573, Marcel Dekker, New York (1969).
206. G. Mamantov and D. L. Manning, *J. Electroanal. Chem.* **18**:309 (1968).
207. B. F. Hitch and C. F. Baes, Jr., *J. Inorg. Nucl. Chem.* **34**:163 (1972).
208. B. F. Hitch and C. F. Baes, Jr., *Inorg. Chem.* **8**:201 (1969).
209. K. A. Romberger and J. Braunstein, *Inorg. Chem.* **9**:1273 (1970).
210. K. A. Romberger, J. Braunstein, and R. E. Thoma, *J. Phys. Chem.* **76**:1154 (1972).
211. P. G. Zambonin and J. Jordan, *Anal. Lett.* **1**:1 (1967); *J. Am. Chem. Soc.* **89**:6365 (1967); **91**:2225 (1969).
212. M. Francini and S. Martini, *Electrochim. Acta* **13**:851 (1968).
213. H. S. Swofford, Jr. and P. G. McCormick, *Anal. Chem.* **37**:970 (1965); P. G. McCormick and H. S. Swofford, *Anal. Chem.* **41**:146 (1969).
214. A. A. El Hosary and A. M. Shams El Din, *Electrochim. Acta* **16**:143 (1971) (and references therein).
215. P. G. Zambonin, *J. Electroanal. Chem.* **24**:365 (1970).
216. P. G. Zambonin, V. L. Cardetta, and G. Signorile, *J. Electroanal. Chem.* **28**:237 (1970).
217. P. G. Zambonin, *Anal. Chem.* **43**:1571 (1971).
218. P. G. Zambonin, *J. Electroanal. Chem.* **32**:A1 (1971).
219. P. G. Zambonin, *J. Electroanal. Chem.* **33**:243 (1971).
220. A. M. Shams El Din and A. A. El Hosary, *Electrochim. Acta* **13**:135 (1968).
221. B. W. Burrows and G. J. Hills, *Electrochim. Acta* **15**:445 (1970).
221a. M. Fredericks, R. B. Temple, and G. W. Thickett, *J. Electroanal. Chem.* **38**:A5 (1972).
222. P. G. Zambonin, *Anal. Chem.* **44**:763 (1972).
223. D. Inman, D. G. Lovering, and R. Narayan, *Trans. Faraday Soc.* **63**:3017 (1967).
224. M. Francini and S. Martini, *Z. Naturforsch.* **23A**:795 (1968).
225. G. Tridot, G. Nowogroki, J. Nicole, M. Wozniak, and J. Canonne, *Compt. Rend. Ser. C* **270**:204 (1970).
226. R. Kalvoda and N. Kh. Tumanova, *Soviet Progr. Chem.* **33**(1):90 (1967).
227. G. Mamantov, J. M. Strong, and F. R. Clayton, Jr., *Anal. Chem.* **40**:488 (1968).
228. S. Casadio, A. Conte, and F. Salvemini, *Electrochim. Acta* **16**:1533 (1971).
229. K. Kawamura, *Electrochim. Acta* **12**:1233 (1967).
230. J. E. L. Bowcott and B. A. Plunkett, *Electrochim. Acta* **14**:363 (1969).
231. S. Sternberg and C. Herdlicka, *Rev. Roum. Chim.* **14**:991 (1969).
232. W. K. Behl and H. C. Gaur, *J. Electroanal. Chem.* **32**:293 (1971).
233. G. V. Shilina and Yu. K. Delimarskii, *Ukr. Khim. Zh.* **36**:781 (1970).
234. C. T. Moynihan and C. A. Angell, *J. Phys. Chem.* **74**:736 (1970).
235. J. Braunstein, L. Orr, A. L. Alvarez-Funes, and H. Braunstein, *J. Electroanal. Chem.* **15**:337 (1967).
236. G. V. Shilina and Yu. K. Delimarskii, *Soviet Progr. Chem.* **33**(3):9 (1967).
237. M. Francini and S. Martini, *Electrochim. Metal.* **3**:132 (1968).

238. H. S. Swofford, Jr. and C. L. Holifield, *Anal. Chem.* **37**:1513 (1965).
239. M. Francini, S. Martini, and C. Monfrini, *Electrochim. Metal.* **2**:325 (1967).
240. G. J. Hills and P. D. Power, *J. Polarog. Soc.* **13**:71 (1967).
241. H. S. Swofford, Jr. and J. H. Propp, *Anal. Chem.* **37**:974 (1965).
242. R. B. Fulton and H. S. Swofford, Jr., *Anal. Chem.* **41**:2027 (1969).
243. W. E. Triaca, H. A. Videla, and A. J. Arvia, *Electrochim. Acta* **16**:1671 (1971).
244. G. A. Sacchetto, G. G. Bombi, and M. Fiorani, *J. Electroanal. Chem.* **20**:89 (1969).
245. G. G. Bombi, G. A. Sacchetto, and G. A. Mazzocchin, *J. Electroanal. Chem.* **24**:23 (1970).
246. Yu. K. Delimarskii and G. V. Shilina, *Electrochim. Acta* **10**:973 (1965).
247. A. A. El Hosary and A. M. Shams El Din, *J. Electroanal. Chem.* **30**:33 (1971).
248. G. A. Mazzocchin, G. G. Bombi, and M. Fiorani, *Ric. Sci.* **36**:338 (1966); *J. Electroanal. Chem.* **17**:95 (1968).
249. Yu. K. Delimarskii and G. V. Shilina, *Ukr. Khim. Zh.* **33**:352 (1967).
250. G. G. Bombi, M. Fiorani, and C. Macca, *Chem. Comm.* **1966**:455.
251. H. Giess, M. Francini, and S. Martini, *Electrochim. Metal.* **4**:17 (1969).
252. R. Marassi, V. Bartocci, P. Cescon, and M. Fiorani, *J. Electroanal. Chem.* **22**:215 (1969).
253. R. Marassi, V. Bartocci, and M. Fiorani, *Chim. Ind. (Milan)* **52**:365 (1970).
254. M. Francini, S. Martini, and H. Giess, *Electrochim. Metal.* **3**:355 (1968).
255. P. Cescon, R. Marassi, V. Bartocci, and M. Fiorani, *J. Electroanal. Chem.* **23**:255 (1969).
256. T. Yanagi, K. Hattori, and M. Shinagawa, *Rev. Polarog. (Kyoto)* **14**:11 (1966).
257. T. Yanagi, T. Nagata, M. Kobayashi, and M. Shinagawa, *Rev. Polarog. (Kyoto)* **15**:102 (1968).
258. Yu. K. Delimarskii, O. G. Zarubitskii, and V. G. Budnik, *J. Appl. Chem., USSR* **42**:2349 (1969); *Soviet Electrochem.* **6**:1563 (1970).
259. J. Goret and B. Tremillon, *Bull. Soc. Chim. Fr.* 2872 (1966); *Electrochim. Acta* **12**:1065 (1967).
260. A. Eluard and B. Tremillon, *J. Electroanal. Chem* **18**:277 (1968); **26**:259 (1970); **27**:117 (1970); **30**:323 (1971); *Anal. Lett.* **2**:221 (1969).
261. L. Martinot and G. Duyckaerts, *Inorg. Nucl. Chem. Lett.* **6**:541 (1970).
261a. G. Schiavon, S. Zecchin, and G. G. Bombi, *J. Electroanal. Chem.* **38**:473 (1972).
262. Yu. K. Delimarskii and N. Kh. Tumanova, *Ukr. Khim. Zh.* **34**:988 (1968).
263. V. V. Kuzmovich and N. Kh. Tumanova, *Soviet Electrochem.* **5**:1003 (1969).
264. P. K. Lorenz and G. J. Janz, *J. Electrochem. Soc.* **118**:1550 (1971).
264a. A. J. Appleby and S. Nicholson, *J. Electroanal. Chem.* **38**:A13 (1972).
265. T. Baak and R. L. Frederick, *J. Am. Ceramic Soc.* **50**:38 (1967).
266. F. Colom and M. De la Cruz, *Electrochim. Acta* **15**:1155 (1970).
267. A. Kisza, *Z. Phys. Chem. (Leipzig)* **237**:97 (1968).
268. G. Picard and J. Vedel, *Bull. Soc. Chim. Fr.* **1969**:2557.
269. B. Bry and B. Tremillon, *J. Electroanal. Chem.* **30**:457 (1971).
270. A. A. Fennin, Jr., L. A. King, and D. A. Seegmiller, *J. Electrochem. Soc.* **119**:801 (1972).
271. L. G. Boxall, H. L. Jones, and R. A. Osteryoung, *J. Electrochem. Soc.* **120**:223 (1973).
272. G. L. Holleck and J. Giner, *J. Electrochem. Soc.* **119**:1161 (1972).

INDEX

Acetates, molten, 238, 245
Acid-base concepts, 76, 200, 218
A. C. polarography, 205, 210
Activated complex, 13, 71
$AlCl_3$ – alkali chlorides, 218, 223, 224
Alkaline earth chlorides, molten, 226
Alkylammonium chloride, molten, 244, 246
Alloy formation, 201
Aluminum, electrochemistry of, 221
Amines, electrochemistry of, 222, 227, 228
Antimony, electrochemistry of, 208, 227, 242, 243, 244
Aromatic hydrocarbons, electrochemistry of, 222

Beryllium, electrochemistry of, 227
Bismuth chloride, molten, 226
Bismuth, electrochemistry of, 204, 208, 222, 226, 227, 236, 242, 243
Bonding in polyatomic molecules, theory of, 103–122
Borax, molten, 244, 246
Bromides, molten, 228
Bromine, electrochemistry of, 237
Bunsen coefficient, 30

Cadmium chloride, molten, 226
Cadmium, electrochemistry of, 204, 205, 214, 219, 220, 225, 226, 235, 236, 238, 242
Calcium chloride–sodium chloride, molten, 225, 227
Carbonate, electrochemistry of, 227, 235
Carbonates, molten, 238, 243, 246
Catalyst, 30, 70, 76
Cerium chloride, molten, 226
Cerium, electrochemistry of, 227, 236
Chlorides, molten, 210, 216, 223, 229, 230

Chlorine, electrochemistry of, 227, 237
Chloroaluminate complexes of cobalt (II), 145–147, 150
 of nickel (II), 167
Chloroaluminates, molten, 218–224
Chromium, electrochemistry of, 204, 205, 206, 212, 236, 243
Chronoamperometry, 236
Chronoamperometry with linear sweep (see linear sweep voltammetry), 201
Chronopotentiometry, 201, 205, 207, 208, 211, 212, 213, 214, 215, 216, 220, 221, 224, 226, 227, 229, 230, 231, 232, 233, 235, 236, 239, 240, 241, 242, 243, 245, 246
Cobalt (II) coordination and spectra, 135–151
Cobalt, electrochemistry of, 204, 205, 212, 219, 235, 242, 243
Complex species, 11, 13, 29, 34, 35, 40, 64, 71
Complexes, molecular orbital theory of, 118
Conductance, electrical, 10, 13, 15, 23
Contamination, atmospheric, 202
Controlled-potential coulometry, 202
Controlled-potential electrolysis, 206
Cooperative motion, 13, 23
Coordination, method of studying at high temperatures, 85
Coordination
 of cobalt (II), 142–151
 of iridium (III), 178, 179
 of iron (II), 132–135
 of iron (III), 130–132
 of nickel (II), 152–172
 of osmium (III), 173–175
 of osmium (IV), 173–175
 of palladium (II), 179, 180

Coordination *(cont'd)*
  of platinum (II), 180, 181
  of rhodium (III), 176, 178
  of ruthenium (III), 172, 173
Copper, electrochemistry of, 204, 205, 212, 213, 219, 227, 235, 236, 242, 243
Corresponding states, 13, 14, 17, 24
Corrosion, 30
Coulomb field, 1, 3, 4, 5, 12, 16, 19, 71
Coulometry, 202, 206, 208, 213, 217, 224, 229, 235
Cryoscopy, 52
Crystal field theory, 87–122
Cyclic voltammetry *(see also* linear sweep voltammetry), 224

Decomposition pressure, 52, 55, 65, 69
Density, 6, 8, 10, 19
Dew point, 52
Differential thermal analysis, 1
Diffusion, 15, 19, 23
Diffusion coefficient, 205, 208, 213, 220, 227, 237, 238
Dihalide molecules, spectra of, 126–128
Dimethylsulfone, molten, 244, 246

Electrode, area, 202
  dropping, 202, 205, 210, 212, 223, 238, 239, 245, 246
  indicator, 202
  quasi-reference, 203
  reference, 203, 243
  rotating, 201, 210, 220, 223, 243, 245, 246
  solid, 202
Electrodes, working, 202, 203
Electrolysis, 68
Electronic configuration for diatomic molecules, spectroscopic terms of, 99
EMF measurements, 220, 221, 226
EMF series, 203
Enthalpy of mixing, 33, 41, 42, 55, 60
  of vaporization, 31
Entropy of fusion, 2, 3
  of mixing, 30, 34, 42, 55
  of vaporization, 32
Equilibrium constant, 29, 33, 36, 61, 77
Erbium, electrochemistry of, 237
Europium, electrochemistry of, 208, 237
Eutectic, 37, 38, 39, 40, 61, 65, 66, 70, 77

Fluorides, molten, 228, 231, 232
Free energy of vaporization, 31
Fugacity, 31, 37

Gallium, electrochemistry of, 209
Glass, sulfate, spectrum of cobalt (II) in, 138, 139
Gold, electrochemistry of, 212, 214

Hafnium, electrochemistry of, 206, 214, 226
Halide gases of group VIII compounds, spectra of, 124, 126–128
Halogens, electrochemistry of, 237, 238, 244
Hard sphere, 7, 8, 10, 16, 19, 23
Hartree-Fock treatment of molecules, 101, 102
Hexachloro coordination of iridium (III), 178, 179
  of osmium (IV), 174
Hexanitrato complex of cobalt (II), 135
Heyrovsky–Ilkovic equation, 201, 205, 208, 214, 225, 238, 244
Holes, 11, 13, 27, 29
Hydrolysis, 202, 204
Hydroxide, electrochemistry of, 227, 234
Hydroxides, molten, 238, 242, 243, 245

Increasing limiting current effect, 201
Indium, electrochemistry of, 209, 238, 242
Intensities of $d-d$ transitions, rules for, 122
Interaction parameter, 42
Iodides, molten, 228
Iodine, electrochemistry of, 237
Iridium (III) coordination and spectra, 178, 179
Iron (II) coordination and spectra, 132–135
Iron (III) coordination and spectra, 130–132
Iron, electrochemistry of, 212, 219, 220, 225, 227, 235, 243, 244
Isoelectronic, 1, 3
Isomers, 1, 9

Kolthoff–Lingane equation, 201, 205, 208, 214, 225, 244

Lanthanides, electrochemistry of, 208, 209, 214, 227, 236
Lanthanum, electrochemistry of, 209, 214
Lead chloride, molten, 226

Index

Lead, electrochemistry of, 204, 205, 212, 214, 219, 220, 225, 226, 235, 236, 238, 242, 243
Lewis acid-base concept, 200, 218
Ligand field theory, 87–122
Linear sweep voltammetry, 201, 204, 205, 206, 210, 214, 216, 220, 224, 231, 232, 233, 235, 236, 237, 239, 240, 241, 242, 243, 245, 246
Lithium chloride–potassium chloride eutectic, 204, 209, 210, 211, 212, 222
Lithium chloride–sodium chloride–potassium chloride, molten, 227
Lux–Flood acid-base concept, 200, 218

Madelung constant, 5
Magnesium chloride–alkali chlorides, molten, 222, 225
Magnesium chloride, molten, 226
Magnesium, electrochemistry of, 226
Manganese, electrochemistry of, 204, 206, 212, 243
Mercury, electrochemistry of, 221, 227, 237, 242, 243
Metal–molten salt solutions, 226
$MgCl_2$–alkali chlorides, 222
Molecular dynamics, 19
Molybdenum, electrochemistry of, 207, 213, 214, 219, 220, 225

Neodymium, electrochemistry of, 214, 227, 236, 237
Neptunium, electrochemistry of, 208, 243
Nickel (II) coordination and spectra, 152–172
Nickel, electrochemistry of, 204, 205, 212, 219, 228, 235, 243
Niobium, electrochemistry of, 206, 207, 213, 220
Nitrate, electrochemistry of, 209
Nitrates, molten, 228, 234–241
Nitrato complexes of nickel (II), 154
Nitrite, electrochemistry of, 209, 228, 234
Noble gases, 27, 28

Onsager relations, 15
Oscillator strength, 119
Oscillographic polarography (*see* linear sweep voltammetry), 201

Osmium (III) coordination and spectra, 173–175
Osmium (IV) coordination and spectra, 173–175
Ostwald solubility, 30
Oxalate, electrochemistry of, 238
Oxidation states of transition metal ions in relation to coordination and spectra, 129–181
Oxygen, electrochemistry of, 209, 227, 228, 234, 235, 243

Packing fraction, 8
Palladium (II) coordination and spectra, 179, 180
Palladium, electrochemistry of, 208, 212
pCl^-, 200, 218, 227, 228
Peroxide, electrochemistry of, 209, 228, 243
Phosphates, molten, 246
Phosphorus, electrochemistry of, 209, 215
Platinum (II) coordination and spectra, 180, 181
Platinum, electrochemistry of, 208, 212, 214
Plutonium, electrochemistry of, 208
Polarography, 235, 237, 239, 241, 245
Potassium, electrochemistry of, 215
Potentiometry, 203, 205, 206, 207, 208, 209, 211, 212, 213, 214, 215, 217, 219, 220, 222, 223, 225, 226, 227, 229, 230, 231, 232, 233, 234, 235, 237, 238, 239, 240, 242, 245, 246
Praseodymium, electrochemistry of, 215, 227, 236, 237
Premelting, 2
Pulse polarography, 201, 237, 239
Purification of melts, 202, 204, 212, 218, 228

Quasi-lattice, 14, 35, 41

Raman spectra, 221
Rare earth chlorides, molten, 227
Rare earths, electrochemistry of, 208, 209, 214, 227, 236
Reference electrode, 203
Reference state, 32, 33
Repulsion, 19, 21, 35, 41
Rhodium (III) coordination and spectra, 176-178

Rotating disc electrode, 201, 220, 223, 243, 245, 246
Ruthenium (III) coordination and spectra, 172, 173

Samarium, electrochemistry of, 237
$SbCl_3$, molten, 227
Scaled particle theory, 7, 9, 28
Scaling parameter, 6
Selection rules for atomic spectra, 96
Selenium, electrochemistry of, 209
Silicates, molten, 243, 246
Silver, electrochemistry of, 204, 205, 212, 214, 219, 226, 236, 242
Sodium chloride–potassium chloride, equimolar, 212–217, 222
Sodium, electrochemistry of, 215
Spectra of atomic ions, theory of, 87–96
  of cobalt (II), 142–151
  of diatomic molecules, theory of, 96–99
  of iridium (III), 178, 179
  of iron (II), 132–135
  of iron (III), 130–132
  of molecules containing group VIII atoms, 123–128
  of molecules, polyatomic, theory of, 103–122, 124, 125
  of osmium (III), 173–175
  of osmium (IV), 173–175
  of palladium (II), 179, 180
  of platinum (II), 180, 181
  of rhodium (III), 176, 178
  of ruthenium (III), 172, 173
  of transition metal ions, theory of, 87–122, 125, 185–191
Standard state, 32, 37
Stationary electrode polarography (*see* linear sweep voltammetry), 201
Stripping, 53
Strontium chloride–sodium chloride–potassium chloride, molten, 226
Sulfates, molten, 246
Sulfur, electrochemistry of, 209, 215
Sulfur heterocycles, electrochemistry of, 222
Superoxide, electrochemistry of, 228, 234, 243
Surface tension, 7, 8, 9, 11, 27
Symmetries of group VIII complexes, general information on, 86, 87, 182–184

Synthesis, chemical, 30, 69, 70, 72

Tantalum, electrochemistry of, 208, 220, 221
Tellurium, electrochemistry of, 209
Term energies, relative, for d^n configurations, 91
Tetrachloro coordination of cobalt (II), 143, 144, 147, 151
  of iron (II) and iron (III), 132
  of ruthenium, 172
Tetracyanato complexes of nickel (II), 169
Tetranitrato complex of cobalt (II), 150
Thallium, electrochemistry of, 204, 205, 214, 225, 235, 236, 238, 242
Theory of spectra of atomic ions, 87–96
  of diatomic molecules, 96–99
  of polyatomic molecules, 103–122, 124, 125
  of transition metal ions, 87–122, 125, 185–191
Thin-layer electrochemistry, 202
Thiocyanates, molten, 238, 242, 245
Thorium, electrochemistry of, 215, 226, 228
Tin, electrochemistry of, 212, 215, 219, 220, 225, 243, 244
Titanium, electrochemistry of, 220, 226, 228, 243
Tracer, 54
Transition time, 202
Transitions, 2, 14, 15, 35, 67
Transpiration, 52
Transuranium elements, electrochemistry of, 208
Tungsten, electrochemistry of, 208, 219

Uranium, electrochemistry of, 208, 228, 242

Valence-bond theory, 100
Vanadium, electrochemistry of, 205, 243
Viscosity, 10, 13, 17
Voltammetry, 201, 205, 206, 208, 209, 210, 213, 214, 215, 216, 219, 220, 221, 223, 224, 225, 226, 227, 229, 234, 235, 236, 237, 238, 240, 242, 243, 245, 246
Volume, 2, 5, 12, 17, 29, 30, 41, 54

Index

Walden product, 10, 11, 16, 17
Water, determination of, 204, 227, 234
Water, removal of, 202, 204, 228
Working electrodes, 202

X-ray diffraction, 221

Ytterbium, electrochemistry of, 208
Yttrium, electrochemistry of, 209

Zinc chloride–indium chloride, molten, 226
Zinc chloride–lead chloride–sodium chloride, molten, 226
Zinc chloride, molten, 226
Zinc chloride–sodium chloride, molten, 225
Zinc, electrochemistry of, 205, 212, 219, 225, 226, 235, 238, 242
Zirconium, electrochemistry of, 206, 213, 226

QD
189
A33
v.2
1973

MAR 29 1977